T0201610

Modelling Mortality with Actuarial Applications

Actuaries have access to a wealth of individual data in pension and insurance portfolios, but rarely use its full potential. This book will pave the way, from methods using aggregate counts to modern developments in survival analysis.

Based on the fundamental concept of the hazard rate, Part One shows how and why to build statistical models based on data at the level of the individual persons in a pension scheme or life insurance portfolio. Extensive use is made of the R statistics package. Smooth models, including regression and spline models in one and two dimensions, are covered in depth in Part Two. Finally, Part Three uses multiple-state models to extend survival models beyond the simple life/death setting, and includes a brief introduction to the modern counting process approach.

Practising actuaries will find this book indispensable and students will find it helpful when preparing for their professional examinations.

ANGUS S. MACDONALD is Professor of Actuarial Mathematics at Heriot-Watt University, Edinburgh. He is an actuary with much experience of modelling mortality and other life histories, particularly in connection with genetics, and as a member of Continuous Mortality Investigation committees.

STEPHEN J. RICHARDS is an actuary and principal of Longevitas Ltd., Edinburgh, a software and consultancy firm that uses many of the models described in this book with life insurance and pension scheme clients worldwide.

IAIN D. CURRIE is an Honorary Research Fellow at Heriot-Watt University, Edinburgh. As a statistician, he was chiefly responsible for the development of the spline models described in this book, and their application to actuarial problems.

INTERNATIONAL SERIES ON ACTUARIAL SCIENCE

MODELLING MORTALITY WITH ACTUARIAL APPLICATIONS

ANGUS S. MACDONALD
Heriot-Watt University, Edinburgh

STEPHEN J. RICHARDS
Longevitas Ltd, Edinburgh

IAIN D. CURRIE
Heriot-Watt University, Edinburgh

CAMBRIDGE
UNIVERSITY PRESS

CAMBRIDGE
UNIVERSITY PRESS

University Printing House, Cambridge CB2 8BS, United Kingdom

One Liberty Plaza, 20th Floor, New York, NY 10006, USA

477 Williamstown Road, Port Melbourne, VIC 3207, Australia

314–321, 3rd Floor, Plot 3, Splendor Forum, Jasola District Centre,
New Delhi – 110025, India

79 Anson Road, #06–04/06, Singapore 079906

Cambridge University Press is part of the University of Cambridge.

It furthers the University's mission by disseminating knowledge in the pursuit of
education, learning, and research at the highest international levels of excellence.

www.cambridge.org
Information on this title: www.cambridge.org/9781107045415
DOI: 10.1017/9781107051386

© Angus S. Macdonald, Stephen J. Richards and Iain D. Currie 2018

First published 2018

Printed in the United Kingdom by Clays, St Ives plc

A catalogue record for this publication is available from the British Library.

ISBN 978-1-107-04541-5 Hardback

Contents

v

Preface

This book brings modern statistical methods to bear on practical problems of mortality and longevity faced by actuaries and analysts in their work for life insurers, reinsurers and pension schemes. It will also be of interest to auditors and regulators of such entities. The following is a list of questions on demographic risks which this book will seek to answer. Practising actuaries will recognise many of them from their daily work.

- Insurance portfolios and pension schemes often contain substantial amounts of individual data on policyholders and beneficiaries. How best can this information be used to manage risk? How do you get the greatest value from your own data?
- Historically, actuarial work modelled mortality rates for grouped data. Does this have drawbacks, and is there a better way of modelling risk? Are there models which recognise that it is individuals who experience insurance events, rather than groups?
- In many markets insurers need to find new risk factors to make their pricing more competitive. How do you know if a risk factor is significant or not? And is a risk factor statistically significant, financially significant, both or neither?
- Even in the very largest portfolio, combinations of some risk factors can be relatively rare. How can you build a model that handles this?
- How do you choose between models?
- Some portfolios are exposed to different modes of exit. An example is a term-insurance portfolio where a policy can lapse, or result in a death claim or a critical-illness claim. How do you build a model when there are competing risks?
- Many modern regulatory frameworks are explicitly statistical in nature, such as the Solvency II regime in the European Union. How do you perform your

analysis in a way that meets the requirements of such regulations? In particular, how do you implement the "events occurring in one year" requirement for risks that are long-term by nature?

- How have mortality rates in your portfolio changed over time? How do you separate time-based trends from changes in the composition of the portfolio?
- After you have built a model, what uncertainty lies over your fitted rates? How do you measure mis-estimation risk in a multi-factor model? And how do you measure mis-estimation risk in terms of the financial consequences for a given portfolio of liabilities?
- The future path of mortality rates is unknown, yet a pension scheme is committed to paying pensions for decades. How do you project mortality rates? How do you acknowledge the uncertainty over the projection?
- How can the analyst or advisor working for a small pension scheme convince a lay audience that a statistical model fits properly?

This book aims to provide answers to these and other questions. While the book is of immediate application to practising actuaries, auditors and regulators, it will also be of interest to university students and provide supplementary reading for actuarial students.

When we refer to "modern" statistical methods, we mean survival modelling. As a sub-discipline of statistics, it developed from the 1950s onwards, mostly with clinical questions in mind, and actuaries paid it little attention. It followed the modern (by then) statistical paradigm, as follows:

- Specify a probabilistic model as a plausible description of the process generating the observations.
- Use that model to deduce the statistical properties of what can be observed.
- Use those statistical properties to test hypotheses, measure goodness-of-fit, choose between alternative models and so on.

Two major themes have emerged in survival modelling in the past 40 years. The first and most obvious is the arrival of cheap computing power and statistics packages. All major statistics packages can now handle the survival models useful in medical research (although not usually those useful to actuaries). In this book we use the R language.

The second development is a thorough examination of the mathematical foundations on which survival modelling rests. This is now also the mathematical basis of the survival models that actuaries use. Few actuaries are aware of it, however. As well as introducing modern statistical methods in a practical setting, which occupies the first two-thirds of this book, we also wish to bring some of the recent work on the foundations of survival models to an actuarial audience. This occupies the last third of the book.

The book is divided into three parts: in Part One, Analysing Portfolio Mortality, we introduce methods of fitting statistical models to mortality data as it is most often presented to actuaries, and assessing the quality of model fit. In Part Two, Regression and Projection Models, we discuss the graduation and forecasting of mortality data in one and two dimensions. In Part Three, Multiple-State Models, we extend the models discussed in Part One to life histories more complex than being alive or dead, and in doing so we introduce some of the modern approach to the foundations of the subject.

Part One consists of nine chapters. Chapter 1 begins by introducing mortality data for individuals and for groups of individuals. Grouped data lead naturally to mortality ratios, defined as a number of deaths divided by a measure of time spent exposed to risk. We give reasons for preferring person-years rather than the number of persons as the denominator of such ratios. Chapter 1 also introduces a case study, a UK pension scheme, which we use as an example to illustrate the fitting of survival models. This chapter discusses the choice of software package for fitting models, and provides a section on the notation that will be followed throughout the book.

Chapter 2 discusses data preparation. We assume that the actuary has data from an insurance or pension portfolio giving details of individual persons. Typically this will contain anomalous entries and duplicates, and we describe how to identify and correct these. Data can then be grouped if an analysis of grouped data is required, or for tests of goodness-of-fit.

In Chapter 3 we introduce the basic probabilistic model that describes the lifetime of an individual person, leading to the key quantity in survival modelling, the hazard rate or force of mortality. Then in Chapter 4 we discuss statistical inference based on the data described by the probabilistic model. We discuss at length the two features that distinguish survival modelling from other branches of statistics, namely left-truncation and right-censoring. Then, in Chapter 5, we focus on parametric models fitted by maximum likelihood, with examples of R code applied to our pension scheme Case Study. We deal with both grouped data and individual data, and show how these analyses are related. By making simplifying assumptions, we obtain the binomial and Poisson models of mortality well known to actuaries. These three chapters form the heart of Part One, and introduce the key model and methodology.

Chapter 6 discusses tests of model goodness-of-fit. We introduce standard statistics such as information criteria that are widely used in statistical practice for model selection, and also the more familiar battery of detailed tests designed to assess the suitability of a fitted model for actuarial use.

One of the main advantages of modelling individual data, rather than grouped data, lies in the possibility of allowing for the effects of covariates by modelling rather than by stratifying the data. Chapter 7 compares stratification and modelling, based on a simple example, and shows how the fitting process in Chapter 5 can be simply extended.

Chapter 8 introduces non-parametric estimates of mortality rates and hazard rates, and their possible use in actuarial practice. These provide a useful, easily visualised presentation of a mortality experience, if individual mortality data are available.

Finally, in Part One, Chapter 9 discusses the role of survival models in risk-based insurance regulation.

Part Two is divided into four chapters. In Chapter 10 we consider regression models of graduation for one-dimensional data. We start with the Gompertz model (Gompertz, 1825) which we fit initially by least squares. This leads to Poisson and binomial models which we describe in a generalised linear model setting. A particular feature of all four chapters is the use of the R language to fit the models; we hope not only that this enables the reader to fit the models but also that the language helps the understanding of the models themselves. Computer code is provided as an online supplement for all four chapters.

In Chapter 11 we discuss smooth models of mortality. We begin with Whittaker's well-known method (Whittaker, 1923) and use this to introduce the general smoothing method of *P*-splines (Eilers and Marx, 1996). We use the R package *MortalitySmooth* (Camarda, 2012) to fit these models.

In Chapter 12 we consider two-dimensional data and model mortality as a function of both age and calendar year. We concentrate on three particular models: the Lee–Carter model (Lee and Carter, 1992), the Cairns–Blake–Dowd model (Cairns et al., 2006) and the smooth two-dimensional model of Currie et al. (2004). We fit the Lee–Carter model with the R package *gnm* (Turner and Firth, 2012), the Cairns–Blake–Dowd model with R's glm() function, and the smooth two-dimensional model with the *MortalitySmooth* package.

In the final chapter of Part Two, Chapter 13, we consider the important question of forecasting. We lay particular emphasis on the importance of the reliability of a forecast. We consider both time-series and penalty methods of forecasting; the former are used for forecasts for the Lee–Carter and Cairns–Blake–Dowd models, while the latter are used for the smooth models.

Part Three explores extensions of the probabilistic model used in Part One, which represent life histories more complicated than being alive or dead. It is divided into four chapters.

The framework we use is that of multiple-state models, in which a person occupies one of a number of "states" at any given time and moves between states at random times governed by the probabilistic model. We think these are now quite familiar to actuaries. They are introduced in Chapter 14, mainly in the Markov setting, in which the future is independent of the past, conditional on the present. They are defined in terms of a set of *transition intensities* between states, a natural generalisation of the actuary's force of mortality. The key to their use is a system of differential equations, the *Kolmogorov equations*, which in turn generalises the well-known equation (3.19) met in Chapter 3.

Chapter 15 discusses inference of the transition intensities of a Markov multiple-state model from suitable life history data. Since life histories consist of transitions between pairs of states at random times, what is observable is the number and times of transitions between each pair of states in the model. These are described by *counting processes*. Once these are defined, inference proceeds along practically the same lines as in Part One.

Chapter 16 applies the multiple-state model in a classical setting, that of *competing risks*. This is familiar to actuaries, for example, in representing the decrements observed in a pension scheme. However, it gives rise to some subtle problems of inference, not so easily discerned in a traditional actuarial approach to multiple decrements, which we compare with our probabilistic approach.

Chapter 17 returns to the topic of counting processes. Pioneering work since the 1970s has placed these at the very heart of survival modelling, and no modern book on the subject would be complete without a look at why this is so. It uses the toolkit of modern stochastic processes – filtrations, martingales, compensators, stochastic integrals – that actuaries now use regularly in financial work but not, so far, in mortality modelling. Our treatment is as elementary as we dare to make it, completely devoid of rigour. It will be enough, we hope, to give access to further literature on survival models, much of which is now written in this language.

PART ONE

ANALYSING PORTFOLIO MORTALITY

1

Introduction

1.1 Survival Data

This part of the book is about the statistical analysis and modelling of survival data. The purpose we usually have in mind is the pricing or valuation of some insurance contract whose payments are *contingent* on the death or survival of an individual. So, our starting point is the question: what form does survival data take?

1.1.1 Examples of Survival Data

Consider the following two examples:

 (i) On 1 January 2014, Mr Brown took out a life insurance policy. The premium he paid took into account his age on that date (he was exactly 31 years and three months old) and the fact that he had never smoked cigarettes. On 19 April 2017 (the date when this is being written) Mr Brown was still alive.

(ii) On 23 September 2013, Ms Green reached her 60th birthday and retired. She used her pensions savings on that date to purchase an annuity. Unfortunately, her health was poor and the annual amount of annuity was higher than normal for that reason. The annuity ceased when she died on 3 April 2016.

These observations, typical of what may be extracted from the files of an insurance company or pension scheme, illustrate the raw material of survival analysis, as actuaries practise it. We can list some features, all of which may be relevant to the subsequent analysis:

- There are three *timescales* in each example, namely age, calendar time and the duration since the life insurance policy or annuity commenced.

- Our observations began only when the insurance policy or annuity commenced. Before that time we had no reason to know of Mr Brown's or Ms Green's existence. All we know now is that they were alive on the relevant commencement dates.
- Observation of Mr Brown ceased when this account was written on 19 April 2017, at which time he was still alive. We know that he will die after 19 April 2017, but we do not know when.
- Observation of Ms Green ceased because she died while under observation (after 23 September 2013 but before 19 April 2017).
- In both cases, additional information was available that influenced the price of the financial contract: age, gender, Mr Brown's non-smoking status and Ms Green's poor health. Clearly, these data influenced the pricing because they tell us something about a person's chances of dying sooner rather than later.

1.1.2 Individual Life Histories

The key features of life history data can be summarised as follows:

- The age at starting observation, the date of starting observation and the reason for starting observation.
- The age at ending observation, the date of ending observation and the reason for ending observation.
- Any additional information, such as gender, benefit amount or health status.

1.1.3 Grouped Survival Data

One main purpose of this book is to describe statistical models of mortality that use, directly, data like the examples above. This is a destination, not a starting point. We will soon introduce the idea of representing the future lifetime of an indvidual as a non-negative random variable T. Ordinary statistical analysis proceeds by observing some number n of observations $t_1, t_2, \ldots, t_n$ drawn from the distribution of T. A key assumption is that these are independent and identically distributed (i.i.d.). In the case of Mr Brown and Ms Green, we have no reason to doubt independence, but they are clearly not identically distributed.

So we take a step back, and ask how we can define statistics derived from the life histories described above that are plausibly i.i.d.. One way is to group data according to qualities that advance homogeneity and reduce heterogeneity. For example, we could group data by the following qualities:

- age
- gender
- policy size (sum assured or annuity payment)
- type of insurance policy
- calendar time
- duration since taking out insurance policy
- smoking status
- occupation
- medical history.

Another way is to propose a statistical model which incorporates directly any important sources of heterogeneity, for example as covariates in a regression model. In Chapter 7 we discuss the relative merits of these two approaches.

1.2 Software

Throughout this book we will illustrate basic model-fitting with the freely available R software package. This is both a programming language and a statistical analysis package, and it has become a standard for academic and scientific work. R is free to download and use; basic instructions for downloading and installing R can be found in Appendix A. Partly because it is free of charge, R comes with no warranties. However, support is available in a number of online forums.

Many actuaries in commerce use Microsoft Excel®, and they may ask why we do not use this (or any other spreadsheet) for model-fitting. The answer is twofold. First, R has many advantages, not least the vast libraries of scientific functions to call upon which mean we can often fit complex models with a few lines of code. Second, there are some important limits to Excel, especially when it comes to fitting projection models like those in Part Two. Some of these limits are rather subtle, so it is important that an analyst is aware of Excel's limitations.

The first issue is that at the time of writing Excel's standard Solver feature will not work with more than 200 variables (that is, parameters which have to be optimised in order to fit the model). This is a problem for a number of important stochastic projection models in Part Two. One option is to use only models with fewer than 200 parameters, but this would allow software limitations to dictate what the analyst can do.

Another issue is that Excel's Solver function will often claim an optimal solution has been found when this is not the case. If the Solver is re-run several times in succession, it often finds a better-fitting set of parameters on the second and third attempts. It is therefore important that the analyst re-runs the Solver a few times until no further change is found. Even then, we have come across examples where R found a better-fitting set of parameters, which the Solver agreed was a better fit, but which the Solver could not find on its own.

One option would be to consider one of the commercially supported alternative plug-ins for Excel's Solver, although analysts would need to check that it was indeed capable of finding the solutions that Excel cannot.

Whatever the analyst does, it is important not to rely uncritically on a single software implementation without some form of checking.

1.3 Grouped Counts

Consider Table 1.1, which shows the mortality-experience data for the UK pension scheme in the Case Study (see Section 1.10 for a fuller description). It shows the number of deaths and time lived in ten-year age bands for males and females combined. The main advantage of the data format in Table 1.1 is its simplicity. The entire human age span is represented by just 11 data points (age bands), and a reasonably well-specified statistical model can be fitted in just four R statements (more on this in Section 1.5). We call the data in Table 1.1 *grouped data*, because there is no information on individuals. (It is likely that information on individuals was collected, but then aggregated. The analyst might not have access to the data originally collected, only to some summarised form.) A natural and intuitive measure of mortality in each age band is the ratio of the number of deaths to the total time lived, which is shown in the last column of Table 1.1. We call quantities of this form *mortality ratios*.

1.4 What Mortality Ratio Should We Analyse?

Suppose in a mortality analysis we want to calculate mortality ratios, as in the rightmost column of Table 1.1. The numerator for the mortality ratio is obvious: it is the number of deaths which have occurred. However, we have two fundamental choices for the denominator:

- the number of lives (which is not shown in Table 1.1), or
- the time lived by those lives.

Table 1.1 *High-level mortality data for Case Study (see Section 1.10); time lived and deaths in 2007–2012.*

Age interval	Time lived, t (years)	Deaths, d	Mortality ratio ($d/t \times 1000$)
[0, 10)	71.9	0	0
[10, 20)	449.0	2	4.5
[20, 30)	163.9	0	0
[30, 40)	121.7	3	24.6
[40, 50)	893.1	6	6.7
[50, 60)	5,079.3	48	9.5
[60, 70)	32,546.7	278	8.5
[70, 80)	21,155.9	510	24.1
[80, 90)	10,606.7	866	81.6
[90, 100)	1,751.5	363	207.3
[100, ∞)	23.1	11	475.7
All ages	72,862.7	2,087	28.6

The distinction arises because some of the individuals in the study may not have been present for the whole period 2007–2012. For example, consider someone who retired on 1 January 2009. Such a person would contribute one to the total number of lives, but a maximum of four out of a possible six years of time lived while a member of the scheme. The methods needed to analyse these alternative formulations will clearly be different.

If we use the number of lives as the denominator, we are calculating the proportion dying. For example, suppose a total of 3,500 individuals were pensioners aged between ages 70 and 80. Then the mortality ratio, which is $510 \div 3500 = 0.1457$, is the proportion of members between ages 70 and 80 who died during the six calendar years 2007–2012. The proportion dying during a single calendar year might then be estimated by $0.1457 \div 6 = 0.0243$. It is natural to suppose that this estimates the probability of dying during a single year. Such probabilities are denoted by q ($0 \le q \le 1$).

As it stands, this may not be a very good or reliable estimate. It takes no account of persons who, as mentioned above, were not under observation throughout all of 2007–2012, or who passed from one age band to the next during 2007–2012. Adjustments would have to be made to allow for these, and other, anomalies. Nevertheless, this analysis of mortality ratios based on "number of lives" has been very common in actuarial work, perhaps motivated by the fact that the probabilities being estimated are precisely the probabilities of the life table.

The alternative, which we will advocate in this book, is to use the time lived as the denominator. In detail, for each individual we record the time at which they entered an age group and the time when they left it, and the difference is the *survival time* during which they were alive and in that age group. Then the sum of all survival times in an age group is the total lime lived, shown in the second column of Table 1.1. Analysis based on time lived has certain advantages. Potentially important from a statistical point of view is that it avoids losing information on who died and when. We will illustrate this in the following example adapted from Richards (2008).

Consider two small groups of pension scheme members, A and B, each with four lives. Over the course of a full calendar year one life dies in each group. The proportion dying is the same in each group: $\hat{q}_A = \hat{q}_B = 1/4$ (we use the circumflex to denote an estimate of some true-but-unknown quantity; thus, $\hat{q}$ is an estimate of q). Analysis of the proportion dying does not distinguish between the mortality experience of groups A and B.

Let us denote mortality ratios based on time lived by m. Suppose that the death in group A occurred at the end of January. The total time lived in group A was therefore $3\frac{1}{12}$ years ($= 1+1+1+\frac{1}{12}$), and the ratio of deaths to time lived is thus $\hat{m}_A = 1 \div 3\frac{1}{12} = \frac{12}{37}$. In contrast, suppose that the death in group B occurred at the end of November. Then the total time lived in group B was $3\frac{11}{12}$ years ($= 1+1+1+\frac{11}{12}$), and the mortality ratio for group B is $\hat{m}_B = 1 \div 3\frac{11}{12} = \frac{12}{47}$. Thus, using the time lived as the denominator enables us to distinguish a genuine difference between the two mortality experiences. Using the number of lives leads us to overlook this difference; the information on the time actually lived is discarded.

We do not need to worry if we need q-type probabilities (that is, a life table) for specific kinds of work. As we will see later, we can derive any actuarial quantity we need having estimated m-type mortality ratios.

Let us develop the example further. Suppose that in group A one of the three surviving individuals leaves the scheme at the end of August. The reason might be resigning from employment (if an active member accruing benefits), or a trivial commutation (if a pensioner member). Using the number of lives, we now have a major problem in calculating $\hat{q}_A$, because we will not know if the departed individual dies or not in the last third of the year. If they did, then we should have $\hat{q}_A = 2/4$; if they did not, then we should have $\hat{q}_A = 1/4$, but we do not know. We will be forced to complicate our analysis on a number-of-lives basis by making some additional assumptions. Unfortunately, the assumptions which are easiest to implement are seldom justified in practice. In contrast, using time lived, the adjustment is trivial and no further assumptions are required.

The total time lived is now simply $2\frac{3}{4}$ years ($= 1 + 1 + \frac{8}{12} + \frac{1}{12}$), and the mortality ratio is $\hat{m}_A = 1 \div 2\frac{3}{4} = \frac{4}{11}$.

This example exhibits the other advantage of using time lived instead of the number of lives – it is better able to handle real-world data where individuals enter and leave observation for various reasons, at times that are not under the control of the analyst.

The mortality ratio q is referred to as the *initial rate of mortality*, while m is referred to as the *central rate of mortality* (see Section 3.6). When used in the denominator, the number of lives is called the *initial exposed-to-risk* (sometimes denoted by E), while the time lived is called the *central exposed-to-risk* (sometimes denoted by E^c). Having set out some reasons for preferring mortality ratios based on time lived, the next section demonstrates how to fit a model to grouped counts.

1.5 Fitting a Model to Grouped Counts

One recurring feature in this book is that quantities closely related to Poisson random variables and, later on, Poisson processes arise naturally in survival models. Why this is so will ultimately be explained in Chapter 17, but for now we shall just accept that the data in Table 1.1 appear to be suitable for modelling as a Poisson random variable from age band (30, 40] upwards. For reasons we explain in Section 1.6, we exclude data below age 30 as having too few observed deaths.

We can build a statistical model for the data in Table 1.1 in just four R commands:

```
vExposures = c(121.7, 893.1, 5079.3, 32546.7, 21155.9,
               10606.7, 1751.5, 23.1)
vDeaths = c(3, 6, 48, 278, 510, 866, 363, 11)
oModelOne = glm(vDeaths ~ 1, offset=log(vExposures),
               family=poisson)
summary(oModelOne)
```

We shall explain what each of these four commands does.

- We first put the times lived and deaths into two separate vectors of equal length. The R function c() concatenates objects (here scalar values) into a vector. It can be handy to begin the variable names with a v as a reminder that they are vectors, not scalars.

- We next fit the Poisson model as a generalised linear model (GLM; see Section 10.7) using R's `glm()` function. We specify the deaths as the response variable, and we have to provide the exposures as an offset. We also specify a distribution for the response variable with the `family` argument. The results of the model are placed in the new model object, `oModelOne`. It can be handy to begin such variable names with an o as a reminder that it is a complex object, rather than a simple scalar or vector.
- Last, we inspect the model object using R's `summary()` function.

Part of what we will see in the output is the following:

```
Coefficients:
            Estimate Std. Error z value Pr(>|z|)
(Intercept)  -3.5444     0.0219  -161.8   <2e-16 ***
```

The above model has fitted a single constant parameter applying across all ages, which R calls the "intercept". The default behaviour in R is to operate on a logarithmic scale, so the parameter labelled (`Intercept`) is $\log \hat{m}$. Thus, the fitted model is $\hat{m} = \exp(-3.5444) = 0.02889$. This is simply the mortality ratio: the total number of deaths (2,085) divided by the total time lived (72,178.0) years). What this tells us is that the mortality ratio is not just an intuitive measure of mortality, but emerges as the estimate in a probabilistic model. In this context it has sampling properties, such as the standard error and the p-value, and these are provided automatically in the R output.

We can do better. Table 1.1 suggests that mortality ratios increase sharply with increasing age. A better model would therefore allow the Poisson parameter to vary by each age group. We can do this by running the following R commands:

```
vExposures = c(121.7, 893.1, 5079.3, 32546.7, 21155.9,
               10606.7, 1751.5, 23.1)
vDeaths = c(3, 6, 48, 278, 510, 866, 363, 11)
vAgeBand = factor(c(35, 45, 55, 65, 75, 85, 95, 105))
oModelTwo = glm(vDeaths ~ -1 + vAgeBand,
                offset=log(vExposures), family=poisson)
summary(oModelTwo)
```

We shall explain the two new features:

- The age bands are labelled with the age at the mid-point of each band. The `factor()` command ensures that the age bands are to be treated as factor levels, rather than values for regression. In other words, a mortality rate will be fitted to each age band separately.

- The -1 in the model specification tells R not to calculate the effect of each group relative to a baseline, but to calculate the effect for each group on a stand-alone basis.

This time we will see in the output that R has calculated estimates of log m for each age group, together with standard errors and p-values:

```
Coefficients:
              Estimate Std. Error z value Pr(>|z|)
vAgeband35    -3.70295    0.57735   -6.414 1.42e-10 ***
vAgeband45    -5.00294    0.40825 -12.255  < 2e-16 ***
vAgeband55    -4.66173    0.14434 -32.297  < 2e-16 ***
vAgeband65    -4.76281    0.05998 -79.412  < 2e-16 ***
vAgeband75    -3.72526    0.04428 -84.128  < 2e-16 ***
vAgeband85    -2.50536    0.03398 -73.727  < 2e-16 ***
vAgeband95    -1.57383    0.05249 -29.985  < 2e-16 ***
vAgeband105   -0.74194    0.30151  -2.461   0.0139 *
```

However, the simplicity of the Poisson model comes at a price. The most obvious drawbacks to using grouped counts are that they lose information – we do not know who died, when they died or what their characteristics were. Also, mortality levels can vary a lot over ten years of age, so the grouping in Table 1.1 is quite a heavy degree of summarisation. Table 1.1 answers the broad question of how mortality varies by age, but that is not nearly enough for actuarial work. We would therefore need to split up the age intervals into ranges where the mortality ratios were approximately constant.

There is another important technical aspect of modelling Poisson counts, which imposes severe limitations on the grouped-count approach for actuarial work. This is the need to have a minimum number of expected deaths for the Poisson assumption to be reasonable. This is explained in the next section.

1.6 Technical Limits for Models for Grouped Data

In Section 1.5 we modelled the data in Table 1.1 at ages over 30 as Poisson random variables. Why did we not model the data below age 30 in the same way?

Part of the answer lies with the testing of the goodness-of-fit of the model where the number of expected events is below 20. This is discussed in detail in

Table 1.2 *Partial probability function for deaths in age interval [103, 104)*
from Table 1.1.

d	$\Pr(D = d)$	$\Pr(D \leq d)$
0	0.19165	0.19165
1	0.31662	0.50827
2	0.26154	0.76981
3	0.14403	0.91385
4	0.05949	0.97333
5	0.01966	0.99299

Section 6.6.1. Another part of the answer lies in a feature of the Poisson model which matters if the number of expected events is very low. In the Case Study, there were 1.21 years of time lived and two deaths in the age interval [103, 104). Under a Poisson model for the number of deaths, we would have an esti-mated Poisson parameter of $\hat{m} = 2/1.21 = 1.6529$. The probability of observ-ing four or more deaths is then 0.08615 $(= 1-0.91385)$. The problem with this is that there were in fact only three individuals contributing 0.20 years, 0.29 years and 0.72 years, respectively, to the 1.21 years lived. The Poisson model therefore attaches a non-zero probability to an impossible event, namely ob-serving four or more deaths among three lives; Table 1.2 shows more details of the probability function.

In technical terms, we say that the Poisson model is not *well specified* in this example. It is the reason why it is important with grouped data to avoid cells with very low numbers of expected deaths. It is also the origin of the rule of thumb to collapse grouped counts across sub-categories until there are at least five expected deaths (say) in each cell. This is why we could not include the data below age 30 in the model in Section 1.5.

It is true that the time lived in the denominator of a mortality ratio will always be contributed by a finite number of individuals. Therefore a Poisson model will always attach non-zero probability to an impossible event, namely observing more deaths than there were individuals. However, if the number of individuals is reasonably large, this probability is so close to zero that it can be ignored. This technical limitation of the Poisson model exposes a fundamental conflict between the desire of the analyst (to analyse as many risk factors as possible) and the requirements of the Poisson model (to minimise the risk-factor combinations to keep the minimum expected deaths above five). We will illustrate this in the next section.

1.7 The Problem with Grouped Counts

To reduce the variability of mortality within a single age band, suppose we split the data into single years of age. Suppose that we know that most of the scheme's liabilities are concentrated among pensioners receiving £10,000 p.a. or more. If these individuals had mortality materially lower than that of other pensioners, there would be important financial consequences. We therefore have good reason to investigate the mortality of this key group separately. To address the questions of age and pension size together, we might sub-divide the data in Table 1.1 a bit further, a process called *stratification*. This is done in Table 1.3, concentrating on the post-retirement ages where we have meaningful volumes of data.

Table 1.3 immediately reveals a problem with stratifying grouped data – even a simple sub-division into two risk factors results in many small numbers of deaths, which we saw in Section 1.6 were problematic. There is a fundamental tension between conflicting aims. On the one hand we need to sub-divide to have relatively homogeneous groups with respect to age and any other risk factors of interest. On the other hand that same sub-division produces too many data points with too few expected deaths. In Table 1.3 we have only stratified by two risk factors; the problem only gets worse when we add more. For example, we might also want to consider gender and early-retirement status as risk factors. This would create many more data points with even fewer deaths, and thus make the Poisson model even more poorly specified.

The conundrum is that Table 1.1 shows that there should be plenty of information: there are over 2,000 deaths and over 70,000 life-years of exposure. However, the problems of stratification mean that using grouped data is not an efficient way of using the information.

There is a solution to this conundrum, and that is to use data on each individual life instead of grouped counts. The modelling of individual lifetimes has four important advantages over grouped data:

(i) Data validation is easier, and more checks can be made, if data on individual lives or policies are available (see Section 2.3).

(ii) No information is lost. The precise dates at which an individual joins and leaves the scheme are known, as is the date of death (if they died) and all the recorded characteristics of that person (such as gender and pension size).

(iii) Using individual-level data better reflects reality. It is individuals who die, not groups.

(iv) Stratification ceases to be an issue. With individual-level data there is no limit to the number of risk factors we can investigate (see Section 7.3).

Table 1.3 *Mortality data for Case Study by age and whether revalued pension size exceeds £10,000 p.a. (ages 60–99 only). Lives are categorised by their revalued pension size at 31 December 2012.*

Age	Pension < £10,000 p.a. Time lived	Deaths	Pension ≥ £10,000 p.a. Time lived	Deaths
60	2,439.2	16	480.5	2
61	2,737.7	15	534.1	0
62	2,881.8	20	543.0	3
63	2,957.6	34	547.9	4
64	2,923.7	20	536.9	8
65	3,181.8	22	512.7	3
66	2,926.2	30	438.8	1
67	2,754.5	26	392.9	4
68	2,625.5	35	363.2	2
69	2,450.9	29	316.5	4
70	2,318.0	29	277.8	3
71	2,239.3	34	257.0	5
72	2,170.7	38	239.4	3
73	2,081.1	40	228.4	4
74	1,933.7	44	213.5	5
75	1,828.6	44	199.8	4
76	1,732.6	52	202.1	6
77	1,635.2	47	194.1	6
78	1,562.9	68	184.9	7
79	1,485.4	63	171.6	8
80	1,380.7	71	160.7	2
81	1,264.2	94	157.5	4
82	1,175.8	79	137.5	7
83	1,101.0	71	121.3	7
84	1,004.3	78	110.8	4
85	909.6	78	110.9	8
86	838.0	83	100.0	12
87	731.8	87	82.1	12
88	605.9	79	67.4	8
89	493.4	72	53.8	10
90	406.4	63	42.2	11
91	325.4	54	28.2	7
92	243.9	54	18.8	3
93	181.4	29	18.5	2
94	156.4	34	14.6	5
95	115.1	30	12.4	2
96	66.7	29	12.2	1
97	46.1	12	7.4	7
98	33.9	7	1.2	1
99	19.2	11	1.5	1
All ages	57,988.1	1,831	8,094.3	197

As it happens, using individual-level data is natural for actuaries, as administration systems for insurance companies and pension schemes usually hold data this way. Indeed, it is usually far easier to ask an IT department to run a simple database query to extract all individual records than it is to ask the same department for a program to be written to perform the aggregation into grouped counts. Individual-level data are often, nowadays, easy to obtain (which would not have been the case before cheap computing power became available).

We shall give an example of how individual-level data improve data validation. Consider the 11 deaths occurring among lives over age 100 in Table 1.1. If we were presented only with the data as in the example in Section 1.3, we would have little choice but to accept its validity. However, if we had individual-level data and could see, say, that many of these individuals had a date of birth of 1 January 1901 (1901-01-01), we would immediately know that something was amiss. Section 2.3 discusses some of the ways to validate data extracts from administration systems, together with some other real-world examples of data issues met in practice.

Collecting data for analysis is therefore best done at the level of the individual or insurance policy. Individual-level data are the "gold standard", and actuarial analysts should always try to obtain it, whatever kind of model they wish to fit.

1.8 Modelling Grouped Counts

We have made a case for using data on individual lives where possible. However, there are many cases where such data are simply not available. National population data collected by the Office for National Statistics in the UK or, at the time of writing, on 39 countries worldwide through the Human Mortality Database are of this nature. This is the kind of data that nearly all mortality modelling used until the advent of survival models and data on individual lives. The need for modelling grouped data and using these models to forecast the future course of mortality remains an important subject for actuaries and indeed for governments. We devote Chapter 13 to this topic.

1.9 Survival Modelling for Actuaries

In Section 1.4 we explored some of the benefits of using time lived, rather than numbers of lives, for calculating mortality ratios. In Sections 1.6 and 1.7 we saw some drawbacks of using grouped data and some benefits of using individual-level data. The combination of individual-level data and analysis based on time lived underlies the subject of *survival analysis*. It is the purpose of Part One of the book to show practising actuaries how to apply survival analysis in their daily work.

Actuaries drove much of the early research into mortality analysis, and, because of this, the actuarial analysis of mortality tended to follow a particular direction. This work mostly predated the development of statistics as a discipline (see the quotation from Haycocks and Perks, 1955, in Section 3.1) and as a result it was, in some technical respects, overtaken by the later work of demographers and medical statisticians. Many of the survival-model methods in this book will be new to actuaries, but were developed several decades ago. These methodologies have great potential for application to actuarial problems.

However, actuaries do need to approach survival models in a different way to statisticians, especially statisticians who use survival models in medical trials. One difference concerns the nature of the data records. In a medical trial there is usually one record per person, as the data have been recorded at the level of the individual. Actuaries, however, often have to contend with the problem of duplicates in the data: the basic unit of administration is the policy or benefit, and people are free to have more than one of these. Indeed, wealthier people tend to have more policies than others (Richards and Currie, 2009). Before fitting any model, actuaries have to deduplicate the data. This is an essential stage of data preparation for actuaries, and Section 2.5 is devoted to it in detail; see also Richards (2008).

A second difference is that statisticians working in medical trials address different, and often simpler, questions to those addressed by actuaries. A statistician often wants to test for a difference between two populations, say with and without a particular treatment. This kind of question can often be answered without detailed modelling of mortality as a function of age, which would usually be of primary importance for an actuary. It is possible that actuaries have, in the past, overlooked survival analysis for this reason.

A third major difference is that statisticians are usually modelling the time lived since treatment began, so their timescales begin at time $t = 0$. In contrast, actuaries always need to deal with missing observation time, because lives enter actuarial investigations at adult ages, as in Table 1.1. The actuary's timescale starts at some age $x > 0$. Technically, the actuary's data have been *left-truncated* (see Section 4.3). Because it is rarely an issue for statisticians, standard software packages such as SAS and R (see Appendix A) do not often allow for left-truncation. For this reason, actuaries typically have to do a modest amount of programming to get software packages to fit survival models to their data.

Survival modelling is a well-established field of statistics, and offers much for actuarial work. However, actuaries' specific requirements are not addressed adequately (or at all) in standard statistical texts or software. This book aims to remedy that gap, and has been written for actuaries wishing to implement survival models in their daily work.

1.10 The Case Study

In Part One of this book we will use a data set to illustrate statistical analysis and model-fitting. The data are from a medium-sized local-authority pension scheme in the United Kingdom. The data comprise 16,043 individual records of pensions in payment with 72,863 life-years lived over the six-year period 2007–2012. There were 2,087 deaths observed in that time. The distributions of exposure time (life-years lived) and deaths are shown in Figure 1.1.

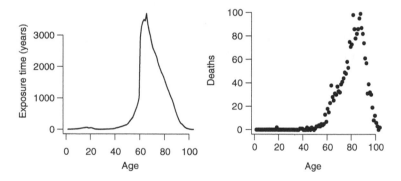

Figure 1.1 Exposure time (left panel) and deaths (right panel) by age for Case Study. The sharp increase in exposure times between ages 60 and 65 reflects the scheme's normal retirement age.

Figure 1.2 shows the log mortality ratios $\log \hat{m}_x$ for single years of age x in the Case Study. This shows two features typical of pension-scheme mortality:

(i) Above age 60 (a typical retirement age) $\log \hat{m}_x$ strongly suggests a linear dependence on age x. This is in fact characteristic of mortality, not just of humans but of many animal species.

(ii) Below age 60 the mortality ratios seem to decrease with age. This pattern is largely confined to pension schemes. It is believed to be associated with the reasons why individuals retire early.

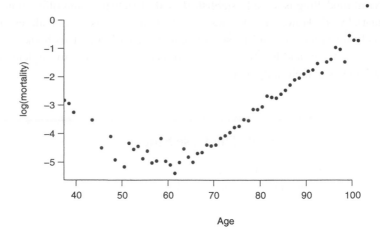

Figure 1.2 Crude log mortality ratios, $\log \hat{m}_x$, for single years of age x for the Case Study.

Table 1.4 shows total annual amounts of pension by decile, revalued at 2.5% per annum up to the end of 2012. The first decile is the smallest pensions, the tenth decile the largest. This shows the concentration of pension scheme liabilities on the better-off pensioners.

Table 1.4 *Pension amounts by decile for the Case Study, revalued at 2.5% per annum.*

Pension decile	Total annual pension (£)	Percentage of total pension	Cumulative percentage of total
1	503,987	0.7	0.7
2	1,181,911	1.6	2.2
3	1,866,303	2.5	4.7
4	2,655,968	3.5	8.2
5	3,693,268	4.9	13.2
6	4,991,669	6.6	19.8
7	6,794,025	9.0	28.8
8	9,435,178	12.5	41.4
9	14,256,324	18.9	60.3
10	29,884,250	39.7	100.0
All	75,262,883	100.0	

1.11 Statistical Notation

We mentioned in Section 1.4 that the circumflex (^) denotes an estimate of some true-but-unknown quantity. Actuaries will be most familiar with the approach of "hatted" quantities being estimates of rates or probabilities at single years of age; thus $\hat{q}_x$ is an estimate of q_x at non-negative integer ages x. Smoothing or graduation of these quantities (see Section 4.7) then proceeds as a second stage once the estimates $\hat{q}_x$ have been found, so that some other notation such as $\dot{q}_x$ is needed to denote the smoothed quantities.

Our approach in this book is to formulate probabilistic models "capable of generating the data we observe" (an idea we begin to develop properly in Chapter 4). The probabilistic model then leads to well-specified estimates such as $\hat{q}_x$, of quantities of interest such as q_x. By "well-specified" we mean that the probabilistic model not only tells us what estimates to use, but also lets us deduce their sampling properties.

In this book "hatted" quantities have a slightly broader interpretation than that described above. This is because many statistical approaches formulate a probabilistic model that treats estimation and smoothing as one, not as two separate stages. Then the estimates we obtain from the model, which we still denote as "hatted" quantities, are already smoothed, and there is no subsequent smoothing or graduation stage. The models we begin to introduce in Chapter 5 are of this kind.

Thus, "hatted" quantities may sometimes be smoothed already, and sometimes not. The context will generally make it clear what "hatted" quantities mean, and this should not lead to confusion.

2

Data Preparation

2.1 Introduction

Data preparation is the foundation on which any statistical model is built. If our data are in any way corrupted, then our model – and any analysis derived from it – is invalid. This chapter describes five essential data-preparation steps before fitting any models, namely extraction, field validation, relationship checking, deduplication and sense-checking. It is possible to automate many of these tasks, but some are unavoidably manual.

2.2 Data Extraction

2.2.1 Data Source

The first question is where to get the data from. For example, it is common for data to be available which have been pre-processed for other business purposes, such as performing a valuation. However, we recommend avoiding using such data if possible, as risk modelling often requires different data items. For example, valuation extracts rarely include the policyholder's surname, whereas this can be very useful for deduplication (see Section 2.5). Furthermore, it is harder to check the validity of data which have already been processed or transformed for another purpose. It is therefore better to extract data directly from the administration or payment systems, which often involves a straightforward SQL query of the database. In addition to being able to extract all the fields directly from the source, this enables detailed checking and validation of the kind described in Section 2.3 onwards.

2.2.2 How Far Back to Extract Records

The second question is how far back to extract the records. The short answer is "as far back as you can go", but there is a minimum required period for modelling to be unaffected by seasonal mortality fluctuations. This is particularly important for modelling pensioner or annuitant mortality, as the elderly experience greater fluctuations due to seasonal mortality drivers such as influenza. This is illustrated in Figure 2.1, which shows both the extent of annual variability and the disproportionate impact on post-retirement ages.

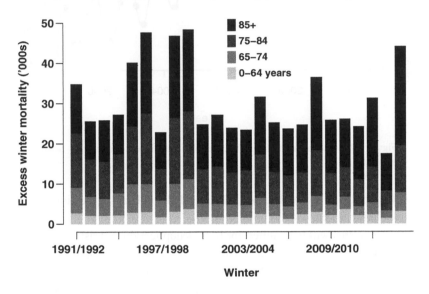

Figure 2.1 Excess winter mortality in UK. Source: own calculations using data from Office for National Statistics.

Figure 2.1 is for population data, but seasonal fluctuations appear in portfolios of annuities and pensions as well, as shown in Figure 2.2. A corollary of Figure 2.1 is that it is important to have the same number of each season in the modelling period.

2.2.3 Data Fields to Extract

The data fields available to the analyst will vary. In some circumstances the analyst can simply specify which fields to extract, whereas in other circumstances a data file will be made available with no opportunity to influence the content.

The essential fields are as follows:

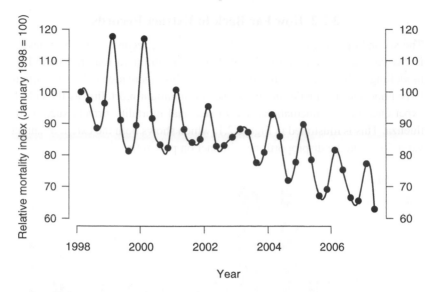

Figure 2.2 Seasonal mortality in UK defined-benefit pension scheme. Source: own calculations.

- date of birth, as this will be needed for both deduplication (Section 2.5) and calculating the exact age
- gender, as this will be required for both deduplication and risk modelling
- on-risk date or commencement date; required mainly for calculating the exact time observed, but also for modelling selection risk
- off-risk date, i.e. the date observation ceased; required for calculating the exact time observed
- status, i.e. whether a claim happened on the off-risk date or not; required for modelling the claim risk.

If any of the above fields are missing, then no mortality modelling can take place. In order for the modelling to be meaningful for financial purposes the following fields are also necessary, or at least highly desirable:

- surname and forename, for deduplication
- postcode, for both deduplication (Section 2.5) and geodemographic profiling (Section 2.8.1)
- pension size, or some other measure of policy value such as sum assured.

In order to correctly calculate the time observed, the analyst will also need to enquire as to whether the policies have been migrated from another administration system at any point. For example, it is not uncommon for pension schemes in the UK to change administrator, or for an insurer to consolidate two or more administration systems. As a general rule, policies which have ceased by the migration date are not carried across to the new administration system, so the analyst will have to ask for the date of migration, which might even be policy-specific. One useful approach is to ask for the date of the earliest transaction for a policy on the administration system – for a pension in payment this could be the earliest date of payment, whereas for term-insurance business it could be the earliest premium-collection date. This "earliest activity date" can then be used to correctly calculate the time observed on a policy-by-policy basis. Another, albeit much rarer, possibility is where business has been migrated to another system but where the only experience data available are from the original system. Such circumstances could arise where part of a portfolio is transferred to another provider. In this case a "latest usable end date" would be required for each policy, which could be the last date of payment for a pension or the last premium-collection date for term-insurance business.

Another possible complication is when the data are sourced from two or more administration systems. This can arise where a history of mergers and acquisitions has resulted in an insurer having several portfolios of the same type of product, or an employer having several pension schemes. Here it is a good idea to record the source of each record and use this as a potential risk factor for modelling.

Beyond these fields the data items to extract depend on what is available. Possible further examples include:

- Early-retirement status. Some pension schemes record a marker for early-retirement cases, which often have higher mortality.
- First life or surviving spouse. Many pension schemes and annuity providers record a marker for whether the benefit is paid to the first life or a surviving spouse. In some cases this status can be deduced from the client identifier.
- Smoker status. In insurance business the smoker status is usually recorded.
- Rating. In insurance business there is often a marker for any underwriting rating which has been applied.
- Product code. Some product types are sold to different classes of life, e.g. group pensions *versus* individual pensions *versus* pensions for the self-employed.
- Distribution channel. This can be very important when modelling lapse risk or persistency.
- Guarantees or options. Annuities arising due to a guaranteed annuity rate (GAR) or guaranteed annuity option can experience higher mortality than other lives. Annuitants freely choosing benefit escalation often experience lighter mortality than lives choosing non-escalating benefits.

The list above covers only a few suggestions. In practice the analyst will have to enquire as to the available fields, which will depend on the administration system and the nature of the business.

2.2.4 File Format

Many potential file formats could be used, but we find that comma-separated value (CSV) files are both simple and convenient. An example CSV input file for survival modelling is shown in Section 5.5.1. Most software packages can read and write CSV files, including the R system we use in this book. Actuaries working across international borders need to remember that CSV files in some countries use the semicolon as a separator instead of the comma.

XML (eXtensible Markup Language) is another useful file format, and has the advantage over CSV files that the data can be validated using a document type definition (DTD). XML files are often marked up with human-readable annotation, but they are generally rather verbose and inefficient for storing very large numbers of records. An XML file will therefore generally be noticeably larger than its CSV equivalent, but XML files can also be read and written using the XML package in R.

Spreadsheets such as Microsoft Excel are also occasionally used, but they are a proprietary binary format which cannot be read by as many software tools as CSV and XML files. Excel files are also a relatively inefficient means of storing large data volumes, and tend to take up a lot more space than the equivalent CSV file. Excel itself is a useful tool for reading and writing both CSV and XML files, although it must be borne in mind that Excel has a limit to the number of data rows (whereas there are no such limits for CSV and XML files). When writing CSV files from Excel, users should beware of Excel's tendency to add empty columns to the CSV file.

We recommend that readers standardise on using CSV files as the most convenient means of storing large data sets in a format that is easy to read for most commonly used software tools.

2.2.5 Date Format

When creating date fields in a CSV file or similar it is important to remember two aspects:

- Four-digit years are essential to avoid ambiguity. This is particularly the case for pensions and annuity business, where it is perfectly possible to have two beneficiary records with dates of birth a hundred or more years apart: very old pensioners may have dates of birth in the 1900s, whereas dependents can have dates of birth in the 2000s. For example, if a date of birth is given as 30/01/07, is it a very old pensioner born in 1907 or a dependent child born in 2007? The answer can usually be inferred manually from other data, but it is simpler and easier to supply four-digit years for all dates.

- US dates are month/day/year, whereas European dates are day/month/year. We recommend standardising on the ISO 8601 date format, which is year-month-day. This format has additional benefits for sorting, as the most significant elements of the date start on the left.

2.3 Field Validation

The first and most basic type of check is that mandatory fields are present: date of birth, gender, commencement date, date of death and status (alive or dead). Beyond this both mandatory and optional fields must be checked for validity:

- Dates must be valid – no "30th Februaries".
- Gender must be M or F – or whatever the local coding is – rather than leaving the field blank or putting in a marker like X or U for "unknown".
- Benefit amounts must be positive, or at least non-negative. Depending on the administration system, a zero-valued benefit may or may not be a sign of some kind of error.

2.4 Relationship Checking

After basic field validation, checks must be performed on the natural relationships between certain fields. Examples are given below for some common data fields:

- The date of birth must lie in the past, and cannot be after the date of extract.
- The policy commencement date or date of retirement must be after the date of birth.
- The date of death must lie in the past, and must be after the later of the date of birth and policy commencement date.

These kinds of errors are not uncommon, and commonly arise in conjunction with other data-integrity issues. In one portfolio of annuities we saw there were thousands of records with invalid or future commencement dates. These records turned out to be dummy records for possible surviving spouses. Exclusion of these records was obviously necessary, but finding this sort of error is harder without individual records extracted directly from the administration system.

2.5 Deduplication

The assumption that events befalling different individuals are independent, in the statistical sense, is key when building models. However, it is usual for administration systems to be orientated around the administration of policies, not people. Some examples of this include the following:

- *Endowment policies*. Where these policies were taken out in conjunction with a mortgage in the UK, it was common to take out a new top-up policy when the policyholder moved house and needed a larger sum assured. Such top-up policies were not always taken out with the same insurer, but when they were, multiple policies for the same life would obviously arise in the same portfolio.
- *Annuities*. It is common in the UK and elsewhere for retirees to have several savings policies, each of which buys a separate annuity. Even where a retiree has a single consolidated fund, some individuals will phase their retirement by buying annuities staggered in time to limit the risk of buying when interest rates are low (known in the UK market as "phased retirement"). Often there are tax or legal restrictions on different types of funds which then force an annuitant to buy separate annuities of different types; one example of this is the "protected-rights" portion of a UK defined-contribution pension fund, which must be used to buy an inflation-linked annuity with a spouse's benefit.
- *Pension schemes*. A pensioner may have more than one benefit record if she has more than one period of service. Similarly, a pensioner may have one benefit record in respect of her own pensionable service, and a second benefit record if she is also a surviving widow of another pensioner in the same scheme.

(In the following, we refer to "policyholder" but assume that the term includes annuitants and pensioners.)

Deduplication is not just important when building models. To have a proper view of overall risk in a portfolio, it is important to know the total liability for each policyholder. In particular, wealthier and longer-lived people are more likely to have multiple contracts, which means that the tendency to have multiple contracts is very likely to be positively correlated with some of the risk factors of interest. This is shown in Figure 2.3, which is an example of this for a portfolio of life-office pension annuities in payment in Richards and Currie (2009). Quite apart from the importance for statistical modelling, Figure 2.3 shows that deduplication can give an insurer a better understanding of its policyholders and their purchasing behaviour.

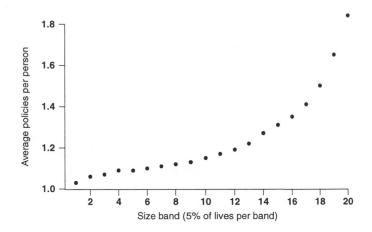

Figure 2.3 Average number of policies by pension size (revealed by deduplication). One individual has 31 policies. Source: Richards and Currie (2009).

When attempting deduplication, decisions have to be made about how to piece together a composite view of the policyholder from the various policy records. For example, it usually makes sense to sum the benefit amounts to give a total risk. Where policies have commenced on different dates, it is usually appropriate to take the earliest commencement date, this being the first date when the policyholder became known. However, deduplication can also throw up reasons for rejection. For example, an annuitant might have two annuities, but be marked as dead on one record and alive on the other. This contradiction would usually result in both records being rejected and flagged for further inspection. Deduplication here offers a potential business benefit, namely releasing unnecessarily held reserves. Deduplication can also help guard against money-laundering or fraud by assisting with the "know your client" requirement in many territories.

The various alternative options for deduplication are described in detail for an annuity portfolio in Richards (2008). In general it makes sense to use the data fields available to create a combined "key" for identifying duplicates. As a rule each deduplication key should involve both the date of birth and the gender code, e.g. a male born on 14 March 1968 would have a key beginning "19680314M". However, on its own this is insufficient to identify duplicates, since in large portfolios many people will share the same date of birth. Some examples of how to extend the deduplication key are given below:

- Add the surname and first initial. If surnames have not been reliably spelled, say due to teleservicing, then metaphone encoding of names might be required to deal with variant spellings; see Philips (1990).

- Add the postcode. In countries with hierarchical postcodes, including the postcode will make for a very powerful deduplication key because postcodes are so specific; see Figure 2.4 for an example with UK postcodes. This may still be useful for other countries with non-hierarchical postcodes, such as Germany.

- Add the administration system's client identifier. If the source administration database has separate tables for client data and policy data, the system's client identifier may be useful. However, this will not help if the administration system has multiple client records for policyholders.

- Add the National Insurance number or tax identifier. For portfolios where names and postcodes are unavailable, the addition of a social security number would make for a powerful deduplication key. However, this is not always wholly reliable, as we have encountered administration systems where dummy data have been entered for some policyholders. Where the same dummy identifier is used, AB123456C for example, this could lead to false duplicates being identified.

There is also no need to restrict deduplication to one single key; a thorough deduplication process might make several passes with a different style of deduplication key each time. It is a good idea to use the most restrictive and reliable deduplication schemes first, e.g. date of birth, gender, surname, first initial and postcode, and then try progressively less restrictive or less reliable schemes thereafter. This minimises the risk of over-deduplication, e.g. in circumstances where the National Insurance field contains some dummy data among mainly valid numbers.

Note that it is always worth deduplicating, even if the data source insists it is unnecessary. In one case known to us a life insurer insisted that deduplication was unnecessary because it had already been done as part of a demutualisation process; the deduplication turned out to have been far from complete. In another example, the deduplication process for a pension scheme uncovered far more duplicates than expected, which turned out to be evidence of a flawed data-extraction process.

2.6 Bias in Rejections

A small number of rejected records is almost inevitable in any portfolio, and this is not always a problem. For example, if 30 records fail validation out of a total of, say, 30,000, then this would not be a major issue. However, one thing to watch for is if there is a bias in rejection. For example, we would feel less comfortable about those 30 failing records if they were all deaths, especially if there were only a few hundred deaths overall.

2.7 Sense-Checking

Many of the data-preparation stages we have described so far can be automated in a computer program, and the program can decide whether records are valid or not. However, not all data issues can be detected by computer and some can only be spotted by a human analyst. After validation and deduplication, we find it useful to tally the five most frequently occurring values in each data field for visual inspection. This often immediately reveals features or issues with the data which a computer cannot detect. Here are some real-life examples based on portfolios of pensions and annuities:

- *Date of birth.* It is perfectly valid to have a date of birth of 1901-01-01; however, it is suspicious when several thousand records share this date of birth, especially when there is only a handful of records with the date of birth 1901-01-02. Suspiciously common dates of birth are often evidence of false dates entered during a system migration, or for policies which do not terminate on the death of a human life. An example of the latter might be a single payment stream in respect of a buy-in policy covering multiple lives in a pension scheme.
- *Surname.* In a large life-office annuity portfolio in the UK we expected the surnames SMITH, JONES and TAYLOR to be among the most frequent. However, when the most common surname turned out to be SPOUSE, it was immediately obvious that some dummy records had been extracted which should not have been included.
- *Commencement date.* One portfolio had a disproportionately large number of annuities commencing in September of each year. Upon enquiry this

Table 2.1 *Five most frequently occurring dates of death in a real annuity portfolio.*

Date of death	Cases
1999-09-09	447
2004-05-14	126
2002-10-16	114
2005-11-11	111
2005-06-27	105

turned out to be perfectly legitimate: the insurer wrote pension annuities for teachers and lecturers, and for their employers it was natural for the policies to begin at the start of the academic year. In other portfolios this can be evidence of the migration of data from another administration system, or the merging of two portfolios.

- *Date of death.* Suspiciously large numbers of deaths on the same date can be evidence that the date given is actually the date of processing, not the date of death itself.

- *Zero-valued pensions.* A small number of pensions with zero value is not necessarily an issue. However, a large number of zero-valued pensions might be indicative of either pension amounts being set to zero on death or trivial commutation. In one portfolio it was the insurer's policy to set the annuity amount to zero on death, which obviously invalidated the use of pension size for modelling.

- *Postcode.* In the UK perhaps around 35 people on average share the same residential postcode (see Section 2.8.1). It was therefore suspicious for one piece of analysis to find over ten thousand records sharing the same postcode in Glasgow. Further investigation revealed that when a pensioner died, their address was changed to that of the administrator as a means of suppressing mailing. While this invalidated the use of geodemographic profiling for modelling, it did not invalidate modelling by age, gender and pension size.

By way of illustration, the five most frequently occurring dates of death for a real annuity portfolio are shown in Table 2.1. Although these are valid dates, and would certainly pass an automated validation stage, it is clear that there is something amiss, and that the date 1999-09-09 is unlikely to be the actual date of death for all 447 annuitants. One possibility would be simply to model mortality from 1 January 2000 onwards, but this presumes that some post-2000 deaths have not been falsely set as 1999-09-09.

Table 2.2 *Five most frequently occurring dates of birth in a real annuity portfolio.*

Date of birth	Cases
1900-01-01	728
1920-01-01	36
1921-01-01	35
1944-07-23	26
1904-12-31	24

Table 2.3 *Five most frequently occurring retirement dates in a real pension scheme.*

Retirement date	Cases
2005-01-31	3,428
1998-01-31	1,737
1994-02-28	1,603
1991-05-31	1,447
1990-04-30	1,300

The same annuity portfolio exhibits a similar problem with the annuitant dates of birth. Table 2.2 suggests that 1 January 1900 is being used as a default date of birth, and that there may be an issue with 1 January 1920 and 1921 as well. These cases would need to be excluded from any mortality model, as age is the most important risk factor and its calculation needs a valid date of birth.

Tables 2.1 and 2.2 show the importance of extracting the underlying data from the payment or administration system. If we had been passed a valuation extract, for example, it might have simply provided the age when the annuity was set up and the age at death. We would therefore have missed the data corruptions and any model based on the valuation-extract data would have been rendered less reliable as a result.

Note, however, that some kinds of date heaping can be perfectly justified. For example, Table 2.3 shows the most commonly occurring dates of retirement in another real pension scheme with just over 40,000 pensioners. In this case the heaping around certain dates is not suspicious, but instead corresponds to restructuring activities of the sponsoring employer where early retirement options have been granted.

Table 2.4 *Age-standardised mortality rates for countries of UK in 2012.*
Source: Office for National Statistics.

Country	SMR
England	523.9
Northern Ireland	567.0
Wales	567.8
Scotland	640.1
UK	538.6

2.8 Derived Fields

Another reason to collect individual records directly from the administration system is that a number of potentially useful fields can be derived from the basic data:

- *Birth year.* Derived from the date of birth, this can be useful if there are strong year-of-birth or cohort effects. See Willets (1999) and Richards et al. (2006) for discussion of year-of-birth mortality patterns.
- *Start year.* Derived from the commencement date, this can be useful for modelling selection effects if the nature of the business written has changed over time. For example, in the UK annuities started to be underwritten from around the year 2000 with the advent of enhanced terms for proposers who could demonstrate a medical impairment. Business which was not under-written therefore became inceasingly strongly anti-selected over time, meaning that the start year of a UK annuity became a useful risk factor.
- *Socio-economic profile.* In many territories the socio-economic status of policyholders can be inferred from their address, or sometimes just their postcode. This is an increasingly important area, and so it is the subject of the following two subsections.

2.8.1 Geodemographics in the UK

Where somebody lives can tell us a lot about their mortality. Table 2.4 shows the age-standardised mortality rates (SMRs) for the four constituent countries of the UK, arranged in ascending order of mortality. As can be seen, Scotland has a markedly higher level of mortality than England.

However, the inhabitants of each country are by no means homogeneous. For example, we can look at some of the SMRs for local council areas within Scotland, as shown in Table 2.5. We can see a dramatic widening of the range

Table 2.5 *Age-standardised mortality rates for selected council areas in Scotland in 2012. Source: Office for National Statistics.*

Council	SMR
East Dunbartonshire	481.9
Glasgow City	827.7
Scotland	640.1

in SMR: from 481.9 for East Dunbartonshire (75% of the Scottish SMR) up to 827.7 in Glasgow City (129% of the Scottish SMR). Further geographical sub-division would reveal still greater extremes, but at the risk of creating too many classifications for practical use (there were 32 Scottish council areas in 2012, for example).

Tables 2.4 and 2.5 use the idea of geography to analyse mortality differentials. Carstairs and Morris (1991) and McLoone (2000) took this further by looking at postcode sectors, which cover several hundred households. A Carstairs index is a measure of deprivation for a postcode sector based on social class, car ownership (or lack of it), overcrowding and unemployment levels. The index can then be used for mortality modelling, with postcode sectors sharing a similar index level grouped together. This works fine as long as the inhabitants of the postcode sector are relatively homogeneous, but this is not always the case. By way of illustration, Figure 2.4 shows the breakdown of a typical UK postcode. The UK uses a hierarchical postcode system, where each successive element from left to right tells you more and more precisely where the address is. In theory a house number and a postcode should be enough to deliver a letter. Other countries which use such hierarchical postcodes include the Netherlands, Canada and the USA (where the postcode is called the zip code).

Richards and Jones (2004) and Richards (2008) introduced the idea of using more granular address data for mortality modelling, i.e. at the level of the postcode, household or street. In the UK a postcode covers around 15 households, or 35 individuals, on average, and on its own a postcode is clearly useless for mortality modelling – there are around 1.6 million residential postcodes in use in the UK, and the number of lives at each postcode is too small. The solution is to use the concept of *geodemographics*, i.e. the grouping of lives according to shared demographic characteristics. The idea is that a lawyer in Edinburgh, Scotland would share many important characteristics with a lawyer in York, England; despite their geographic separation, they are likely to share similar

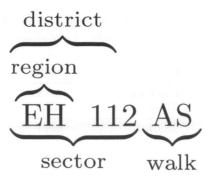

Figure 2.4 Anatomy of a UK postcode. Source: Longevitas Ltd.

levels of education, relative wealth and income and other health-related attributes (such as the propensity to smoke). In the UK, a hierarchical postcode such as in Figure 2.4 can be mapped onto one of the geodemographic type codes illustrated in Figure 2.5 for the Mosaic® classification system from Experian Ltd. This mapped code can then be used for mortality analysis instead of the actual postcode. A number of other classification systems are available for the UK, including the Acorn® classification from CACI Ltd.

Geodemographic profiles such as Mosaic and Acorn were originally developed for marketing purposes. However, their application has spread more widely, including to the pricing of general insurance and life insurance. The market for transfer of longevity risk in pension schemes, as either bulk annuities or longevity swaps, is characterised by heavy use of postcodes for pricing. Two features of such profiles make them much more powerful predictors of mortality than Carstairs scores: first, they operate at a more granular level, thus avoiding issues with heterogenous populations over larger areas; second, the profiles are commonly derived from detailed individual-level data on wealth, income and credit history. Geodemographic profiles work well in addition to knowledge of, say, pension size, because the latter often gives only an incomplete picture of an individual's wealth and income. For a more detailed discussion, see Richards (2008) and Madrigal et al. (2011).

The importance of both geodemographic profiles and deduplication (Section 2.5) is shown in Table 2.6 for a large portfolio of life-office annuitants in the UK. Average pension size is strongly correlated with the postcode-driven geodemographic profile, but so is the average number of policies per life.

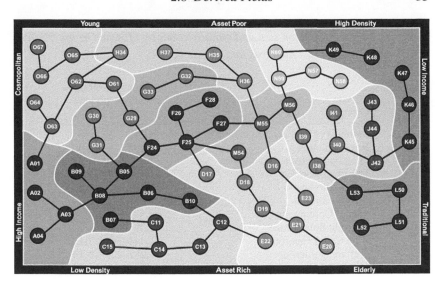

Figure 2.5 The Mosaic classification for UK residents. Source: © Experian Ltd.

2.8.2 Geodemographics in Other Territories

Other developed countries have similar-seeming postal codes, e.g. the *code postal* in France and the *Postleitzahl* in Germany. However, such codes do not have the granularity of hierarchical postcodes in the UK, USA, Canada and the Netherlands. For example, the German *Postleitzahl* 89079 covers exactly 100 residential streets, and thus is more akin to the granularity of the UK postcode district, rather than the postcode sector used by Carstairs and Morris (1991) and McLoone (2000) (see Figure 2.4).

However, this does not mean that geodemographic profiling is impossible for such countries. On the contrary, such profiling can be done if the entire address is used, thus enabling the profiling of each individual household. This usually requires specialist software for address matching, and this is typically territory-specific. Actuaries working for life insurers should contact their marketing department, since such functions will typically already have a customer-profiling solution in-house and this can often be reused at minimal cost for geodemographic modelling.

Note that codes like the *Postleitzahl* would still be very much usable as part of a deduplication key, as in Section 2.5.

Table 2.6 *Average pension size and annuity count by Mosaic Group. Source: Richards and Currie (2009).*

Mosaic Group	Average annuity (£ p.a.)	Average policies per life
Symbols of Success	4,348	1.33
Rural Isolation	3,405	1.30
Grey Perspectives	2,708	1.29
Suburban Comfort	2,203	1.24
Urban Intelligence	2,489	1.22
Happy Families	1,856	1.19
Ties of Community	1,592	1.19
Twilight Subsistence	1,394	1.17
Blue Collar Enterprise	1,444	1.16
Welfare Borderline	1,281	1.14
Municipal Dependency	1,093	1.12
Unmatched or unrecognised postcodes	2,619	1.17
Commercial addresses	4,365	1.35
All lives	2,663	1.24

2.8.3 Bias in Geodemographic Profiling

One issue to watch for in geodemographic profiling is whether the profiler includes historic address or postcode data. It is not uncommon for administration systems to hold out-of-date postcodes for deaths; after all, the customer-service department will see little point in keeping the address data for dead people up to date. In a country like the UK, where the Post Office retires old postcodes and replaces them with new ones, it is important that the profiling system has profiles for formerly valid postcodes, not just currently valid postcodes. If not, deaths will have a greater likelihood of not being profiled at all, leading to a bias in profiles between deaths and survivors.

2.9 Preparing Data for Modelling and Analysis

Once the data have been validated, deduplicated and sense-checked, the file needs to be prepared for modelling. This means that each individual record needs to have the dates turned into ages and times observed. If pension size is to be used in the modelling, then special consideration needs to be given to the pension amounts for deceased individuals or other early exits from the portfolio.

2.9.1 Transforming Records for Modelling and Analysis

The basic data from most administration systems contain dates: dates of birth, commencement dates, dates of death or other dates of exit. However, for modelling purposes we need to convert these data items into ages and times observed for each individual. Furthermore, we will very likely also want to select a particular age range for modelling; for example, if we wanted to fit a log-linear model to the data set in Figure 1.2, we would want to ignore deaths and time lived below age 60. In addition to selecting an age range, we would also want to specify a calendar period for the investigation. For example, in pensions and annuities work it is common to discard the most recent few months of deaths and time lived to minimise the impact of delays in death reporting. Similarly, Figure 2.2 shows that we must balance the numbers of seasons, so we need to specify exact start and end dates for the modelling period.

The process of preparing the data for a survival analysis under these restrictions is relatively straightforward. We define the following for the ith individual:

- *dateofbirth$_i$*, the date of birth of the ith individual.
- *commencementdate$_i$*, the date of commencement for the ith individual, that is, when they entered observation. In a pensioner portfolio this would be the date of retirement, for example.
- *enddate$_i$*, the end date of observation for the ith individual, i.e. when they ceased observation. This would either be the date of death or the date the life was last observed to be alive (often the date of extract).

In addition to the above individual-level data items, we also need to define the following for the entire data set:

- *modelminage*, the lower age from which to model the data. This might be 50 or 60 in a pensioner data set.
- *modelmaxage*, the upper age to which to model the data. If all the data were to be used, this might be set to an artificially high value like 200.
- *modelstartdate*, the model start date. This would be the earliest calendar date where the data were still felt to be relevant and free from data-quality concerns.

- *modelenddate*, the model end date. This would be the latest calendar date where the data were felt not to be materially affected by reporting delays, i.e. it would be earlier than the extract date.

We need to standardise to prepare the data, and we have the option of driving data preparation by either dates or ages. If we use the age-based approach, then we would calculate the following for the ith individual:

- *entryage$_i$*, the entry age of the individual at the date of commencement. This would be calculated as the number of years between *dateofbirth$_i$* and *commencementdate$_i$*.
- *exitage$_i$*, the individual's age at the end of the observation period. This would be calculated as the number of years between *dateofbirth$_i$* and *enddate$_i$*.
- *ageatmodelstartdate$_i$*, the individual's age at the model start date. This would be calculated as the number of years between *dateofbirth$_i$* and *modelstartdate*.
- *ageatmodelenddate$_i$*, the individual's age at the model end date. This would be calculated as the number of years between *dateofbirth$_i$* and *modelenddate*.

When calculating the number of years between two dates, we count the exact number of days and divide by the average number of days per year (365.242 to allow for leap years). The resulting age would therefore be a real number, rather than an integer. For the purposes of modelling we can then calculate the following:

(i) $x_i = \max(entryage_i, modelminage, ageatmodelstartdate_i)$, that is, the age when modelling can commence for individual i after considering the model minimum age and modelling period

(ii) $x_i + t_i = \min(exitage_i, modelmaxage, ageatmodelenddate_i)$, that is, the first age when modelling has to stop after considering the model maximum age and the modelling period

(iii) If $x_t + t_i \leq x_i$ then the life does not contribute to the model

(iv) $d_i = 0$ if the life was alive at *enddate$_i$* or if $x_i + t_i < exitage_i$; otherwise $d_i = 1$.

Although the data-preparation steps above sound complicated, they lend themselves particularly well to the column-wise calculations in a simple spreadsheet. Table 2.7 shows some examples. The formulae in steps (i) and (ii) above are extensible, too. For example, imagine that an administration system contains annuities transferred from another system in batches over time such that each annuitant has a transfer date, *transferdate$_i$*. As is common under such circumstances, records for deceased cases are not transferred, so modelling for an individual cannot begin until after the transfer date. In this example we would calculate *ageattransferdate$_i$* as the number of years between *dateofbirth$_i$* and *transferdate$_i$* and change the calculation in step (i) to:

$$x_i = \max(\textit{entryage}_i, \textit{modelminage}, \textit{ageatmodelstartdate}_i, \textit{ageattransferdate}_i).$$

The data-preparation steps in (i) to (iv) above are for models based on time lived, such as those discussed in Chapter 4. Moreover, the resulting data for the ith individual describe their entire life history during the period of the investigation, usually a period of several years. This is our preferred approach, but an alternative, which we will discuss in later chapters, has been to divide each individual's life history into separate years of age and calculate mortality ratios (based on time lived) at single years of age. Then the data preparation proceeds as above, but with the added complication that each year of age must be treated as if it fell into a separate investigation.

The data-preparation steps for calculating mortality ratios based on the number of lives bring yet more complications, as they require consideration not just of complete years of time lived but also of *potential* complete years. This would involve discarding records which did not have the potential to complete a full year, thus leading to information loss. However, since such models offer us less flexibility than survival models based on time lived, we do not concern ourselves with the extra data-preparation steps for them.

2.9.2 Revaluation of Pensions to Deceased Cases

In pension schemes in the UK it is common for annual pension increases to be granted. This poses a potential bias problem if pension amount is to be used in modelling. Consider two pensioners with the pension amount of 100 per month at the start of a five-year investigation period. If the first pensioner survives to the end of the five-year period, and if the annual rate of pension increases is 2%, then her pension amount in the data extract will be 110.41 ($=100 \times 1.02^5$). However, if the second pensioner dies before the first pension increase, his pension amount in the data extract will be 100. Naturally, the data record for the deceased pensioner will show the pension payable at the time of death, not what the pension would have been had he survived. This discrepancy is important, as it can lead to bias when modelling mortality by pension size.

One solution is to revalue the pension amounts of deceased cases to the end of the observation period using the actual scheme increases. However, this information is not always available, or is not simple to use; indeed, in UK pension schemes, typically different parts of the total pension receive different

Table 2.7 *Examples of data-preparation calculations. The model start age is 50 and the model end age is 105. The model start date is 2000-01-01 and the model end date is 2004-12-31.*

Date of birth	Commencement date	End date	Status at end date	Age at:				x_i	$x_i + t_i$	d_i	Comments
				(i) Comm. date	(ii) End date	(iii) Model start date	(iv) Model end date				
1968-03-14	1999-12-01	2005-02-01	Alive	31.72	36.89	31.80	36.80	50.00	36.80	0	Excluded as no time lived to contribute to model ($x_i < x_i + t_i$)
1938-03-14	1998-03-14	2005-02-01	Alive	60.00	66.89	61.80	66.80	61.80	66.80	0	5.00 years of time lived
1908-03-14	1973-05-21	2002-08-29	Dead	65.19	94.46	91.80	96.80	91.80	94.46	1	2.66 years of time lived
1908-03-14	1973-05-21	2005-10-02	Dead	65.19	97.55	91.80	96.80	91.80	96.80	0	5.00 years of time lived; $d_i = 0$ as $x_i + t_i <$ exitage$_i$

increases. Often a simple *ad hoc* adjustment will suffice. In the Case Study used in this book we increased pension amounts from the date of death to the end of the observation period by 2.5% per annum.

2.10 Exploratory Data Plots

On the assumption that the data have passed the validation and consistency checks, further useful checks come from plotting the data. The first thing to establish is the time interval over which mortality modelling can take place. Figure 2.6 shows an annuity portfolio in the UK which has grown steadily in the number of lives receiving an annuity. The time lived dips in the final year because the data extract was taken in July and so each life can contribute at most half a year of time lived. Because of delays in the reporting of deaths, it is advisable to discard the most recent three or six months (say) of time lived and deaths data in order for the analysis not to be biased; for this portfolio it would mean setting the end date for modelling at 2006-12-31 (*modelenddate* in Section 2.9.1). The bottom panel of Figure 2.6 shows an interesting discontinuity in the number of deaths between 1997 and 1998. Since this discontinuity is not mirrored in either the number of lives or deaths, it would seem that a large number of deaths have been archived in late 1997 or early 1998. The fact that there are still deaths prior to this is most likely due to late-reported deaths being processed after the date of archival. Since deaths prior to 1998 are incomplete, we would only commence modelling from 1998-01-01 (*modelstartdate* in Section 2.9.1). Since the time interval (1998-01-01, 2006-12-31) includes equal numbers of seasons (see Section 2.2.2) we could commence mortality modelling; if need be we could have shortened or lengthened the period discarded for late-reported deaths, but one should normally discard the experience of the most recent three or six months prior to the extract date.

After selecting a time interval for modelling, the next thing to do is check the quality of the mortality data. One of the simplest checks is to plot the crude mortality ratios on a logarithmic scale, as in Figure 2.7. The left-hand panel shows a typical pattern of increasing mortality by age; in contrast, the right-hand panel suggests that the mortality data have been corrupted.

Another useful check for pension schemes and annuity portfolios is to plot the empirical survivor function. For a modern population under normal circumstances, we would expect a clear difference in the survival rates for males and females, as in the left-hand panel in Figure 2.8. However, in some administration systems the same policy record is used for paying benefits to a surviving spouse after death of the first life, and this can lead to corruption in the mor-

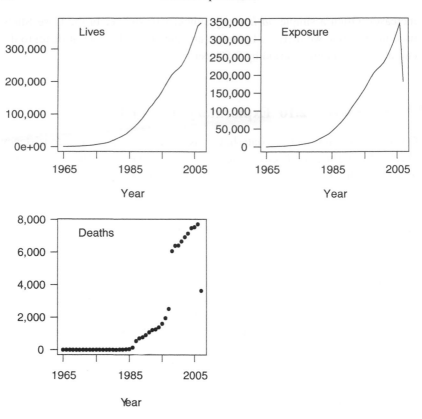

Figure 2.6 Lives (top left), time lived (top right) and deaths (bottom left) in UK annuity portfolio. Source: Richards (2008).

tality data by gender. This manifests itself in a confused picture for gender differentials, as in the right-hand panel of Figure 2.8, where the survival rates for males and females are essentially the same between ages 60 and 80. This sort of data problem cannot be picked up by a traditional actuarial analysis comparing mortality to a published table.

However, often an analyst might only be presented with crude aggregate mortality rates for five-year age bands. This is a nuisance, because such summarisation loses important details in the data. What can you do when you don't have the detailed individual information to build a full survival curve? One option is to use the crude five-year mortality rates to compute approximate survival curves for males and females and compare the survival differential with some other benchmarks. An example of this is shown in Table 2.8.

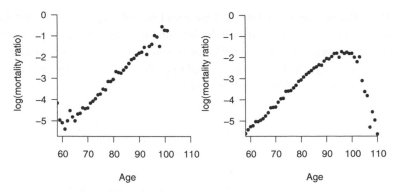

Figure 2.7 log mortality ratio as data-quality check. The left-hand panel shows the data for the Case Study, while the right-hand panel shows the data for a different UK annuity portfolio. The right-hand panel suggests the data have been badly corrupted and cannot immediately be used for mortality modelling.

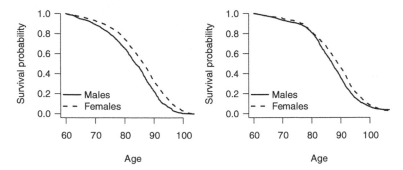

Figure 2.8 The empirical survival function as a data-quality check; Kaplan–Meier survival function (see Section 8.3) for a pension scheme (left) and annuity portfolio (right). The right-hand panel suggests that the data have been corrupted with respect to gender.

Table 2.8 *Difference in male–female survival rates from age 60 for various mortality tables and portfolios (female survival rate minus male survival rate). Source: own calculations using lives-weighted mortality.*

Survival from 60 to age	SAPS table S2PL	Interim life tables 2009–2011	Bulk-annuity portfolio A	Bulk-annuity portfolio B
70	4.3%	4.2%	3.0%	4.8%
75	7.9%	7.5%	5.2%	5.8%
80	11.9%	11.0%	8.4%	4.4%
85	14.5%	13.3%	11.5%	8.9%

The different groups in Table 2.8 have widely differing levels of mortality, from the population mortality of the interim life tables to the SAPS (self-administered pension scheme) table for private pensioners. The calculations also apply to slightly different periods of time. Nevertheless, there is a degree of consistency in the differential survival rates between males and females for the first three columns. We can see that over the 20-year range from age 60 to age 80 there should be a differential of between 8% and 12%. This makes the differential of 4.4% for Portfolio B look rather odd. Furthermore, the differential widens steadily with age for the first three columns, whereas it does not for Portfolio B. This makes us rather suspicious about the data for Portfolio B, and it raises questions about any mortality basis derived from it.

As with the Kaplan–Meier results (see Chapter 8) in Figure 2.8, this data problem could not be detected with simple comparison of the mortality experience against a published table, such as described in Section 8.2. However, here the published tables have actually proved useful *indirectly*. By (i) calculating the survival rates under the published tables and (ii) comparing the excess female survival rate to that of the portfolio in question, we can see that there is something wrong with the data for Portfolio B.

3

The Basic Mathematical Model

3.1 Introduction

Chapter 1 introduced intuitive estimates of the proportion of people of a given age (and perhaps other characteristics), alive at a given time, that should die in the following period of time. The estimates were of the form:

$$\frac{\text{Number of deaths observed}}{\text{Number of person-years of life observed}}. \tag{3.1}$$

In Chapter 1 these were called "mortality ratios". Another name commonly seen is "occurrence-exposure rate". Our ultimate aim will be to use these estimates in actuarial calculations involving financial contracts, principally life insurance and pensions contracts. We therefore wish to know how "good" these estimates are of the "true" quantitities they purport to represent.

A further point concerns the progression of these estimates with age. It is common in actuarial practice to use quantities that are functions of age x, and to assume that these functions are reasonably smooth. If we calculate mortality ratios from equation (3.1) at different ages, for example:

$$\frac{\text{Number of deaths observed between ages } x \text{ and } x + 1}{\text{Number of person-years of life observed between ages } x \text{ and } x + 1} \tag{3.2}$$

for integer ages $x = 0, 1, \ldots$, they are likely to progress irregularly because of sampling variation. To be used in practice, we must somehow smooth them.

Both of these requirements – the "quality" of the raw estimates and the possible need to smooth them – are met by specifying a probabilistic model of human mortality. The model then gives us a plausible representation of the mechanism that is generating the data, leading to the numerator and denominator of equation (3.1). The probabilistic model then performs three essential functions:

- It tells us the statistical properties that the data should possess, so we can assess the "quality" of the raw mortality ratios in terms of their sampling distributions.
- It tells us exactly what quantities the mortality ratios are estimating.
- If we want the quantities estimated by the raw mortality ratios to be smooth functions of age, then the model may tell us how to smooth the raw mortality ratios in a way that is in some sense optimal.

We begin, therefore, with the following basic definition.

Definition 3.1   Let (x) represent a person who is currently alive and age exactly x, where $x \geq 0$. Then define T_x to be a random variable, with continuous distribution on $[0, \infty)$, representing the remaining future lifetime of (x).

We will assume $P[T_x < t] = P[T_x \leq t]$ for all $t > 0$ and there are no probability "masses" concentrated on any single age (technically, the distribution of T_x is assumed to be absolutely continuous). This accords with most observation.

Thus, we assume that (x) will die at some age $x + T_x$, which is unknown but can be modelled as a suitable random variable. Putting this to practical use will depend on being able to say something about the distribution of T_x. That is the subject of much of this book. Our subject goes under several names: actuaries used to call it "analysis of mortality", statisticians call it "survival analysis", engineers call it "reliability analysis", others such as economists call it "event history analysis". A very recent application in the financial markets has been modelling the risk that a bond will default. Our preferred name is "survival modelling".

Much of the subject's early history can be traced to life insurance applications. It so happens that the analysis of mortality, in healthy populations, does not depend much on the methods employed. For nearly 200 years actuaries were able to use simple, intuitive approaches that were robust and good enough in practice. In 1955, Haycocks and Perks could introduce a standard textbook on the subject in the following terms:

Our subject is thus a severely practical one, and the methods used are such as are sufficient for the practical purposes to be served. Elaborate theoretical development would be inappropriate for our purpose; utilitarianism is the keynote and approximation pervades the whole subject. The modern developments of mathematical statistics have made hardly any impact in this field . . . (Haycocks and Perks, 1955)

While life insurance is a very old activity, insurers now offer many more complex products. For example, term assurance premiums allowing for smoking habits, or postcode-driven pricing for individual and bulk annuities, are now the norm (Richards and Jones, 2004). Other contracts may cover events not so clear-cut as death. Modern insurance contracts may cover time spent off work because of illness (disability insurance), or the onset of a serious disease (critical illness insurance), or the need for nursing care in old age (long-term care insurance), or many other contingencies. These features bring into play many more risk factors than were considered relevant to life insurance management in 1955. Their sound financial management needs new tools, which do, in fact, come from the modern developments of mathematical statistics (mostly developed since 1955, so Haycocks and Perks were not being unduly harsh at the time).

The main topic of Part One is the question of how to infer the distribution of T_x from suitable data. We will then be able to extend and generalise inference in many ways in Parts Two and Three. This is the source of the tools needed for the management of more complex long-term insurance products.

3.2 Random Future Lifetimes

The definition of the random future lifetime T_x of (x) requires some thought if it is to be useful, arising from the fact that we are interested in the life or death of a single person, as long as a financial contract is in force. That person's age is continually changing – by definition, they grow older at a rate of one year per annum.

- A year ago, (x) was exact age $x - 1$, and we were interested in T_{x-1} and its distribution.
- In a year's time, if (x) is then alive and our insurance contract is still in force, (x) will be exact age $x + 1$, and we will be interested in T_{x+1} and its distribution.

In fact we see that we have defined not just a single random variable T_x, but a family of random variables $\{T_x\}_{x \geq 0}$.

Definition 3.2   Define $F_x(t) = P[T_x \leq t]$ to be the distribution function of T_x, and define $S_x(t) = 1 - F_x(t) = P[T_x > t]$ to be the *survival function* of T_x.

If this family of random variables is to represent the lifetime of the same person as they progress from birth to death, it is intuitively clear that they must be related to each other in some way. We therefore make the following assumption.

Assumption 3.3   For all $x \geq 0$ and $t \geq 0$ the following holds:

$$F_x(t) = P[T_x \leq t] = P[T_0 \leq x + t \mid T_0 > x]. \tag{3.3}$$

In words, if we consider a person just born, and ask what will be their future lifetime beyond age x, conditional on their surviving to age x, it is the same as T_x. We call this the *consistency condition*.

International actuarial notation includes compact symbols for these distribution and survival functions, as follows:

Definition 3.4

$$\text{Distribution function}: \ _tq_x = F_x(t) \tag{3.4}$$

$$\text{Survival function}: \ _tp_x = S_x(t). \tag{3.5}$$

We will use either notation freely. By convention, if the time period t is one year it may be omitted from the actuarial symbols, so $_1p_x$ can be written p_x and $_1q_x$ can be written q_x. The probability q_x is often called the annual rate of mortality, an old but perhaps unfortunate terminology since "rate" should really be reserved for use in connection with continuous change. In modern parlance q_x is a probability, not a rate.

An important relationship is the following *multiplicative property* of survival probabilities:

$$
\begin{aligned}
S_x(t + s) &= P[T > x + t + s \mid T > x] \\
&= P[T > x + t + s \mid T > x + t]P[T > x + t \mid T > x] \\
&= S_{x+t}(s)S_x(t). \tag{3.6}
\end{aligned}
$$

That is, the probability of surviving for $s + t$ years is equal to the probability of surviving for t years and then for a further s years from age $x + t$.

With these functions, we can answer most of the questions about probabilities that are needed for simple insurance contracts. For example, the probability that a person now age 40 will die between ages 50 and 60 is:

$$S_{40}(10)F_{50}(10) = S_{40}(10)(1 - S_{50}(10)) = S_{40}(10) - S_{40}(20) \tag{3.7}$$

or, in the equivalent actuarial notation:

$$_{10}p_{40}\,_{10}q_{50} = {}_{10}p_{40}(1 - {}_{10}p_{50}) = {}_{10}p_{40} - {}_{20}p_{40}. \tag{3.8}$$

3.3 The Life Table

A short calculation using equation (3.6) leads to:

$$S_x(t) = \frac{S_0(x+t)}{S_0(x)}. \tag{3.9}$$

So, if we know $S_0(t)$, we can compute any $S_x(t)$. In pre-computer days this meant that a two-dimensional table could be replaced by a one-dimensional table and a simple calculation. Traditionally, $S_0(t)$ would be represented in the form of a *life table*. Choose a large number, denoted l_0, to be the *radix* of the life table; for example $l_0 = 100,000$. The radix represents a large number of identical lives aged 0. Then at any age $x > 0$, define l_x as follows:

Definition 3.5

$$l_x = l_0 S_0(x). \tag{3.10}$$

If we suppose that the future lifetimes of all l_0 lives are independent, identically distributed (i.i.d.) random variables, we see that l_x represents the expected value of the number of the lives who will be alive at age x. It is usual to tabulate l_x at integer ages x; the function l_x so tabulated is called a life table. It allows any relevant probability involving integer ages and time periods to be calculated; for example for non-negative integers x and t:

$$S_x(t) = \frac{S_0(x+t)}{S_0(x)} = \frac{l_0 S_0(x+t)}{l_0 S_0(x)} = \frac{l_{x+t}}{l_x}. \tag{3.11}$$

Table 3.1 shows the lowest and highest values of l_x from the English Life Table No.16 (Males), which has a radix $l_0 = 100,000$ and under which it is assumed that no-one lives past age 110. It represents the mortality of the male population of England and Wales in the calendar years 2000–2002.

The reasons for defining l_x in this way are historical. The life table is much older than formal probabilistic models. It was often presented as a tabulation, at integer ages x, of the *number* who would be alive at age x in a cohort of l_0 births. With a large enough radix, l_x could be rounded to whole numbers, as in Table 3.1, avoiding difficult ideas such as fractions of people being alive. This interpretation is sometimes called the deterministic model of mortality. Of course, the mathematicians such as Halley, Euler and Maclaurin who used

Table 3.1 *Extract from the English Life Table No.16 (Males) which has a radix $l_0 = 100,000$ and it is assumed that $_{110}p_0 = 0$.*

Age x	l_x
0	100,000
1	99,402
2	99,358
3	99,333
4	99,316
...	...
...	...
105	25
106	12
107	5
108	2
109	1
110	0

life tables knew quite well the nature of l_x, but in times when knowledge of mathematical probability was scarce, it was a convenient device.

Sometimes actuarial interest is confined to a subset of the whole lifespan. For example, actuaries pricing endowment assurances may have no need for the life table below about age 16, while actuaries pricing retirement annuities may have no need for the life table below about age 50 (and indeed, no data). It is common in such cases to define the radix of the life table at the lowest age of interest. For example, a life table for use in pricing retirement annuities might have radix $l_{50} = 100,000$, and the tabulation of l_x for $x \geq 50$ then allows us to compute $S_{50}(t)$ for integers $t \geq 0$, and hence $S_x(t)$ for $x \geq 50$ and $t \geq 0$.

3.4 The Hazard Rate, or Force of Mortality

From this point on, we will assume that we are interested in some person alive at age $x > 0$, as is usual in actuarial practice, although everything we do will be perfectly valid if $x = 0$. This will keep the context and the notation closer to those familiar to actuaries. Thus, the objects of interest to us are the random variables $\{T_x\}_{x \geq 0}$ and their distribution functions $F_x(t)$.

The random variable T_x is assumed to have a continuous distribution with no positive mass of probability at any single age. Many problems will require us to ask the question: what is the probability of death between ages $x + t$ and $x + t + dt$, where dt is small? The distribution function $F_x(t)$ is of no help directly, since our assumption that there is no probability mass at any single age means that, for all $t \geq 0$:

$$\lim_{dt \to 0^+} P[T_x \leq t + dt \mid T_x > t] = 0 \qquad (3.12)$$

(recall that all distribution functions are right-continuous). By analogy with the familiar derivative of ordinary calculus, we define the *hazard rate* or *force of mortality* at age $x + t$, denoted $\mu_{x,t}$, as follows.

Definition 3.6 The hazard rate or force of mortality at age $x + t$ associated with the random lifetime T_x is:

$$\mu_{x,t} = \lim_{dt \to 0^+} \frac{P[T_x \leq t + dt \mid T_x > t]}{dt}, \qquad (3.13)$$

assuming the limit exists. (Do not confuse this with the notation $\mu_{x,y}$ used in Part Two to represent a hazard rate at age x in calendar year y.) The hazard rate has the same time unit as we use to measure age. Usually it is a rate per annum (just as the speed of a car – the rate of change of distance in a defined direction – might be expressed as a rate per hour). The term "hazard rate" is usual among statisticians, who commonly denote it by λ; the (much older) term "force of mortality", and the use of μ to denote it, is usual among actuaries and demographers. Probabilists would often call the hazard rate a *transition intensity*, and engineers might call it a *failure rate*. We may use any of these terms interchangeably. Although we adopt the actuarial μ notation, our preference is for the name "hazard rate".

Equation (3.13) defines $\mu_{x,t}$ in terms of the random lifetime T_x. Suppose y is another non-negative age, and that $s > 0$, such that $y + s = x + t$. We can define a hazard rate at age $y + s$ in terms of the random lifetime T_y, denoted here by $\mu_{y,s}$:

$$\mu_{y,s} = \lim_{ds \to 0^+} \frac{P[T_y \leq s + ds \mid T_y > s]}{ds}. \qquad (3.14)$$

It is easily shown, using the consistency condition, that $\mu_{y,s} = \mu_{x,t}$, so they are consistent and we can write μ_{y+s} or μ_{x+t}. For example, the pairs $x = 30$, $t = 20$ and $y = 40$, $s = 10$ both define identical hazard rates at age 50, $\mu_{30,20} = \mu_{40,10}$, so we can write μ_{50} without any ambiguity arising.

The hazard rate can be interpreted through the approximate relationship $_{dt}q_{x+t} = F_{x+t}(dt) \approx \mu_{x+t}dt$, for small dt. It will help later on if we make this more precise. A function $g(t)$ is said to be $o(t)$ ("little-oh-of-t") if:

$$\lim_{t \to 0} \frac{g(t)}{t} = 0 \tag{3.15}$$

or, in other words, if $g(t)$ tends to zero sufficiently faster than t itself. It is easy to see that the sum of a finite number of $o(t)$ functions is again $o(t)$, as is the product of any $o(t)$ function and a bounded function. Then we can show from the definition of μ_{x+t} that:

$$_{dt}q_{x+t} = F_{x+t}(dt) = \mu_{x+t}dt + g(dt), \tag{3.16}$$

where $g(t)$ is some function which is $o(t)$. We usually just write the right-hand side as $\mu_{x+t}dt + o(dt)$ since the precise form of $g(t)$ is of no concern.

We can now find the density function of T_x, denoted $f_x(t)$:

$$
\begin{aligned}
f_x(t) = \frac{\partial}{\partial t} F_x(t) &= \lim_{dt \to 0^+} \frac{F_x(t + dt) - F_x(t)}{dt} \\
&= \lim_{dt \to 0^+} \frac{P[T_x \le t + dt] - P[T_x \le t]}{dt} \\
&= \lim_{dt \to 0^+} \frac{P[T_x \le t] + P[T_x > t]P[T_x \le t + dt \mid T_x > t] - P[T_x \le t]}{dt} \\
&= P[T_x > t] \lim_{dt \to 0^+} \frac{P[T_x \le t + dt \mid T_x > t]}{dt} \tag{3.17} \\
&= S_x(t)\mu_{x+t} = {}_tp_x\mu_{x+t}.
\end{aligned}
$$

From the fact that $F_x(t) = \int_o^t f_x(s)\,ds$, we obtain the important relationship:

$$_tq_x = \int_0^t {}_sp_x\mu_{x+s}\,ds, \tag{3.18}$$

or, in words, the probability that someone age x will die before age $x + t$ is the probability that they live to an intermediate age $x + s$ and then die, summed (integrated) over all intermediate ages. Since $F_x(t) = 1 - {}_tp_x$, the differential equation in (3.17) can be rewritten as:

$$\frac{\partial}{\partial t} {}_tp_x = -{}_tp_x\mu_{x+t}. \tag{3.19}$$

The density function ${}_tp_x\mu_{x+t}$ was known to earlier actuarial researchers as the *curve of deaths*; see for example Beard (1959).

Equation (3.17) (and equation (3.19)) is written using a partial derivative, recognising the fact that the function being differentiated is a function of x and

t. However, since no derivatives in respect of x appear, it is in fact an ordinary differential equation (ODE). We can solve it with boundary condition $_0p_x = 1$ as follows, where c is a constant of integration:

$$\frac{\partial}{\partial t} {}_t p_x = -{}_t p_x \mu_{x+t} \tag{3.20}$$

$$\Rightarrow \frac{\partial}{\partial t} \log {}_t p_x = -\mu_{x+t}$$

$$\Rightarrow \int_0^t \left(\frac{\partial}{\partial s} \log {}_s p_x \right) ds = - \int_0^t \mu_{x+s} ds + c$$

$$\Rightarrow \log {}_t p_x = - \int_0^t \mu_{x+s} ds + c.$$

The boundary condition implies $c = 0$, so we have the extremely important result:

$$S_x(t) = {}_t p_x = \exp\left(- \int_0^t \mu_{x+s} ds \right). \tag{3.21}$$

If $t = 1$, and we approximate $\mu_{x+s} \approx \mu_{x+1/2}$ for $0 \leq s < 1$, we have the important approximation:

$$q_x \approx 1 - e^{-\mu_{x+1/2}}. \tag{3.22}$$

The *integrated hazard* $\int_0^t \mu_{x+s} ds$ that appears in the right-hand side of equation (3.21) is important in its own right, and we denote it by $\Lambda_{x,t}$:

Definition 3.7

$$\Lambda_{x,t} = \int_0^t \mu_{x+s} ds. \tag{3.23}$$

3.5 An Alternative Formulation

Equation (3.19) allows us to calculate μ_{x+t}, given the probability function $_t p_x$. Equation (3.21) allows us to do the reverse. If μ_{x+t} or $_t p_x$ are particularly simple functions, we may be able to do so analytically, but sometimes we have to use numerical approximations. Given modern personal computers, this is straightforward, for example using the R function `integrate()`. Perhaps more im-

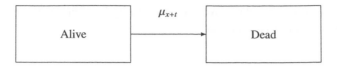

Figure 3.1 A two-state model of mortality.

portantly, equation (3.21) offers an alternative formulation of the model that has many advantages when we consider more complex insurance contracts.

Suppose we have an insurance contract taken out by (x). Our model above was specified in terms of T_x and its distribution $F_x(t)$, from which we derived the hazard rate μ_{x+t}. Alternatively, we could formulate a model in which (x) at any future time t occupies one of two states, alive or dead, as in Figure 3.1. It is given that (x) starts in the alive state and cannot return to the alive state from the dead state. Transitions from the alive state to the dead state are governed by the hazard rate μ_{x+t}, which we therefore take to be the fundamental model quantity instead of $F_x(t)$. From equation (3.21) we can calculate $F_x(t)$ for $t \geq 0$ and define the random future lifetime T_x to be the continuous non-negative random variable with that distribution function.

This is the simplest example of a *Markov multiple-state model*. In this context, the hazard rate would usually be called a *transition intensity*. More complex lifetimes, for example involving periods of good health and sickness, can be formulated in term of states and transitions between states, governed by transition intensities in an analogous fashion. This turns out to be the easiest way, in many respects, to extend the simple model of a random future lifetime to more complicated life histories. This will be the topic of Chapter 14.

3.6 The Central Rate of Mortality

The life table l_x is sometimes interpreted as a deterministic model of a stable population, in which lives are born continuously at rate l_0 per year, and die so that the "number" alive at age x is always l_x. Then $l_0 \, q_x$ is the "number" of deaths between ages x and $x + 1$, while $l_0 \, {}_t p_x \, dt$ is the total "number" of years lived by all members of the population between ages $x + t$ and $x + t + dt$. This leads to the following definition.

Definition 3.8   The *central rate of mortality* at age x, denoted m_x, is:

$$m_x = \frac{q_x}{\int_0^1 {}_tp_x\, dt} = \frac{l_0\, q_x}{l_0 \int_0^1 {}_tp_x\, dt}. \tag{3.24}$$

The central rate of mortality m_x can be interpreted as a death rate per person-year of life. This idea is useful in some applications, such as population projection. From equation (3.18) we have:

$$m_x = \frac{\int_0^1 {}_tp_x\, \mu_{x+t}\, dt}{\int_0^1 {}_tp_x\, dt}, \tag{3.25}$$

so the central rate of mortality can be interpreted as a probability-weighted average of the hazard rate over the year of age.

We may note that the central rate of mortality explains the notation used for the mortality ratios defined in Section 1.4.

3.7 Application to Life Insurance and Annuities

Our main interest is in estimation, but it is useful motivation to mention briefly how simply the model may be applied. The chief tool of life insurance mathematics is the expected present value (EPV) of a series of future cash-flows. Given a deterministic force of interest δ per annum, the present value of a cash-flow of 1 at time t in the future is $\exp(-\delta t)$. If the cash-flow is to be paid at a random time such as on the date of death of a person now age x, that is, at time T_x, its present value $\exp(-\delta T_x)$ is also a random variable. The most basic life insurance pricing and reserving values a random future cash-flow as its EPV, which can be justified by invoking the law of large numbers. Thus, the EPV of the random cash-flow contingent on death is denoted by $\overline{A}_x$ and is by, equation (3.17):

$$\overline{A}_x = \int_0^\infty e^{-\delta t}\, {}_tp_x \mu_{x+t}\, dt. \tag{3.26}$$

EPVs of annuities can be as easily written down. Life insurance mathematics based on random lifetimes is described fully in Bowers et al. (1986), Gerber (1990) and Dickson et al. (2013).

4

Statistical Inference with Mortality Data

4.1 Introduction

In statistics, it is routine to model some observable quantity as a random variable X, to obtain a sample $x_1, x_2, \ldots, x_n$ of observed values of X and then to draw conclusions about the distribution of X. This is statistical inference.

If the random variable in question is T_x, obtaining a sample of observed values means fixing a population of persons aged x (homogeneous in respect of any important characteristics) and observing them until they die. If the age x is relatively low, say below age 65, it would take anything from 50 to 100 years to complete the necessary observations. This is clearly impractical.

The available observations (data) almost always take the following form:

- We observe n persons in total.
- The ith individual ($1 \leq i \leq n$) first comes under observation at age x_i. For example, they take out life insurance at age x_i or become a pensioner at age x_i. We know nothing about their previous life history, except that they were alive at age x_i.
- We observe the ith individual for t_i years. By "observing" we mean knowing for sure that the individual is alive. Time spent while under observation may be referred to as *exposure*. Observation ends after time t_i years because either: (a) the individual dies at age $x_i + t_i$; or (b) although the individual is still alive at age $x_i + t_i$, we stop observing them at that age.

Additional information may also be available, for example data relating to health obtained as part of medical underwriting. Thus, instead of observing n individuals and obtaining a sample of n values of the random variable T_x, we are most likely to observe n individuals, of whom a small proportion will actually die while we are able to observe them, and most will still be alive when we cease to observe them.

56

The fact that the ith individual is observed from age $x_i > 0$ rather than from birth means that observation of that lifetime is *left-truncated*. This observation cannot contribute anything to inference of human mortality at ages below x_i (equivalently, the distribution function $F_0(t)$ for $t \leq x_i$).

If observation of the ith individual ends at age $x_i + t_i$ when that individual is still alive, then observation of that lifetime is *right-censored*. We know that death must happen at some age greater than $x_i + t_i$ but we are unable to observe it.

Left-truncation and right-censoring are the key features of the mortality data that actuaries must analyse. (Left-truncation is a bigger issue for actuaries than for other users of survival models such as medical statisticians.) They matter because they imply that any probabilistic model capable of generating the data we actually observe cannot just be the family of random future lifetimes $\{T_x\}_{x \geq 0}$. The model must be expanded to allow for:

- a mechanism that accounts for individuals entering observation; and

- a mechanism that explains the right-censoring of observation of a person's lifetime.

We may call this "the full model". Although we may be interested only in that part of the full model that describes mortality, namely the distributions of the family $\{T_x\}_{x \geq 0}$ (or, given the consistency condition, the distribution of T_0), we cannot escape the fact that we may not sample values of T_x at will, but only values of T_x in the presence of left-truncation and right-censoring. Strictly, inference should proceed on the basis of an expanded probabilistic model capable of generating *all* the features of what we can actually observe. Figure 4.1 illustrates how left-truncation and right-censoring limit us to observing the density function of T_0 over a restricted age range, for a given individual who enters observation at age 64 and is last observed to be alive at age 89.

Intuitively, the task of inference about the lifetimes alone may be simplified if those parts of the full model accounting for left-truncation and right-censoring are, in some way to be made precise later, "independent" of the lifetimes $\{T_x\}_{x \geq 0}$. Then we may be able to devise methods of estimating the distributions of the family $\{T_x\}_{x \geq 0}$ without the need to estimate any properties of the left-truncation or right-censoring mechanisms. We next consider when it may be reasonable to assume such "independence", first in the case of right-censoring.

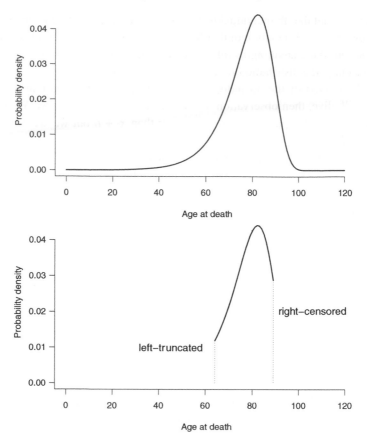

Figure 4.1 A stylised example of the full probability density function of a random lifetime T_0 from birth (top) and that part of the density function observable in respect of an individual observed from age 64 (left-truncated) to age 89 (right-censored) (bottom).

4.2 Right-Censoring

At its simplest, we have a sample of n newborn people (so we can ignore left-truncation), and in respect of the ith person we observe either T_0^i (their lifetime) or C^i (the age at which observation is censored). Exactly how we should handle the censoring depends on how it comes about. Some of the commonest assumptions are these (they are not all mutually exclusive):

- *Type I censoring.* If the censoring times are known in advance then the mechanism is called Type I censoring. An example might be a medical study in which subjects are followed up for at most five years; in actuarial studies a more common example would the ending of an investigation on a fixed calendar date, or the termination date of an insurance contract.

- *Type II censoring.* If observation is continued until a predetermined number of deaths have been observed, then Type II censoring is said to be present. This can simplify the analysis, because then the number of deaths is non-random.

- *Random censoring.* If censoring is random, then the time C^i (say) at which observation of the ith lifetime is censored is modelled as a random variable. The observation will be censored if $C^i < T_0^i$. The case in which the censoring mechanism is a second decrement of interest (for example, lapsing a life insurance policy) gives rise to multiple-decrement models, also known as competing risks models; see Chapter 16.

- *Non-informative censoring.* Censoring is non-informative if it gives no information about the lifetimes T_0^i. If we assume that T_0^i and C^i are independent random variables, then censoring is non-informative. Informative censoring is difficult to analyse.

Most actuarial analysis of survival data proceeds on the assumption that right-censoring is non-informative. So, for example, if the full model accounting for all that may be observed from birth is assumed to be a bivariate random variable (T_0, C_0), where T_0 is the random time until death and C_0 is the random time until censoring, with T_0 and C_0 independent, we can estimate the distribution of T_0 alone without having to estimate any of the properties of C_0. If T_0 and C_0 could not be assumed to be independent, we might have to estimate the joint distribution of the pair (T_0, C_0). Clearly this is much harder.

Right-censoring is not the only kind of censoring. Data are left-censored if the censoring mechanism prevents us from knowing when entry into the state which we wish to observe took place. Many medical studies involve survival since onset of a disease, but discovery of that disease at a medical examination tells us only that the onset fell in the period since the patient was last examined; the time since onset has been left-censored. Data are interval-censored if we know only that the event of interest fell within some interval of time. For example, we might know only the calendar year of death. Censoring might also depend on the results of the observations to date; for example, if strong enough evidence accumulates during the course of a medical experiment, the investigation might be ended prematurely, so that the better treatment can be extended to all the subjects under study, or the inferior treatment withdrawn. Andersen et al. (1993) give a comprehensive account of censoring schemes.

How censoring arises depends on how we collect the data (or quite often how someone else has already collected the data). We call this an *observational plan.* As with any study that will generate data for statistical analysis, it is best to consider what probabilistic models and estimation methods are appropriate before devising an observational plan and collecting the data. In actuarial studies this rarely happens, and data are usually obtained from files maintained for business rather than for statistical reasons.

4.3 Left-Truncation

In simple settings, survival data have a natural and well-defined starting point. For human lifetimes, birth is the natural starting point. For a medical trial, the administration of the treatment being tested is the natural starting point. Data are left-truncated if observation begins some time after the natural starting point. For example, a person buying a retirement annuity at age 65 may enter observation at that age, so is not observed from birth. The lifetime of a participant in a medical trial, who is not observed until one month after the treatment was administered, is left-truncated at duration one month. Note that in these examples, the age or duration at which left-truncation takes place may be assumed to be known and non-random.

Left-truncation has two consequences, one concerning the validity of statistical inference, the other concerning methodology:

- By definition, a person who does not survive until they might have come under observation is never observed. If there is a material difference between people who survive to reach observation and those who do not, and this is not allowed for, bias may be introduced.
- If lifetime data are left-truncated, we must then choose methods of inference that can handle such data. All the methods discussed in this book can do so, unless specifically mentioned otherwise.

Suppose we are investigating the mortality experience of a portfolio of term-insurance policies. Consider two persons with term-insurance policies, both age 30. One took out their policy at age 20, the other at age 29, so observation of their lifetimes was left-truncated at ages 20 and 29, respectively. For the purpose of inference, can we assume that at age 30 we may treat their future lifetimes as independent, identically distributed random variables? Any actuary would say not identically distributed, because at the point of left-truncation such persons would just have undergone medical underwriting. The investigation would most likely lead to the adoption of a select life table.

In the example above, an event in a person's life history prior to entry into observation influences the distribution of their future lifetime. In this case, the event (medical underwriting) is known and can be allowed for (construct a select life table). Much more difficult to deal with are unknown events prior to entry that may be influential. For example, if retirement annuities are not underwritten, the actuary may not know the reason for a person retiring and entering observation. It may be impossible to distinguish between a person in good health retiring at age 60 because that is what they always planned to do, and a person retiring at age 60 in poor health, who wanted to work for longer. Studies of annuitants' mortality do indeed show anomalies at ages close to the usual range of retirement ages; see for example Continuous Mortality Investigation (2007) and other earlier reports by the CMI. It has long been conjectured, by the CMI, that undisclosed ill-health retirements are the main cause of these anomalies, but this cannot be confirmed from the available data. Richards et al. (2013) model ill-health retirement mortality.

When an insurance policy is medically underwritten at the point of issue, the insurer learns about a range of risk factors, selected for their proven relevance. In a sense this looks back at the applicant's past life history. Modelling approaches incorporating these risk factors or *covariates* are then a very useful way to mitigate the limitations of left-truncated lifetimes; see Chapter 7.

Left-truncation is just one of many ways in which unobserved heterogeneity may confound a statistical study, but it is particular to actuarial survival analysis. It is often not catered for in standard statistical software packages, so the actuary may have some programming to carry out.

4.4 Choice of Estimation Approaches

We have to abandon the simple idea of observing a complete, uncensored cohort of lives. What alternatives are realistically open to us? A useful visual aid in this context is a *Lexis diagram*, an example of which is shown in Figure 4.2. A Lexis diagram displays age along one axis and calendar time along the other axis. An individual's observed lifetime can then be plotted as a vector of age against calendar time, referred to as a *lifeline*. Observations ending in death or by right-censoring are indicated by lifelines terminating in solid or open circles, respectively. Figure 4.2 shows two such lifelines, one of an individual entering observation at age $x_i = 60.67$ years on 1 May 2005 and leaving observation by right-censoring at exact age $x_i' = 63.67$ years on 1 May 2008, and the other of an individual entering observation at age $x_j = 60.25$ years on 1 July 2006 and leaving observation by death at exact age $x_j' = 62.5$ years on 1 October 2008. Below, we use the first of these individuals to illustrate several alternative approaches to handling exposures.

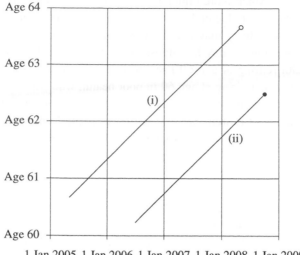

Figure 4.2 A Lexis diagram illustrating: (i) the lifetime of individual i entering observation at age $x_i = 60.67$ years on 1 May 2005 and leaving observation by censoring at exact age $x'_i = 63.67$ years on 1 May 2008; and (ii) the lifetime of individual j entering observation at age $x_j = 60.25$ years on 1 July 2006 and leaving observation by death at exact age $x'_j = 62.5$ years on 1 October 2008.

- The simplest approach is to note that this individual entered observation at age 60.67 and was observed for three years. Therefore, we can develop models that use this data directly. We call this modelling *complete observed lifetimes*. We will develop two such approaches in this book. In Chapter 5, we use maximum likelihood methods to fit parametric models to the hazard rate, based on complete observed lifetimes, and in Chapter 7 we develop this approach to allow additional information about each individual to be modelled.

- The classical approach, which we have already seen in Chapter 1, is to divide the age range into manageable intervals, and to estimate mortality in each age interval separately. Age ranges of one year, five years or ten years are often used. Thus, if an age range of one year is to be used, the censored observation shown uppermost in Figure 4.2 contributes to four separate estimates:

  – it contributes 1/3 years of exposure between ages 60 and 61;

- it contributes 1 year of exposure between ages 61 and 62;
- it contributes 1 year of exposure between ages 62 and 63; and
- it contributes 2/3 years of exposure between ages 63 and 64.

Historically, this approach came naturally because of the central role of the life table in life insurance practice. If the objective was to produce a table of q_x at integer ages x, what could be simpler and more direct than to obtain estimates $\hat{q}_x$ of each q_x at integer ages x? It remains an important, but limiting, mode of actuarial thought today. One purpose of this book is to introduce actuaries to a broader perspective.

We will define our models using complete lifetimes. The word "complete" is, however, qualified by the fact that observation is unavoidably right-censored by the date of the analysis (or the date when records are extracted for analysis). It is then easily seen that analysis of mortality ratios, for single years of age, is just a special case, where right-censoring of survivors takes place at age $x + 1$. Once we have defined an appropriate probabilistic model representing complete lifetimes, adapting it to single years of age is trivial.

- In Chapter 8 we adapt the empirical distribution function of a random variable to allow for left-truncation and right-censoring. The latter is a *non-parametric* approach, and it leads to several possible estimators.

- It is also common to collect data over a relatively short period of calendar time, called the *period of investigation*. One reason is that mortality (or morbidity, or anything else we are likely to study) changes over time. For example, in developed countries during the twentieth century, longevity improved, meaning that lifetimes lengthened dramatically, because of medical and economic advances. Figure 4.3 shows the period expectation of life at birth in Japan from 1947 to 2012. So even if we had observed (say) the cohort of lives born in 1900, it would now tell us almost nothing about what mortality to expect at younger or middle ages in the early twenty-first century. For any but historical applications, estimates of future mortality are usually desired.

The choice of how long a period to use will be a compromise. Too short a period might be unrepresentative if it does not encompass a fair sample of seasonal variation (for example, mild and severe winters). Too long a period might encompass significant systematic changes. Figure 2.2 showed clearly how mortality fluctuates with the seasons in any one calendar year, but also suggested there might be a systematic trend over calendar time.

As a simple example, suppose it has been decided to limit the investigation to the calendar years 2006 and 2007. Then if we analyse complete

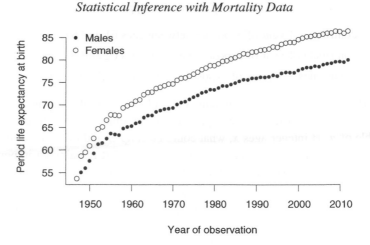

Figure 4.3 Period life expectancies at birth in Japan, 1947–2012; an illustration of lifetimes lengthening in recent times. Source: Human Mortality Database.

observed lifetimes, the uppermost (censored) lifeline in Figure 4.2 enters observation at age 61.33 and is observed for two years. If we divide the observations into single years of age, then:

– it contributes 2/3 years of exposure between ages 61 and 62;
– it contributes 1 year of exposure between ages 62 and 63; and
– it contributes 1/3 years of exposure between ages 63 and 64.

If the data have yet to be collected, a shorter period of investigation gets the results in more quickly. In practice, periods of three or four calendar years are common in actuarial and demographic work. Studies based on data collected during short calendar periods are called *secular* or *cross-sectional* studies.

4.5 A Probabilistic Model for Complete Lifetimes

We now define a simple probabilistic model capable of generating the observed data on complete lifetimes described above – a candidate for the full model.

Definition 4.1 Define the following quantities:

$$x_i = \text{the age at which the } i\text{th individual enters observation} \qquad (4.1)$$

$x_i' =$ the age at which the ith individual leaves observation   (4.2)

$t_i = x_i' - x_i$ (4.3)

= the time for which the ith individual is under observation.

Next, let us define random variables to represent these observations.

Definition 4.2   Define the random variable T_i to be the length of time for which we observe the ith individual after age x_i.

Definition 4.3   Define the random variable D_i to be an indicator random variable, taking the value 1 if the ith individual was observed to die at age $x_i + T_i$, or the value 0 otherwise.

We regard the observed data (d_i, t_i) as being sampled from the distribution of the bivariate random variable (D_i, T_i). This is consistent with the notational convention that random variables are denoted by upper-case letters, and sample values by the corresponding lower-case letters.

It is important to realise that D_i and T_i are not independent; they are the components of a bivariate random variable (D_i, T_i). Their joint distribution depends on the maximum length of time for which we may observe the ith individual's lifetime. For example, suppose we collect data in respect of the calendar years 2008–2011. Suppose the ith individual enters observation at age x_i on 29 October 2010. Then T_i has a maximum value of 1 year and 63 days (1.173 years), and will attain it if and only if the ith individual is alive and under observation on 31 December 2011. In that case, also, $d_i = 0$. Suppose for the moment that no other form of right-censoring is present. Then the only other possible observation arises if the ith individual dies some random time $T_i < 1.173$ years after 29 October 2010. In that case, $t_i < 1.173$ years and $d_i = 1$. Since $t_i < 1.173$ if and only if $d_i = 1$, it is clear that D_i and T_i are dependent. Motivated by this example, we make the following additional definition.

Definition 4.4   Define b_i to be the maximum possible value of T_i, according to whatever constraints are imposed by the manner of collecting the data (the observational plan).

In the example above, $b_i = 1.173$ years.

Together with the age x_i and the maximum time of observation b_i, the bivariate random variable (D_i, T_i) accounts for all possible observations of the ith individual. This is what we mean by a probabilistic model "capable of generating the observed data". Moreover, all the events possible under the model are, in principle, capable of being observed. There is no possibility of the model assigning positive probability to an event that cannot actually happen (unlike the example seen in Section 1.6).

The joint distribution of (D_i, T_i) will be determined by that portion of the density function $f_0(t)$ of the random lifetime T_0 on the age interval x_i to $x_i + b_i$. See the lower graph in Figure 4.1, and consider left-truncation at age x_i and right-censoring at age $x_i + b_i$ (at most). Therefore, if we have a candidate density function $f_0(t)$ of T_0, we can write down the joint probability of each observed (d_i, t_i).

However, we have to recognise and dispose of one potentially awkward feature of this model. We assumed above that the only possible observations of the ith individual were survival to age $x_i + b_i$, or death before age $x_i + b_i$. We excluded right-censoring before age $x_i + b_i$, but this will usually be present in actuarial studies. The probabilistic model described above then does not account completely for all possible observations of the ith individual, because it does not describe how right-censoring may occur before age $x_i + b_i$. Strictly, we ought to define an indicator random variable R_i of the event that observation of the ith individual's lifetime was right-censored before age $x_i + b_i$, and then work with the joint distribution of the multivariate random variable (D_i, R_i, T_i).

For example, one reason for right-censoring of annuitants' lifetimes is that the insurer may commute very small annuities into lump sums to save running costs. If there is an association between mortality and amount of annuity, then D_i and R_i would not be independent.

Working with (D_i, R_i, T_i) is very difficult or impossible in most studies encountered in practice. We almost always assume that any probabilistic model describing right-censoring before age $x_i + b_i$ is "independent" of the probabilistic model defined above, describing mortality and right-censoring only at age $x_i + b_i$. Loosely speaking, this "independence" means that, conditional on (D_i, T_i), it is irrelevant what form any right-censoring takes. This allows us to use just the data (d_i, t_i) without specifying, in the model, whether any right-censoring occurred before age $x_i + b_i$ or at age $x_i + b_i$. In some circumstances, this may be a strong or even implausible assumption, but we can rarely make any headway without it.

We summarise below the probabilistic models we have introduced in this chapter:

- We began in Chapter 3 with a probabilistic model of future lifetimes, the family of random variables $\{T_x\}_{x \geq 0}$, and obtained a complete description in terms of distribution, survival or density functions, or the hazard rate.
- The story would have ended there had we been able actually to observe complete human lifetimes, but this is rarely possible because of left-truncation

and right-censoring. So we had to introduce a probabilistic model for the observations we actually can make, the model (D_i, T_i). Clearly its properties will be derived from those of the more basic model, the random variables $\{T_x\}_{x \geq 0}$. We should not overlook the fact that left-truncation and right-censoring introduce some delicate questions.

4.6 Data for Estimation of Mortality Ratios

Our main focus is on the analysis of complete lifetimes, but estimation of mortality ratios at single years of age is still important, for several reasons:

- It is the approach most familiar to actuaries.
- It may be the only possible approach if only grouped data for defined age intervals are available.
- Mortality ratios are needed for checking the goodness-of-fit of models estimated using complete lifetime data.
- Mortality ratios lead naturally to regression models (see Part Two) which in turn lead naturally to methods of forecasting future longevity.

We will adapt the observations and the corresponding probabilistic model of Section 4.5 to the case that we have grouped data at single years of age. The same approach may be used for other age intervals, for example five or ten years. Our aim will be to estimate the "raw" mortality ratios in equation (3.1).

Consider calculation of the mortality ratio between integer ages x and $x + 1$. We suppose that during a defined calendar period we observe n individuals for at least part of the year of age x to $x + 1$. Consider the ith individual. We have $x_i = x$ if they attain their xth birthday while being observed, and $x < x_i < x + 1$ otherwise. We have $x'_i = x + 1$ if they are alive and under observation at age $x + 1$, and $x < x'_i < x + 1$ otherwise. The time spent under observation is $t_i = x'_i - x_i$ and $0 \leq t_i \leq 1$. We also define an "indicator of death" in respect of the ith individual; if they are observed to die between age x and $x + 1$ (which, if it happens, must be at age x'_i), define $d_i = 1$, otherwise define $d_i = 0$.

From the data above, we can find the total number of deaths observed between ages x and $x + 1$; it is $\sum_{i=1}^{i=n} d_i$. Denote this total by d_x. (The R function `splitExperienceByAge()` described in Appendix H.1 will do this.) We can also calculate the total amount of time spent alive and under observation between ages x and $x + 1$ by all the individuals in the study; it is $\sum_{i=1}^{i=n} t_i$. Denote this total by E_x^c, and call it the *central exposed-to-risk*. Recall the options considered in Section 1.4. We have chosen "time lived" rather than "number of lives". This is simplest because it does not matter whether an individual is observed for the whole year of age or not.

It is important to note that underlying the aggregated quantities d_x and E_x^c are the probabilistic models (D_i, T_i) $(i = 1, 2, \ldots, n)$ generating the data in respect of the n individuals we observe at age x. This remains true even if d_x and E_x^c are the only items of information we have (for example, if we are working with data from the Human Mortality Database). Individual people die, not groups of people, and this is fundamental to the statistical analysis of survival data.

Then the "raw" mortality ratio for the age interval x to $x + 1$, which we denote by $\hat{r}_x$, is defined as:

$$\hat{r}_x = \frac{d_x}{E_x^c}. \tag{4.4}$$

The notation d_x for deaths is self-explanatory. The notation E for the denominator stands for "exposed-to-risk" since this is the conventional actuarial name for the denominator in a "raw" mortality ratio. The superscript c further identifies this quantity as a "central exposed-to-risk", which is the conventional actuarial name for exposure based on "time lived" rather than "number of lives" (Section 1.4). The choice of r is to remind us that we are estimating a mortality ratio, and the "hat" adorning r on the left-hand side of this equation is the usual way of indicating that $\hat{r}_x$ is a statistical estimate of some quantity based on data.

This leads to the following question. The quantity $\hat{r}_x$ is a statistical estimate of something, but what is that something? We shall see in Section 4.8 that under some reasonable assumptions $\hat{r}_x$ may be taken to be an estimate of the hazard rate at some (unspecified) age between x and $x+1$. In actuarial practice, we usually call such a "raw" estimate a *crude* hazard rate. We might therefore rewrite equation (4.4) as:

$$\hat{r}_x = \hat{\mu}_{x+s} = \frac{d_x}{E_x^c}, \tag{4.5}$$

but since s is undetermined this would not be helpful. In some texts, this mortality ratio is denoted by $\hat{\mu}_x$, with the proviso that it does not actually estimate μ_x, but we reject this as possibly confusing. If the hazard rate is not changing too rapidly between ages x and $x + 1$, which is usually true except at very high ages, it is reasonable to assume $s \approx 1/2$, and that $\hat{r}_x$ can reasonably be taken to estimate $\mu_{x+1/2}$.

Part of the "art" of statistical modelling is to specify a plausible probabilistic model of how the observed data are generated, in which reasonably simple statistics (a "statistic" is just a function of the observed data) can be shown to be estimates of the important model quantities. We may then hope to use the features of the model to describe the sampling distributions of the estimates. In Chapter 10 we will specify a model under which $\hat{r}_x$ is approximately normally distributed with mean $\mu_{x+1/2}$ and variance $\mu_{x+1/2}/E_x^c$, which is very simple to use.

Older actuarial texts would have assumed the mortality ratio in equation (4.5) to be $\hat{m}_x$, an estimate of the central rate of mortality m_x (see Section 3.6). In Section 4.9 we outline why this falls outside the statistical modelling framework described above.

Note that we allow for left-truncation here essentially by ignoring it, therefore assuming that mortality depends on the current age but not also on the age on entering observation. Whether or not this is reasonable depends on the circumstances. We have not ignored right-censoring – it is present in some of the observed exposures, and the indicators d_i tell us which ones – but we have assumed that it is non-informative so we have said nothing about how it may arise.

4.7 Graduation of Mortality Ratios

Having calculated the crude hazard rates at single ages, we may consider them as a function of age; for example they may include:

$$\ldots, \hat{r}_{60}, \hat{r}_{61}, \hat{r}_{62}, \hat{r}_{63}, \hat{r}_{64}, \ldots \tag{4.6}$$

Because these are each based on finite samples, we may expect their sequence to be irregular. Experience suggests, however, that the hazard rate displays a strong and consistent pattern over the range of human ages. We may suppose (or imagine) that there is a "true" underlying hazard rate, which is a smooth function of age, and that the irregularity of the sequence of crude estimates is just a consequence of random sampling.

Figure 4.4 shows crude hazard rates as in equation (3.1) for single years of age, based on data for ages 65 and over in the Case Study. Data for males and females have been combined. The points on the graphs are values of $\log \hat{r}_x$, which we take to be estimates of $\log \mu_{x+1/2}$. The logarithmic scale shows clearly a linear pattern with increasing age, suggesting that an exponential function might be a candidate model for the hazard rate. On the left, the crude

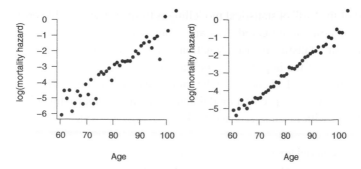

Figure 4.4 Logarithm of the crude mortality hazards for single years of age, for the Case Study, males and females combined. The left panel shows the experience data for 2012 only, showing a roughly linear pattern by age on a log scale. The right panel shows the experience data for 2007–2012, showing how a greater volume of data brings out a clearer linear pattern.

rates are calculated using the data from 2012 only, and on the right using the data from 2007–2012. Comparing the two, we see that the larger amount of data results in estimates $\log \hat{r}_x$ that lie much closer to a straight line.

Then we may attempt to bring together the consequences of all our assumptions:

- From our idea that the "true" hazard rate ought to be a smooth function of age, we may propose a smooth function that we think has the right shape.

- From the probabilistic model for the observations, we calculate the sampling distributions of the crude estimates of the hazard rates.

Putting these together, we obtain a smooth function μ_x of age x that is consistent with the observed crude estimates $\hat{r}_x$ being estimates of $\mu_{x+1/2}$. Moreover, we may try to find a smooth function μ_x that is in some sense (to be determined) optimal among all such smooth functions that may be said to be consistent with the observations. Actuaries call this process *graduation* of mortality data. In the sense of fitting a curve (the smooth function) to some data points (the crude estimates), graduation has been around for nearly two hundred years. Modern developments, made possible by computers, mean that this is now just one of many ways to model the mortality data described in this chapter. This is the subject of later chapters.

4.8 Examples: the Binomial and Poisson Models

In our approach above, the fundamental idea is that of the probabilistic model at the level of the individual, (D_i, T_i). But the observations described by this model may be aggregated in useful ways, for example as in Section 4.6 to obtain the numerator and denominator of equation (4.5). Further progress, for example graduation of the crude hazard rates, requires us to specify a distribution for the probabilistic model, in order to to obtain sampling properties of the data generated. We illustrate this process with two classical examples.

Example *The Bernoulli/binomial model.* The simplest model is obtained at the expense of putting some quite strict limits on the form that the data in Section 4.6 can take. Specifically, for all $i = 1, 2, \ldots, n$ we assume that $x_i = x$, that $b_i = 1$ and that there is no right-censoring between ages x and $x + 1$.

Then the ith individual is observed between ages x and $x + 1$ and if they die between those ages their death is observed. So $d_i = 1$ with probability q_x and $d_i = 0$ with probability $p_x = 1 - q_x$. In other words, D_i has a Bernoulli(q_x) distribution. So we know that $E[D_i] = q_x$ and $\text{Var}[D_i] = q_x(1 - q_x)$.

Defining $D_x = D_1 + D_2 + \cdots + D_n$, it follows that D_x has a binomial(n, q_x) distribution, since it is the sum of n independent Bernoulli(q_x) random variables. So we know that $E[D_x] = nq_x$ and $\text{Var}[D_x] = nq_x(1 - q_x)$. This knowledge of the sampling distribution of D_x is enough to let us graduate the data using a variety of methods, in all of which the ratio D_x/n can be shown to be an estimator of q_x, with sampling mean q_x and sampling variance $q_x(1 - q_x)/n$.

The binomial model is the prototype of analysis based on "number of lives". It is so simple that it has often been taken as the starting point in the actuarial analysis of mortality data. This is to overlook the fact that death is an event that befalls individuals, and that is how the data truly are generated. It also has the drawback that the strict limits on the form of the observations are rarely met in practice. Some consequences of this are discussed in Section 5.6.

Example *The Poisson model.* An alternative approach does not impose any limits on the form of the data, other than those in Section 4.6. This is therefore an approach based on "time lived". We assume instead that the hazard rate is constant between ages x and $x + 1$, and for brevity in this example we denote it by μ. We then proceed by analogy with the theory of Poisson processes.

Consider a Poisson process with constant parameter μ. If it is observed at time t for a short interval of time of length dt, the probability that it is observed to jump in that time interval is $\mu\, dt + o(dt)$ (see Cox and Miller, 1987, pp.153–154). This is exactly analogous to equation (3.16). However, a Poisson process can jump any positive integer number of times, while an individual can die just once. So our observation of the ith individual takes the form of observing a Poisson process with parameter μ between ages x_i and $x_i + b_i$, with the stipulation that if death or right-censoring occurs between these ages, observation of the Poisson process ceases at once.

This form of observation does not conform to any of the standard univariate random variables, so it does not have a name in the literature. Its natural setting is in the theory of counting processes, which is now the usual foundation of survival analysis (see Andersen et al., 1993, and Chapters 15 and 17).

What distribution does the aggregated data D_x then have? D_x is no longer the sum of identically distributed random variables. However, if the number of individuals n is reasonably large, and μ is small, we can argue that D_x should be, approximately, the number of jumps we would see if we were to observe a Poisson process with parameter μ for total time E_x^c. Thus, D_x is distributed approximately as a Poisson($E_x^c \mu$) random variable.

It is obvious that this can be only an approximation, since a Poisson random variable can take any non-negative integer value, whereas $D_x \leq n$. When the expected number of deaths is very small, it cannot be relied upon; see Table 1.2 and the ensuing discussion for an example. In most cases, however, this approximation is very good indeed, so this Poisson model is also often taken as the starting point in the actuarial analysis of mortality data. As in the case of the binomial model, this overlooks the fact that the data are generated at the level of the individual.

In this book we will make much use of the Poisson model and models closely related to it because of the form of their likelihood functions. Here, we just note that if we assume the Poisson model to hold, then $E[D_x] = \text{Var}[D_x] = E_x^c \mu$ and the mortality ratio $\hat{r}_x = d_x / E_x^c$ is an estimate of the parameter μ, and we usually assume that μ is in fact $\mu_{x+1/2}$.

4.9 Estimating the Central Rate of Mortality?

In some treatments, $\hat{r}_x = d_x / E_x^c$ would be taken to estimate m_x rather than $\mu_{x+1/2}$. This appears reasonable, because, from equation (3.25), m_x is a weighted average of the hazard rate between ages x and $x + 1$, so should not be very different from $\mu_{x+1/2}$.

However, there is a big conceptual difference between the two approaches. The central rate of mortality is a derived quantity in a deterministic model, rather than a parameter in a probabilistic model. We should ask, what is the underlying probabilistic model, in which m_x emerges as the quantity estimated by d_x / E_x^c? Such a model would be quite difficult to write down, because m_x is a complicated object compared with the hazard rate. In a probabilistic frame-work, the hazard is the fundamental model quantity, and m_x can be derived from estimates of the hazard if required.

4.10 Census Formulae for E_x^c

If census data are available, counting the numbers alive at stated ages on a given date, it is usually possible to estimate total time lived, giving E_x^c. If we also have total numbers of deaths that can be matched up exactly with the rel-evant exposures, we can calculate crude estimates $\hat{r}_x$ as before. It is essential that the deaths do match the exposures, which is called the *principle of corre-spondence*. The quantities d_x and E_x^c match if and only if, were an individual counted in E_x^c to die, they would appear in d_x.

Two particular examples of using census data to estimate E_x^c are the follow-ing:

• If we have P_x individuals age x under observation at the start of a calendar year, and P_x' at the end of the calendar year, then:

$$E_x^c \approx \frac{P_x + P_x'}{2} \tag{4.7}$$

in respect of that calendar year. The CMI until recently collected "sched-uled" census data of this form from UK life offices and aggregated the re-sults over four-year periods to estimate $\hat{r}_x$.
• If there is a national census, often every ten years, and reliable registration data for births, deaths and migration, then the population in years after a census can be estimated by applying the "movements" to the census totals.

4.11 Two Approaches

In Section 4.5 we defined a probabilistic model capable of generating complete lifetime data. Then in Sections 4.6 to 4.10 we adapted this model to single years of age, with the estimation of mortality ratios in mind. We introduced the idea of graduating "raw" mortality ratios (also called crude hazard rates) and two models, binomial and Poisson, applicable to grouped mortality data, and some ancillary points.

Given the number of the preceding pages devoted to estimating mortality ratios at single years of age, it may appear that this is our main focus. It is not, although it is still important for the reasons given at the start of Section 4.6. Our main focus is on building models based on complete lifetime data. We regard the fact that we were able to set out all the definitions needed for that in one short section as indicating the relative simplicity of that approach.

5

Fitting a Parametric Survival Model

5.1 Introduction

In his landmark paper (Gompertz, 1825), Benjamin Gompertz made the following remarkable observation:

This law of geometrical progression pervades, in an approximate degree, large portions of different tables of mortality; during which portions the number of persons living at a series of ages in arithmetical progression, will be nearly in geometrical progression.

Gompertz gave a number of numerical examples to support this statement and then stated his famous law of mortality:

... at the age x [man's] power to avoid death, or the intensity of his mortality might be denoted by aq^x, a and q being constant quantities.

(Note that Gompertz's q is entirely different to the familiar q_x of the life table.) In modern terminology, the Gompertz law defines a *parametric model*, in which the hazard rate is some parametric function of age. In the Gompertz model, the hazard rate is an exponential function of age.

Experience shows that the hazard rate representing human mortality is, over long age ranges, very close to an exponential function of age. Departures from this pattern occur at ages below about 50, and above about 90. As a result:

- the Gompertz model is often a good model of human mortality; and
- virtually all parametric models that have been used to represent human mortality contain a recognizable Gompertz or Gompertz-like term.

To allow a more systematic notation for parametric models to be developed later (see Section 5.9) we will reformulate the Gompertz law as:

$$\mu_x = \exp(\alpha + \beta x). \tag{5.1}$$

75

That is, rewrite $a = e^\alpha$ and $q = e^\beta$. The term in parentheses is essentially a linear model, in which α is the intercept and β is the slope or age coefficient, and we will sometimes use these terms.

In this chapter our purpose is to show how to fit parametric models to survival data, using maximum likelihood. We will use a Gompertz model for simplicity, but our methods will usually be applicable to any parametric models. We discuss some of the more commonly used parametric models in Section 5.9.

- First we fit a Gompertz model to the complete lifetime data from the Case Study. We write down the likelihood for the Gompertz model, based on the probabilistic model from Chapter 4, paying more attention to right-censoring. We show how it may be fitted, and all its sampling properties obtained, using the nlm() function in R.

- Then we consider the case that complete individual lifetime data are not available, and the task is to graduate the crude mortality ratios $\hat{r}_x$ from Chapter 4. We show that if the deaths D_x are assumed to have a Poisson($E_x^c \mu_{x+1/2}$) distribution, as in Section 4.8, then the likelihood is almost identical to that for complete lifetime data, and the process of graduating the crude $\hat{r}_x$ in R differs from fitting a survival model only in trivial ways.

- The fact that these two approaches yield near-identical results for data from the same experience should not cause us to overlook that they really are different. The first approach does not involve traditional graduation, and the second approach does not amount to fitting a model.

5.2 Probabilities of the Observed Data

In Section 4.5 we defined a probabilistic model (D_i, T_i) capable of generating the observations of complete lifetimes. We can write down the joint distribution of (D_i, T_i) in the special case that the only right-censoring that occurs is upon survival to age $x_i + b_i$, and no right-censoring is possible before that age:

$$P[(0, b_i)] = {}_{b_i}p_{x_i} \tag{5.2}$$

$$P[(1, t_i)] = {}_{t_i}p_{x_i}\,\mu_{x_i+t_i}\,dt_i \qquad (0 < t_i < b_i). \tag{5.3}$$

The random variable T_i has a mixed distribution; it has a continuous distribution on $(0, b_i)$ and a probability mass at b_i. The fact that some combinations of D_i and T_i are excluded by our assumptions (for example, pairs $(0, t_i)$ with

$t_i < b_i$) immediately shows again that D_i and T_i are not independent random variables.

In Section 4.5 we discussed the potential difficulty of introducing right-censoring at ages before $x_i + b_i$ into the model, because the probabilistic model is incomplete if it does not specify a mechanism for generating such right-censored observations. Consideration of probabilities helps us to understand this. If right-censoring is possible only at age $x_i + b_i$ then it is true that:

$$P[\text{died before time } b_i] + P[\text{survived to time } b_i] = 1 \qquad (5.4)$$

because all possible events are accounted for. If right-censoring can happen before age $x_i + b_i$ then equation (5.4) is false, and instead we have:

$$P[\text{died before time } b_i] + P[\text{censored before time } b_i] + P[\text{survived to time } b_i] = 1 \qquad (5.5)$$

and our probabilistic model above does not now account for all possible events.

One approach, which we will use only to motivate what follows, is to use the idea of random censoring introduced in Section 4.2. Its analogue in this model is to suppose that, as well as being subject to the mortality hazard μ_x, the ith individual is also subject to a *censoring hazard* denoted v_x at age x. Thus, at age $x_i + t$, the individual risks being removed from the population during the next small time interval of length dt, by death with approximate probability $\mu_{x_i+t}dt$, or by right-censoring with approximate probability $v_{x_i+t}dt$. If the two hazards act additively the probability of being removed from the population by either means is approximately $(\mu_{x_i+t} + v_{x_i+t})dt$ and, by the same arguments as led to equation (3.21), the probability of still being alive and under observation at age $x_i + t$, denoted by $_tp^*_{x_i}$, is:

$$
\begin{aligned}
_tp^*_{x_i} &= \exp\left(-\int_0^t (\mu_{x_i+s} + v_{x_i+s})\, ds\right)\\
&= \exp\left(-\int_0^t \mu_{x_i+s}\, ds\right)\exp\left(-\int_0^t v_{x_i+s}\, ds\right)\\
&= {_tp_{x_i}}\, \exp\left(-\int_0^t v_{x_i+s}\, ds\right).
\end{aligned}
\qquad (5.6)
$$

The first factor on the right-hand side of equation (5.6) is just the survival probability arising in our probabilistic model. The second factor is a function of v_{x_i+s} alone. If the parameters of interest, defining the function μ_x, do not have any role in the definition of v_x, then the second factor can be dropped everywhere it appears in the likelihood. Thus we have:

$$P[(0, b_i)] = {}_{b_i}p^*_{x_i} \qquad \propto {}_{b_i}p_{x_i} \tag{5.7}$$

$$P[(0, t_i)] = {}_{t_i}p^*_{x_i} \, v_{x_i+t_i} \, dt_i \propto {}_{t_i}p_{x_i} \qquad (0 < t_i < b_i) \tag{5.8}$$

$$P[(1, t_i)] = {}_{t_i}p^*_{x_i} \, \mu_{x_i+t_i} \, dt_i \propto {}_{t_i}p_{x_i} \, \mu_{x_i+t_i} \qquad (0 < t_i < b_i). \tag{5.9}$$

The expanded probabilistic model is complete (the "full model"), in the sense that it accounts for all possible observations, but everything to do with right-censoring can be ignored in the likelihood.

The argument above, which was not at all rigorous, was helpful in getting rid of right-censoring from the likelihood when it was a nuisance factor. Sometimes more than one decrement is of interest; for example we may be interested in the rate at which life insurance policyholders lapse or surrender their policies, as well as in their mortality. Then a similar argument leads to *multiple-decrement models*, which we will meet in Chapters 14 and 16.

5.3 Likelihoods for Survival Data

If we use the value of d_i as an indicator variable then we obtain the following neat expression unifying equations (5.7) to (5.9):

$$P[(d_i, t_i)] \propto {}_{t_i}p_{x_i} \, \mu^{d_i}_{x_i+t_i}. \tag{5.10}$$

Using equation (3.21), this can be expressed entirely in terms of the hazard function μ_{x_i+t} as follows:

$$P[(d_i, t_i)] \propto \exp\left(-\int_0^{t_i} \mu_{x_i+s} \, ds\right) \mu^{d_i}_{x_i+t_i}. \tag{5.11}$$

Suppose there are n individuals in the sample, and we assume that the observations we make on each of them are mutually independent of the observations on all the others (so that the pairs (D_i, T_i) are mutually independent for $i = 1, 2, \ldots, n$). Then the total likelihood, which we write rather loosely as $L(\mu)$, is:

$$L(\mu) \propto \prod_{i=1}^{n} P[(d_i, t_i)] \propto \prod_{i=1}^{n} \exp\left(-\int_0^{t_i} \mu_{x_i+s} \, ds\right) \mu^{d_i}_{x_i+t_i}. \tag{5.12}$$

The log-likelihood, denoted $\ell(\mu)$ is:

$$\ell(\mu) = -\sum_{i=1}^{n} \int_{0}^{t_i} \mu_{x_i+s}\, ds + \sum_{i=1}^{n} d_i \, \log(\mu_{x_i+t_i}). \qquad (5.13)$$

The log-likelihood can also be written in the convenient form:

$$\ell(\mu) = -\sum_{i=1}^{n} \int_{0}^{t_i} \mu_{x_i+s}\, ds + \sum_{d_i=1} \log(\mu_{x_i+t_i}), \qquad (5.14)$$

where the second summation is over all i such that $d_i = 1$ (observed deaths).

5.4 Example: a Gompertz Model

The likelihood approach is completed by expressing μ_{x_i+t} as a parametric function and maximising the likelihood with respect to those parameters. The question of what parametric function to choose is then, of course, the key decision to be made in the entire analysis, and is a large subject in its own right. We list some candidate functions relevant to actuarial work in Section 5.9. In Section 5.1 we noted the persistent effectiveness of the Gompertz model (Gompertz, 1825), positing an exponential hazard rate, which in older texts would have been called the Gompertz law of mortality. In our notation, introduced in Section 5.1, the Gompertz model is:

$$\mu_{x+t} = \exp(\alpha + \beta(x+t)).$$

Using the Gompertz model, the likelihood and log-likelihood, as functions of α and β, are, from equation (5.12):

$$L(\alpha,\beta) \propto \prod_{i=1}^{n} \exp\left(-\int_{0}^{t_i} \exp(\alpha + \beta(x_i + t))\, dt\right) \exp(\alpha + \beta(x_i + t_i))^{d_i} \quad (5.15)$$

$$\ell(\alpha,\beta) = -\sum_{i=1}^{n} \int_{0}^{t_i} \exp(\alpha + \beta(x_i + t))\, dt + \sum_{i=1}^{n} d_i\,(\alpha + \beta(x_i + t_i)). \qquad (5.16)$$

Taking the partial derivatives, the equations to be solved are:

$$\frac{\partial \ell}{\partial \alpha} = -\sum_{i=1}^{n} \int_{0}^{t_i} \exp(\alpha + \beta(x_i + t))\, dt + \sum_{i=1}^{n} d_i = 0 \qquad (5.17)$$

$$\frac{\partial \ell}{\partial \beta} = -\sum_{i=1}^{n} \int_{0}^{t_i} \exp(\alpha + \beta(x_i + t))\,(x_i + t)\, dt + \sum_{i=1}^{n} d_i\,(x_i + t_i) = 0. \, (5.18)$$

Equation (5.17), in a simplified setting, takes us back to mortality ratios. If the range of ages is short enough that the hazard rate may be approximated by a constant, denoted by r, equation (5.17) is:

$$-\sum_{i=1}^{n} \int_{0}^{t_i} r \, dt + \sum_{i=1}^{n} d_i = 0, \qquad (5.19)$$

which immediately gives the estimate:

$$\hat{r} = \frac{\sum_{i=1}^{n} d_i}{\sum_{i=1}^{n} t_i}, \qquad (5.20)$$

and this is the mortality ratio from Chapter 4.

We will now discuss in some detail the steps needed to fit the model illustrated above using R.

5.5 Fitting the Gompertz Model

In this section we fit the Gompertz model to the data from the medium-sized UK pension scheme in the Case Study (see Section 1.10 and Figure 4.4). Here we focus on the main steps that must be coded, and the standard R functions (included in most R implementations) that are used to carry them out.

There are four steps:

(i) Read in the data from a suitably formatted source file. We assume that data are held in comma-separated values (CSV) files with a .csv extension. CSV files can be read by most data-handling software. Data held in other formats (such as Excel spreadsheets) can easily be converted into a CSV format. The standard R function to read a CSV file is read.csv().

(ii) Define a function that takes the data and computes the log-likelihood. Since R does not supply standard likelihood functions for the parametric survival models most likely to be used in actuarial work, we must supply this function ourselves. Since we will make use of R's standard minimisation function, we actually compute the negative log-likelihood, and define the function NegLogL() to calculate the negative log-likelihood for a single individual, and FullNegLogL() to sum the contributions from all individuals.

(iii) Minimise the negative log-likelihood, using the standard R function for non-linear minimisation, nlm(). The main inputs required are the function FullNegLogL() and some suitable starting values for α and β, which tell nlm() where to start searching. The outputs from nlm() include the parameter estimates $\hat{\alpha}$ and $\hat{\beta}$, and also the matrix of second derivatives of the (negative) log-likelihood evaluated at the minimum, called the Hessian matrix or just Hessian. The value of the log-likelihood attained at its maximum can be used to compute quantities helpful in choosing between different models, such as Akaike's Information Criterion (see Section 6.4.1).

(iv) Invert the Hessian matrix to obtain an estimate of the covariance matrix of the parameter estimates (see Appendix B) using the standard R function solve(). Most statistics packages will report the standard deviation of parameter estimates automatically, but the full covariance matrix is needed for some useful applications, such as finding sampling distributions of premium rates or policy values by parametric bootstrapping (see Forfar et al., 1988 or Richards, 2016).

As can be seen, there is relatively little coding for the analyst to do; most of the work is done by standard functions supplied by R and its libraries. We comment further on these four steps below.

5.5.1 R's read.csv() Function

We discussed file formats in Section 2.2.4, but we explain the operation of R's read.csv() function here. The full command to read the data file is:

```
gData=read.csv(file="SurvivalInput_40276.csv", header=TRUE)
```

There are several parameter options to control this function's behaviour, but the two most important are as follows:

- file, which tells R where the CSV file is stored. This might be on your local computer, on a network or even on the Internet. Note that the backslash character "\" in any filename needs to be escaped. Thus, if you would normally specify a filename as C:\Research\filename.csv, then in the R script you have to type C:\\Research\\filename.csv. Alternatively, type C:/Research/filename.csv, which works under both Windows and Linux.
- header, which tells R if there is a header row in the CSV file. If so, R will use the names in the header row. In general it is advisable to store your data with header rows to make things easier to follow and understand.

We show below a few lines extracted from the CSV file being read in (to fit the data illustrated in the left-hand plot in Figure 4.4):

```
Id,EntryAge,EntryYear,TimeObserved,Status,Benefit,Gender
...
231202760,95.114968,2012,0.999337,0,576.1,F
231202761,95.084851,2012,0.999337,0,2017.18,F
231202763,95.065685,2012,0.999337,0,1659.67,F
231202764,95.038306,2012,0.999337,0,771.97,F
231218105,94.9945,2012,0.208081,DEATH,1352.05,F
231205353,94.931528,2012,0.999337,0,2157.85,F
231205354,94.92879,2012,0.999337,0,8169.81,F
231202765,94.86308,2012,0.999337,0,933.14,F
231204767,94.860342,2012,0.509251,DEATH,7459.05,F
231204768,94.857604,2012,0.23546,DEATH,1970.47,F
...
```

There is one row for each life, and seven data columns, of which we will use only three in this chapter, namely the entry age, the time observed and the status (which are the quantities x_i, t_i and d_i from Section 4.5). (Note that the time observed for lives exposed for the whole year is 0.999337 years, not exactly one year. This is because, to allow for leap years, of which there are 47 every 400 years, a standard year of 365.242 days has been adopted. Other conventions may be used.) The file also contains data for males, although the extract above does not. We use the R command gData = read.csv() to read the contents of the CSV file into the object gData, which is created by issuing this command. If a header row has been provided, as here, then the various data columns can be accessed using the $ operator: for example, after reading the above file, we can access the vector of entry ages with gData$EntryAge.

For more details of the read.csv() function, type help(read.csv) at the R prompt.

5.5.2 R's nlm() Function

Most of the work in parameter estimation is done by R's nlm() function (non-linear minimisation). There are numerous options to control this function's behaviour, of which the four most important are as follows:

- f, the function to be minimised. Since we seek to maximise the log-likelihood function ℓ, we have to supply the formula for $-\ell$, which here is called FullNegLogL().
- p, the vector of initial parameter estimates. Sensible initial estimates speed the optimisation, and generally we find that values in the range $(-14, -9)$ work well for the intercept α and values in the range $(0.08, 0.13)$ work well

for the age coefficient β. For parametric models other than the Gompertz (see Section 5.9) we recommend an initial value of -5 for the Makeham parameter ϵ and zero for all other parameters.

- `typsize`, the vector of parameter sizes, i.e. the approximate order of the final optimised values. Often it is most practical to simply reuse the value specified for `p`, the vector of initial parameter estimates. `typsize` is an important option to specify, as without it it is all too easy for `nlm()` to return values which are not optimal.
- `gradtol`, the value below which a gradient is sufficiently close to zero to stop the optimisation. Smaller values of `gradtol` will produce more accurate parameter estimates, but at the cost of more iterations. Depending on the structure of the log-likelihood function, you may need to use larger (or coarser) values to make the algorithm work.
- `hessian`, a boolean indicator as to whether to approximate the Hessian matrix at the end. This is necessary to approximate the variance-covariance matrix of the parameter estimates.

The command to use `nlm()` is:

```
Estimates = nlm(FullNegLogL, p=c(-11.0, 0.1), typsize
     =c(-11.0, 0.1), gradtol=1e-10, hessian=T)
```

We remind the reader that the `c()` function takes as inputs any number of R objects and concatenates them into a vector. Thus, the command `p=c(-11.0, 0.1)` above creates a vector of length two with components -11.0 and 0.1, and assigns it to the object named `p` which is recognised by `nlm()` as a vector of starting values. The length of the vector `p` tells `nlm()` the number of parameters in the model.

The `nlm()` function returns an object, here given the name `Estimates`, with a variety of components, of which the four most important are as follows:

- `estimate`, the vector of maximum-likelihood parameter estimates; so, for example, $\hat{\alpha}$ can be accessed as `Estimates$estimate[1]` and $\hat{\beta}$ as `Estimates$estimate[2]`
- `minimum`, the value of $-\ell$ at the maximum-likelihood estimates
- `hessian`, the matrix of approximate second derivatives of $-\ell$
- `code`, an integer stating why the optimisation process terminated.

The value of `code` tells you whether the optimisation is likely to have been successful or not. It is critical to check the value of `code` after each optimisation, as there are plenty of circumstances when it can fail. Users will find that

```
nlm() return code 1.
Data cover 13085 lives and 365 deaths.
Total exposure time is 12439.95 years.

Iterations=17
Log-likelihood=-1410.26
AIC=2824.53

Parameter     Estimate Std. error
------------  -------- ----------
Intercept     -12.9721    0.466539
Age           0.122874  0.00563525
```

Figure 5.1 Output from fitting Gompertz model to 2012 data from the Case Study.

it is sometimes necessary to try various alternative starting values, or else to change some of the other parameters controlling the behaviour of the nlm() function call. Even where the code value signals convergence, it is nevertheless advisable to try different starting values in the p option. This is because some log-likelihoods may have more than one local maximum.

For more details of the nlm() function, type help(nlm) at the R prompt. Useful additional options include the gradient attribute for the likelihood function for calculating the formulaic first partial derivatives. Without this the nlm() function uses numerical approximations of those derivatives.

5.5.3 R's solve() Function

To estimate the variance-covariance matrix of the parameter estimates, we need to invert the Hessian matrix for the negative log-likelihood function, $-\ell$. For this we use R's solve() function. The function call solve(A, B) will solve the system $AX = B$, where A and B are known matrices, and X is the solution to the equation. More simply, if A is an invertible square matrix, solve(A) will return A^{-1}.

For more details of the solve() function, type help(solve) at the R prompt.

5.5.4 R Outputs for the Basic Gompertz Model

If we carry out the steps described above on the data from the Case Study, for the calendar year 2012 only, and suitably format the key outputs, we get the results shown in Figure 5.1.

```
nlm() return code 1.
Data cover 14773 lives and 2028 deaths.
Total exposure time is 66082.41 years.

Iterations=17
Log-likelihood=-7902.07
AIC=15808.13

Parameter     Estimate Std. error
------------  -------- ----------
Intercept     -12.4408    0.199095
Age           0.116941 0.00242424
```

Figure 5.2 Output from fitting Gompertz model to 2007–2012 data from the Case Study.

The meaning of `nlm() return code is` 1 is "relative gradient is close to zero, current iterate is probably solution" (R documentation). The other outputs are self-explanatory, except for `AIC=2824.53`. AIC is an acronym of Akaike's information criterion, which is defined and discussed in Section 6.4.1. A smaller AIC value generally represents a better model. The AIC may also be used in smooth models; see Chapter 11, and Section 11.8 in particular.

Fitting a model to the mortality experience data from a single year is of course inefficient if more data are available. This was illustrated in the right-hand plot in Figure 4.4, where mortality ratios for single years of age were calculated using the Case Study data from 2007–2012. Fitting the Gompertz model to this larger data set is simply a matter of supplying a larger file containing the data. The outputs of the R code are shown in Figure 5.2. There are two noteworthy features:

- The relative error has reduced, i.e. the standard errors are much smaller compared to the parameter estimates. This shows the increased estimation power from being able to use more experience data, in this case by being able to span multiple years.

- The AIC has increased massively compared with Figure 5.1. This is because the data have changed: there are more lives, the exposure time is longer and there are more deaths. Since the data have changed, the AICs in Figures 5.1 and 5.2 are not comparable.

The use of the AIC and other information criteria in choosing between different possible models is discussed in Section 6.4.

5.5.5 A Comment on Maximisation Routines

Any maximisation routine written to handle quite general functions whose form may not be known *a priori* can sometimes give erroneous results, without any warning of error. R's nlm() function is no exception.

- If the function being maximised has two or more local maxima, then the routine may find any of them. Sometimes it may be known *a priori* that there can be only one maximum (for example, a bivariate normal distribution with a positive covariance between the components). Otherwise it is prudent to fit the model with different starting values within a reasonable range of the parameter space.
- Routines based on a Newton–Raphson-type algorithm (including R's nlm() function) are known to be poor at finding global maxima if given bad starting values. If necessary, a global maximum can be checked as described above using hessian=F in nlm() and then fitted with good starting values using hessian=T.
- Routines may behave unpredictably in areas where the function being maximised is relatively flat in one or more dimensions.
- If the Hessian matrix is far from diagonal then there may be significant relationships between the components of the function being maximised. Performance may be improved by transforming the parameter space so that the components of the objective function are orthogonal at the global maximum.

No general-purpose maximising function is immune from these and other pitfalls. They must be used with care.

5.6 Data for Single Years of Age

Suppose now that complete lifetime data are not available, and the data we have are deaths and exposures for single years of age.

We first consider why data of this form were used in the past. Many reasons stem from the fact that, during the two centuries or so of actuarial practice predating cheap computing power, the life table was at the centre of everything the actuary did. For that or other reasons, the target of actuarial mortality analysis for a long time tended to be the estimation of the probabilities q_x. Some of the consequences are listed below:

- Data collection and analysis were most naturally based on single years of age, because q_x applies to a single year of age. The same individual might be observed for several years in a mortality study, but this observation would be chopped up into separate contributions to several integer ages, for each of which q_x would be estimated independently. (The reader may notice a troubling consequence. If the same individual contributes to the estimation of q_x at several adjacent ages then those estimated $\hat{q}_x$ cannot be statistically independent. The same anomaly afflicts the estimates $\hat{r}_x$.)

- Methods of estimating q_x were motivated by the binomial distribution, introduced in Section 4.8. Given E_x individuals age x, each of whom may die before age $x + 1$ with probability q_x, the number of deaths D_x has a binomial(E_x, q_x) distribution, and the obvious estimate of q_x is $\hat{q}_x = d_x/E_x$. This is, in fact, the moment estimate, the least squares estimate and the maximum likelihood estimate, so it is an attractive target. However, because we have bypassed the Bernoulli model of the mortality of a single individual, we have introduced the idea that mortality analysis is all about groups, not individuals. This idea became very deep-rooted in actuaries' thinking. While it was a useful shortcut, it became a barrier to adopting more modern methods of analysis.

- The estimate $\hat{q}_x = d_x/E_x$ is a mortality ratio along the lines of equation (3.1) in which the denominator is based on "number of lives" rather than "time lived" (in terms discussed in Section 1.4). This would be easy in the absence of left-truncation (everyone is observed from age x) and right-censoring (no-one leaves before age $x + 1$ except by dying). However, left-truncation and right-censoring are the defining features of actuarial mortality data. Their presence led to elaborate schemes to compute an appropriate denominator for $\hat{q}_x$, usually denoted E_x as above, and called an *initial exposed-to-risk*. See Benjamin and Pollard (1980) and its predecessors for examples.

- Based on the assumption that deaths occur, on average, half-way through the year of age, we have the approximate relationship $E_x \approx E_x^c + d_x/2$. Since E_x^c is often relatively simple to calculate, this relationship has sometimes been used to estimate E_x for the purpose of estimating q_x in the setting of a binomial model. However, we have seen that using E_x^c directly as the denominator of a mortality ratio gives an acceptable estimate of $\mu_{x+1/2}$ in the setting of a Poisson model. Approximating E_x in terms of E_x^c just in order to estimate q_x would be, today, rather pointless.

- The inability to base estimation on individual lives, in a binomial model based on q_x, has serious consequences for actuarial methodology. It means that additional information about individuals, such as gender, smoking status, occupation or place of residence, cannot be taken into account except by *stratifying* the sample according to these risk factors and analysing each cell separately. This quickly reaches a practical limit, mentioned briefly in Section 1.6. For example, if we have two genders, three smoking statuses, six occupation groups and four places of residence, we need $2 \times 3 \times 6 \times 4 = 144$ separate analyses to take all risk factors into account, even before considering individual ages. Even quite large data sets will quickly lose statistical relevance if stratification is the only way to model the effect of risk factors. We will see an example of this in Section 7.2. Richards et al. (2013) gives an example of how quickly even a large data set can be exhausted by stratification.

With the exception of Section 10.5, we will make no further reference to the binomial model based on "numbers of lives" and in which q_x is the target of estimation, except to give historical context. We will use "time lived" with the hazard rate as the target of estimation. If probabilities q_x are required they can be obtained from μ_x using:

$$q_x = 1 - \exp\left(-\int_0^1 \mu_{x+s}\, ds\right) \tag{5.21}$$

(see equation (3.21)). In the UK, this has been the approach used by the CMI for many years.

5.7 The Likelihood for the Poisson Model

In Chapter 4 we considered mortality ratios based on single years of age, using person-years as denominators. This led naturally to a Poisson model in which the mortality ratio $\hat{r}_x$ was a natural estimate of $\mu_{x+1/2}$. We may regard it as a modern version of estimation using mortality ratios, replacing the binomial model with the Poisson model to some advantage. Forfar et al. (1988) is a useful source on this approach.

First, we obtain the log-likelihood for the data d_x and E_x^c. Let D_x be the random variable with $D_x \sim \text{Poisson}(E_x^c \mu_{x+1/2})$. The probability that D_x takes the observed value d_x is:

$$P[D_x = d_x] = \exp(-E_x^c \mu_{x+1/2})(E_x^c \mu_{x+1/2})^{d_x}/d_x!. \tag{5.22}$$

The likelihood $L(\mu)$ is proportional to $P[D_x = d_x]$:

$$L(\mu) \propto \exp(-E_x^c \mu_{x+1/2})\mu_{x+1/2}^{d_x} \tag{5.23}$$

so the log-likelihood is:

$$\ell(\mu) = -E_x^c \mu_{x+1/2} + d_x \log(\mu_{x+1/2}).$$ (5.24)

Note that, just as the likelihood in equation (5.23) is defined up to a constant of proportionality, so the log-likelihood in equation (5.24) is defined up to an additive constant. We will adopt the convention that this additive constant is omitted. Summing over all ages, the total log-likelihood is:

$$\ell^*(\mu) = -\sum_{x=0}^{\infty} E_x^c \mu_{x+1/2} + \sum_{x=0}^{\infty} d_x \log(\mu_{x+1/2}).$$ (5.25)

Now compare this Poisson log-likelihood in (5.25) with that in equation 5.13, which for convenience is repeated below:

$$\ell(\mu) = -\sum_{i=1}^{n} \int_0^{t_i} \mu_{x_i+t}\, dt + \sum_{i=1}^{n} d_i \, \log(\mu_{x_i+t_i}).$$ (5.26)

The likelihood in equation (5.25) is a sum over all ages, while that in equation (5.26) is a sum over all individuals in the portfolio. However, we can show that they are, to a close approximation, the same.

The second term in the log-likelihood (5.25) can be seen to be an approximation to the second term in the log-likelihood (5.26), replacing the hazard rate at the exact age at death of each individual observed to die, with the hazard rate at the mid-point of the year of age x to $x + 1$ in which death occurred. To see that the first term in the log-likelihood (5.25) is an approximation to the first term in the log-likelihood (5.26), define an indicator function in respect of the ith individual as follows:

$Y_i(x) = 1$ if the ith individual is alive and under observation at age x; or

$Y_i(x) = 0$ otherwise. (5.27)

Then we can write:

$$\sum_{i=1}^{n} \int_0^{t_i} \mu_{x_i+t}\, dt = \sum_{i=1}^{n} \int_0^{\infty} Y_i(t)\, \mu_t\, dt \qquad (5.28)$$

$$= \sum_{i=1}^{n} \sum_{x=0}^{\infty} \int_0^1 Y_i(x+t)\, \mu_{x+t}\, dt \qquad (5.29)$$

$$\approx \sum_{x=0}^{\infty} \sum_{i=1}^{n} \int_0^1 Y_i(x+t)\, \mu_{x+1/2}\, dt \qquad (5.30)$$

$$= \sum_{x=0}^{\infty} \left(\sum_{i=1}^{n} \int_0^1 Y_i(x+t)\, dt \right) \mu_{x+1/2} \qquad (5.31)$$

$$= \sum_{x=0}^{\infty} E_x^c \mu_{x+1/2}, \qquad (5.32)$$

noting that the term in parentheses in equation (5.31) is the sum, over all n individuals, of the time spent by each individual under observation between ages x and $x + 1$, namely E_x^c.

This explains why using the Poisson model for crude hazards at single years of age and then graduating by curve-fitting gives results very similar to those obtained by fitting a survival model to individual lifetime data. Note that, although we use a Gompertz model in the examples discussed in this chapter, neither of the log-likelihoods (5.25) or (5.26) assumed any particular parametric model.

What we have shown is that the log-likelihood (5.26) is functionally similar to the likelihood for a set of Poisson-distributed observations. This is a feature that recurs in survival analysis; we will see it again in Chapter 15.

Thus fitting the Poisson model in R requires just a trivial change to the function `FullNegLogL()` supplied to the `nlm()` function. Wherever `FullNegLogL()` evaluates μ_x using $\mu_x = \exp(\alpha + \beta x)$, we replace that with a piecewise-constant version of the Gompertz formula, taking the value:

$$\mu_{\lfloor x \rfloor + 1/2} = \exp(\alpha + \beta(\lfloor x \rfloor + 1/2)) \qquad (5.33)$$

on the age interval $[x, x + 1)$, where $\lfloor x \rfloor$ is the integer part of x.

5.8 Single Ages *versus* Complete Lifetimes

To illustrate the fitted Gompertz models, Figure 5.3 reproduces Figure 4.4 with the Gompertz models fitted in Section 5.5 added. The points shown are

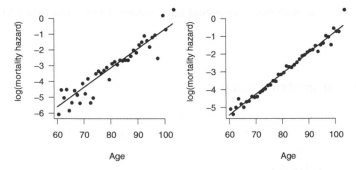

Figure 5.3 Logarithm of the crude mortality hazards for single years of age, for the Case Study, males and females combined, from Figure 4.4, with the Gompertz models fitted in this chapter superimposed. The left panel shows the experience data for 2012 only, the right panel for 2007–2012. The crude hazards were not used to fit the models, and serve only to give a visual impression of goodness-of-fit.

precisely the crude mortality ratios $\hat{r}_x$ that we would have fitted the model to, had we used single years of age. Had we done so, we would have obtained a figure practically identical to Figure 4.4. But the two figures, although almost identical, would be different in an important way, which is one of the main messages of this book.

The difference is that the crude hazards at single ages in Figure 5.3 play no part whatsoever in the fitting of the Gompertz functions displayed there. These functions were obtained directly from the raw data by maximising the log-likelihood in equation (5.16). The crude mortality ratios added from Figure 4.4 are there purely to allow the quality of the fit to be visualised.

Having reached this point, individual lifetime data may appear to offer little by way of *statistical* advantage over data at single years of age, although we did note some practical advantages in Section 1.6. However, it is still the "gold standard" for survival-modelling work:

- Data at single ages can be extracted from complete lifetime data if required. Indeed, that was how the crude mortality ratios in Figure 4.4 were calculated, and they are also useful in checking goodness-of-fit (see Chapter 6).

- As we shall see in Chapter 7, individual lifetime data allow us to model features of the experience in ways that are inaccessible if all we have is grouped data at single ages. They lead the actuary to a much deeper understanding of the mortality risks in any portfolio.

- In case only grouped data at single years of age are available, we have the comfort of knowing that, although what we can do is limited, the results should agree with those we would have obtained using individual lifetime data.

5.9 Parametric Functions Representing the Hazard Rate

Table 5.1 shows a selection of parametric functions for μ_x at adult ages that have been found useful in insurance and pensions practice. All of them contain a Gompertz term, or some transformation of a Gompertz term, and all can be integrated analytically.

In the UK, the CMI in recent years has made much use of the "Gompertz–Makeham family" of functions, in which:

$$\mu_x = \text{polynomial}_1(x) + \exp(\text{polynomial}_2(x)), \tag{5.34}$$

or the "logit Gompertz–Makeham" family, in which:

$$\mu_x = \frac{\text{polynomial}_1(x) + \exp(\text{polynomial}_2(x))}{1 + \text{polynomial}_1(x) + \exp(\text{polynomial}_2(x))}, \tag{5.35}$$

and the polynomials are expected to be of fairly low order. The exponentiated polynomial obviously represents a Gompertz-like term. See Forfar et al. (1988) for details.

Note that mortality in the first decades of life has features that are not represented by any of these functions. In particular:

- mortality falls steeply during the first year of life as infants with congenital conditions die; and
- there is often a levelling off or even a hump at around age 20, attributed to the excesses of young adulthood and widely known as the "accident hump". (This is a development of the additive constant in Makeham's formula (see Table 5.1), which was introduced to represent accidental deaths independent of age.)

These are perhaps more important in demography than in actuarial work. The Heligman–Pollard formulae, such as:

$$q_x/p_x = A^{(x+B)^C} + D\exp(-E(\log x - \log F)^2) + GH^x \tag{5.36}$$

(see Benjamin and Pollard, 1980), attempt to represent mortality over the whole human age range. Parameters A, B and C define a steeply falling curve over the first year or so of life. Parameters D, E and F define a normal-like

Table 5.1 *Some parametric functions for μ_x that are candidates to describe human mortality ("mortality laws"), and the corresponding integrated hazards $\int_0^t \mu_{x+s}ds$. Source: Richards (2008).*

Mortality "law"	μ_x	$\int_0^t \mu_{x+s}ds$
Constant hazard	e^α	te^α
Gompertz (1825)	$e^{\alpha+\beta x}$	$\dfrac{e^{\beta t}-1}{\beta}e^{\alpha+\beta x}$
Makeham (1860)	$e^\epsilon + e^{\alpha+\beta x}$	$te^\epsilon + \dfrac{e^{\beta t}-1}{\beta}e^{\alpha+\beta x}$
Perks (1932)	$\dfrac{e^{\alpha+\beta x}}{1+e^{\alpha+\beta x}}$	$\dfrac{1}{\beta}\log\left(\dfrac{1+e^{\alpha+\beta(x+t)}}{1+e^{\alpha+\beta x}}\right)$
Beard (1959)	$\dfrac{e^{\alpha+\beta x}}{1+e^{\alpha+\rho+\beta x}}$	$\dfrac{e^{-\rho}}{\beta}\log\left(\dfrac{1+e^{\alpha+\rho+\beta(x+t)}}{1+e^{\alpha+\rho+\beta x}}\right)$
Makeham–Perks (Perks, 1932)	$\dfrac{e^\epsilon + e^{\alpha+\beta x}}{1+e^{\alpha+\beta x}}$	$te^\epsilon + \dfrac{1-e^\epsilon}{\beta}\log\left(\dfrac{1+e^{\alpha+\beta(x+t)}}{1+e^{\alpha+\beta x}}\right)$
Makeham–Beard (Perks, 1932)	$\dfrac{e^\epsilon + e^{\alpha+\beta x}}{1+e^{\alpha+\rho+\beta x}}$	$te^\epsilon + \dfrac{e^{-\rho}-e^\epsilon}{\beta}\log\left(\dfrac{1+e^{\alpha+\rho+\beta(x+t)}}{1+e^{\alpha+\rho+\beta x}}\right)$

"bell curve" representing the accident hump. Parameters G and H define a Gompertz term which dominates at later ages. Note that the left-hand side of equation (5.36) is the odds ratio of q_x:

$$\text{odds ratio} = \frac{q_x}{1-q_x}, \tag{5.37}$$

which causes q_x to increase less than exponentially at the highest ages. A similar feature can be seen in Table 5.1 in the Perks, Beard, Makeham–Perks and Makeham–Beard formulae, and it will appear again in the binomial regression model in Section 10.5. The shape of q_x or μ_x at the highest ages, 95–100 and over, has been much discussed in the face of questionable data; see Thatcher et al. (1998), for example.

6

Model Comparison and Tests of Fit

6.1 Introduction

In this chapter we will look at three related topics:

- how to compare models statistically
- how to formally test the fit of a model
- how to test the suitability of a model for financial purposes.

The first two of these involve standard statistical methodologies documented elsewhere. However, the third is specific to actuarial work: relying solely on statistical tests is not sufficient for financial work, and models must be tested for financial suitability before they can be used.

This chapter concerns the concepts of deviance, information criteria and degrees of freedom, all used in the context of assessing the goodness-of-fit of a model after it has been fitted. Chapter 11 also uses these same concepts, but from a perspective of smoothing parameters during the model-fitting process.

6.2 Comparing Models

We have seen how to fit a given parametric model. A major decision is: what parametric model should we fit? Section 5.9 described several models that have been found to be useful in practice, and was by no means exhaustive. Some of these may be classified as families of models, in which more complex models are created by systematically adding more terms. The Gompertz–Makeham family is an example of this. We would like to have some systematic and quantitative basis for choosing which model to fit.

There are numerous metrics for comparing models. So not only do we have to choose a model, we have to choose what metric to use to choose a model.

Some of the most useful rely simply on the log-likelihood function. These include the following:

- The *model deviance* (see Section 6.3). The deviance is a well-understood statistical measure of goodness-of-fit. However, one important drawback of the deviance is that it takes no account of the number of parameters used in the model. Thus, while the deviance can tell us how well a model fits, and whether one model fits better than another, it says nothing about parsimony, that is, whether additional parameters are worth keeping. Model deviance is also covered in Section 11.7.

- An *information criterion* (see Section 6.4). An information criterion is a function of the log-likelihood and the number of parameters. Generally speaking, they favour better-fitting models and penalise large numbers of parameters. Information criteria are therefore useful for comparing models with different numbers of parameters. In particular, they help the analyst to decide whether an improved fit justifies the additional complexity of more parameters. Information criteria are also covered in Section 11.8.

6.3 Deviance

The deviance of a model is a statistic measuring the current fit against the best possible fit. The deviance is twice the difference between the log-likelihood of a model and the log-likelihood of a model with one parameter for each observation. The deviance, Dev, is defined as:

$$\text{Dev} = 2(\ell_1 - \ell_2) \tag{6.1}$$

where ℓ_1 is the log-likelihood with a single parameter for each observation (referred to as the "full model" or "saturated model") and ℓ_2 is the log-likelihood for the model under consideration. The full model is, by definition, the best possible fit to the data, so the deviance measures how far the model under consideration is from a perfect fit. The deviance is an analogue of the χ^2 goodness-of-fit statistic.

6.3.1 Poisson Deviance

If the number of events, denoted by D, has a Poisson distribution with parameter λ, the likelihood of observing d events is as follows:

$$L \propto e^{-\lambda} \lambda^d, \tag{6.2}$$

and so the log-likelihood, ℓ, is:

$$\ell = -\lambda + d \log \lambda. \tag{6.3}$$

The deviance for a single observation, Dev_j, is therefore:

$$\text{Dev}_j = 2\left[d \log\left(\frac{d}{\hat{\lambda}}\right) - (d - \hat{\lambda})\right], \tag{6.4}$$

where $\hat{\lambda}$ is the estimate of the Poisson parameter λ. This is the same definition as given in McCullagh and Nelder (1989, p.34). Note that $d \log d \to 0$ as $d \to 0$, so $0 \log 0$ is taken to be zero. The total deviance for a model is the sum over all observations: $\text{Dev} = \sum_j \text{Dev}_j$. Note the absence of any terms involving the number of model parameters; see also Section 11.7 for a fuller derivation of the Poisson deviance.

When we have fitted a parametric survival model and have estimated the hazard *function* $\hat{\mu}_x$ we replace the Poisson parameter $\hat{\lambda}$ in equation (6.4) with the hazard function integrated over all the observed exposures for the relevant year of age, denoted by Λ_j:

$$\text{Dev}_j = 2\left[d \log\left(\frac{d}{\hat{\Lambda}_j}\right) - (d - \hat{\Lambda}_j)\right]. \tag{6.5}$$

This makes it particularly useful to work with parametric hazard functions that have an explicit expression for the integrated hazard, such as those in Table 5.1. The R function `calculateDevianceResiduals()` described in Appendix H.2 gives an example of calculating Poisson deviance residuals for a Gompertz model.

6.3.2 Binomial Deviance

If the number of events, denoted by D, has a binomial(m, q) distribution, the likelihood of observing d events is as follows:

$$L \propto (1 - q)^{m-d} q^d, \tag{6.6}$$

and so the log-likelihood, ℓ, is:

$$\ell = (m - d) \log(1 - q) + d \log q. \tag{6.7}$$

The deviance for a single observation, Dev_j, is therefore:

$$\text{Dev}_j = 2\left[d \log\left(\frac{d}{m\hat{q}}\right) + (m - d) \log\left(\frac{m - d}{m - m\hat{q}}\right)\right],$$

where $\hat{q}$ is the estimate of q. This is the same definition as given in McCullagh and Nelder (1989, p.118). As with the Poisson model, the total deviance for a model is the sum over all observations: Dev $= \sum_j \text{Dev}_j$.

6.3.3 Analysis of Deviance

The deviance is applicable only to models fitted to single years of age, as is evident from the appearance above of the Poisson and binomial distributions. It is not applicable to models fitted to complete individual lifetimes. If we have two nested models (that is, one includes all the parameters in the other) with p and $q > p$ parameters, respectively, and wish to test the hypothesis that the "true" model is that with p parameters, then the difference between the two deviances is approximately χ^2 with $q - p$ degrees of freedom. McCullagh and Nelder (1989) describe some of the practical difficulties that arise with non-normal models, and remark that the χ^2 approximation is often not very good, even asymptotically.

6.4 Information Criteria

An information criterion balances the goodness-of-fit of a model against its complexity, akin to the philosophy of Occam's Razor. The aim is to provide a single statistic which allows model comparison and selection. An early attempt at this balancing act was from Whittaker (1923); see Section 11.2. Information criteria are a development of Whittaker's balancing statistic to permit formal statistical tests. As a result, information criteria can be used to compare any two models which are based on exactly the same underlying data, whether or not they are nested. As a general rule, the smaller the value of the information criterion, the better the model.

One important point to note when comparing two models is that it is change or difference in an information criterion which is important, not the absolute value. Larger data sets tend to have larger absolute values of a given information criterion, so model selection is based on changes in the criterion value.

There are several different kinds of information criterion, including Akaike's Information Criterion, the Bayesian Information Criterion, the Deviance Information Criterion and the Hannan–Quinn Information Criterion. These are described in the following sections.

Table 6.1 *Difference between AIC and AIC^c for various sample sizes and parameter counts.*

n	\multicolumn{7}{c}{Number of parameters, k}
	1	2	5	10	20	30	50
1,000	0.004	0.012	0.060	0.222	0.858	1.920	5.374
5,000	$< 10^{-3}$	0.002	0.012	0.044	0.169	0.374	1.031
20,000	$< 10^{-3}$	$< 10^{-3}$	0.003	0.011	0.042	0.093	0.256
50,000	$< 10^{-4}$	$< 10^{-3}$	0.001	0.004	0.017	0.037	0.102
100,000	$< 10^{-4}$	$< 10^{-3}$	$< 10^{-3}$	0.002	0.008	0.019	0.051
500,000	$< 10^{-5}$	$< 10^{-4}$	$< 10^{-3}$	$< 10^{-3}$	0.002	0.004	0.010
1,000,000	$< 10^{-5}$	$< 10^{-4}$	$< 10^{-4}$	$< 10^{-3}$	$< 10^{-3}$	0.002	0.005

6.4.1 Akaike's Information Criterion

Akaike (1987) proposed a simple information criterion based on the log-likelihood, ℓ, and the number of parameters, k, as follows:

$$\text{AIC} = -2\ell + 2k. \tag{6.8}$$

Akaike's Information Criterion (AIC) is rather "forgiving" of extra parameters, so Hurvich and Tsai (1989) proposed a correction to the AIC for small sample sizes; the small-sample version, AIC^c, is defined as follows:

$$\text{AIC}^c = \text{AIC} + \frac{2k(k+1)}{n-k-1}, \tag{6.9}$$

where n is the number of independent observations. The difference between the AIC and AIC^c in equation (6.9) is tabulated in Table 6.1 for various values of n and k. Table 6.1 shows that the difference between the AIC and AIC^c is very small unless there are fewer than 20,000 independent observations and there are many parameters. For most survival-modelling work the size of data sets means that the AIC^c leads to the same conclusions as when using the AIC. For projections work, however, the AIC^c is a useful alternative to the AIC because the number of observations is typically smaller than 5,000 and the number of parameters is usually in excess of 50.

Another question is what sort of difference in AIC should be regarded as significant. A rule of thumb is that a difference of four or more AIC units (or AIC^c units) would be regarded as significant. This does require a degree of judgement: normally if there were two models with AICs within 4 units of each other, we would pick the more parsimonious, that is, the one with fewer parameters. However, with actuarial work there are additional considerations. For example, we know from long experience of analysing insurer portfolios that almost every risk factor interacts with age; specifically, mortality differentials reduce with age at a rate proportional to the strength of the initial differential. This would give us grounds for erring on the side of the more complex model if the additional complexity came from age interactions.

6.4.2 The Bayesian Information Criterion

Similar to the AIC, the Bayesian Information Criterion (BIC) also makes use of the number of independent observations, n, as follows:

$$\text{BIC} = -2\ell + k \log n. \tag{6.10}$$

The factor $\log n$, applied to the number of parameters k, will be higher than the factor of 2 used in the AIC if there are eight or more independent observations ($\log 8 = 2.079$). Since mortality models are typically built with over a thousand times more observations than this, selecting models using the BIC has the potential to produce a simpler end-model than when using the AIC. The BIC is also sometimes known as the Schwarz Information Criterion (SIC) after Schwarz (1978).

6.4.3 Other Information Criteria

For completeness we mention two other information criteria, although they generally have no advantages over the AIC or BIC for modelling survival data:

- *Deviance Information Criterion.* The Deviance Information Criterion (DIC) is based on the model deviance as a measure of the model fit, together with the number of parameters. The definition is as follows:

$$\text{DIC} = \text{Dev} + k, \tag{6.11}$$

where k is the number of parameters and Dev is the model deviance defined in Section 6.3. The DIC is particularly suited to Bayesian models estimated by Markov Chain Monte Carlo (MCMC) methods, which we shall not be using.

- *Hannan–Quinn Information Criterion.* The Hannan–Quinn Information Criterion (HQIC) is rarely used, but we include it here for completeness. The HQIC is defined as follows:

$$\text{HQIC} = -2\ell + k \log \log n, \tag{6.12}$$

where ℓ is the log-likelihood, k is the number of parameters and n is the number of independent observations. See Hannan and Quinn (1979).

6.5 Tests of Fit Based on Residuals

The guidance on goodness-of-fit provided by the values of an information criterion is at a very high level indeed, based on the entire log-likelihood. It tells us very little about how well a model fits the data at a lower level, for example at individual ages or over age ranges. Goodness-of-fit at this level is, of course, vital for a model which is to be used in actuarial work. In this section, we describe a battery of tests of more detailed goodness-of-fit, mainly based on the *residuals*, that is, the difference between the observed and fitted values.

6.5.1 Individual *versus* Grouped Tests

In standard survival-model work many tests are based around residuals at the level of the individual. In medical-trials work the relatively small number of observations makes this practical; an example of this is given in Collett (2003, p.238) for 26 ovarian-cancer patients.

However, actuarial work is very different because there are usually many thousands of observations. In Richards et al. (2013) there were over quarter of a million observations, but many data sets are even larger. Actuaries therefore need to use a different set of tools for analysing the residuals in their survival models.

The approach we use is to divide the data into categories, defined by values of the variable against which we want to investigate the residuals, and examine the exposure times and deaths which occurred in each category. This is most useful for variables which are either continuous or categories with large numbers of levels. For example:

- In the case of age, we would use non-overlapping age bands, for example [60, 61), [61, 62), ...
- For residuals against pension size we would divide into non-overlapping bands which contained roughly equal numbers of lives.

As an example, consider the age interval [60,61). Define x_i to be the earliest age and $x_i + t_i$ to be the latest age at which the ith individual was observed to be in that age interval. The vital point is that x_i and $x_i + t_i$ are calculated according to the principles in Section 2.9.1, therefore excluding any contribution from individuals who were never observed in that age interval (in the particular case of testing a model fitted to data at single years of age, the job is already done).

To proceed we make use of a result from Cox and Miller (1987), namely that the number of deaths in each sub-group has a Poisson distribution with parameter equal to the sum of the integrated hazard functions. Continuing the example above, let d_{60} be the number of deaths observed in the age interval $[60,61)$, and define Y_i to be an indicator, equal to 1 if the ith individual contributed to the exposure in that age interval and 0 otherwise. The relevant Poisson parameter, which we denote λ_{60}, is then defined as follows:

$$\lambda_{60} = \sum_{i=1}^{n} Y_i \Lambda_{x_i,t_i}, \qquad (6.13)$$

summing over all individuals, where $\Lambda_{x,t}$ is the integrated hazard function defined in equation (3.23).

Denote by D_{60} the random variable for which d_{60} is the observed value. Then, according to Cox and Miller (1987, Section 4.2), D_{60} will have a Poisson distribution with parameter λ_{60}. (In fact, this is only approximate for survival-model data, because Cox and Miller treat Poisson processes in the absence of right-censoring, but for reasons we will see in Chapter 17 it is an excellent approximation.) In general for any subgroup j we would have d_j deaths and a Poisson parameter λ_j built from the sum over all contributing lives of the integrated hazard functions over the relevant interval. The most appropriate type of residual here is the deviance residual, since the distribution is non-normal (McCullagh and Nelder, 1989, p.38). Since $E[D_j] = \lambda_j$ the Poisson deviance residual (see Section 6.3.1), r_j, is:

$$r_j = \operatorname{sign}(d_j - \hat{\lambda}_j) \sqrt{2 \left[d_j \log \frac{d_j}{\hat{\lambda}_j} - (d_j - \hat{\lambda}_j) \right]}, \qquad (6.14)$$

where $d_j \log d_j \to 0$ as $d_j \to 0$. Also, λ_j is non-zero by definition, as we cannot test the model fit for a sub-group which has no exposure. If the Poisson parameters λ_j are not too small, and if the model is a good fit, then the deviance residuals are assumed to be i.i.d. $N(0,1)$ (this assumption is explored in Section 6.6.1). We can then use these residuals to perform a series of tests of fit. An example set of deviance residuals is shown in Figure 6.1 and tabulated in Table 6.2. The reasons for using the deviance residual in place of the better-known Pearson residual are explored in Section 6.6.1.

(Note that r_j is distinct from $\hat{r}_x$, which we have used since Chapter 4 to denote a raw mortality ratio. There should be no risk of confusion, as r_j the residual is never seen with a "hat" and $\hat{r}_x$ the mortality ratio is never seen without one.)

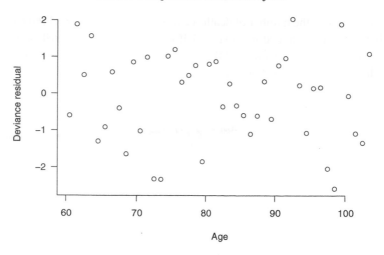

Figure 6.1 Deviance residuals by age calculated using equation (6.14) and the model fit from Figure 5.1.

6.6 Statistical Tests of Fit

When testing the normality of a set of residuals, we can test a number of features:

- *Overall fit.* We want the residuals to be small enough that they are consistent with random variation, i.e. the model should have no large deviations in fit from the observed values. For this we can use the χ^2 test or the standardised-deviations test.
- *Bias.* We want the residuals to be reasonably balanced between positive and negative values. That is, there should be no overall substantial bias towards over- or underestimation.
- *Under- or overestimation over ranges of ages.* We do not want the model to systematically under- or overestimate over ranges of ages. For this we can use the runs test or the lag-1 autocorrelation test. Note that this is different to the bias point above – a model may pass a simple bias test even if all the residuals below the median age are positive and all those above are negative.

6.6.1 The χ^2 Test

The χ^2 test statistic is simply the sum of squared deviance residuals:

$$\tilde{\chi}^2 = \sum_j r_j^2. \tag{6.15}$$

Table 6.2 *Deviance residuals calculated using equation (6.14) from the model fit in Figure 5.1; also shown in Figure 6.1.*

Age	r	r^2	Age	r	r^2
60	−0.598	0.358	82	−0.367	0.135
61	1.884	3.548	83	0.265	0.070
62	0.501	0.251	84	−0.336	0.113
63	1.559	2.431	85	−0.594	0.353
64	−1.303	1.698	86	−1.102	1.215
65	−0.918	0.843	87	−0.607	0.369
66	0.576	0.332	88	0.330	0.109
67	−0.405	0.164	89	−0.691	0.478
68	−1.645	2.705	90	0.764	0.584
69	0.847	0.718	91	0.963	0.927
70	−1.023	1.047	92	2.032	4.131
71	0.981	0.962	93	0.224	0.050
72	−2.319	5.377	94	−1.070	1.146
73	−2.331	5.436	95	0.141	0.020
74	1.010	1.020	96	0.163	0.027
75	1.193	1.424	97	−2.039	4.158
76	0.301	0.091	98	−2.578	6.645
77	0.488	0.238	99	1.883	3.544
78	0.760	0.578	100	−0.076	0.006
79	−1.852	3.430	101	−1.096	1.200
80	0.797	0.635	102	−1.345	1.810
81	0.868	0.753	103	1.073	1.151
Sum				−4.692	62.279

In Table 6.2 we can see that $\tilde{\chi}^2 = 62.279$. The number of degrees of freedom to test against is far from simple to decide. In many instances a pragmatic decision will be made as to the number of degrees of freedom, and often it is simply the number of residuals. In this case, our test statistic of 62.279 has a p-value of 0.036 on $n = 44$ degrees of freedom, so the model fit would fail the χ^2 test.

If the basic model structure is not a good match to the shape of the data, then the χ^2 test statistic will be high, but it may be high for other reasons. One is that major sources of heterogeneity have not been accounted for in the model. Another is that there is a very large amount of data. The χ^2 test is basically testing a hypothesis, which is a conjecture about an unknown "true" model. If enough data are collected, any hypothesis will be rejected.

It is worth noting some important underlying assumptions behind equation (6.15). If the model is correct, the residuals are usually assumed to be i.i.d. $N(0,1)$. This in turn means that $\{r_j^2\}$ are values drawn from the χ_1^2 distribution, and thus that $\tilde{\chi}^2$ is a value drawn from the χ^2_{n-k} distribution (k is the number of constraints or parameters estimated in the model). Comparing $\tilde{\chi}^2$ against the appropriate percentage point of the χ^2_{n-k} distribution then gives us our test of goodness-of-fit.

So, we have our testing edifice, but how sound is the foundation? In particular, how sound is the assumption that $\{r_j\}$ are N(0,1)? We can test this via the simulation of Poisson variates using a known methodology, and then looking at the normal quantile-quantile plot of the two alternative definitions of residuals, i.e. Pearson and deviance. The quantile-quantile plot is a graph of the quantiles of the residuals against the quantiles of the N(0,1) distribution; if the plotted points form a straight line through the original with slope 1, the residuals are plausibly N(0,1). There is a function in R for this, qqnorm(), and we can use this to test the nature of the various definitions of residual.

The Pearson residual arises from the assumption of normality, i.e. subtracting the mean and dividing by the standard error normalises a variate to N(0,1). The Pearson residual is therefore defined as follows for a Poisson random variable with parameter λ and observed number of events d:

$$r = \frac{d - \hat{\lambda}}{\sqrt{\hat{\lambda}}}. \tag{6.16}$$

Figure 6.2 shows two features of Pearson residuals:

(i) The quantile-quantile plots are only passably linear when λ approaches 100, i.e. the number of expected deaths in each count needs to be over 50 (say).
(ii) The N(0,1) distribution is continuous, whereas the quantile-quantile plots show pronounced banding up to $\lambda = 20$.

These observations suggest a rule-of-thumb for deviances and goodness-of-fit tests with models for Poisson counts, that there should be at least 20 expected deaths per cell.

The deviance residual is defined in equation (6.14). The simulated Poisson counts in Figure 6.2 are reused to calculate the quantile-quantile plots of deviance residuals in Figure 6.3.

Figure 6.3 shows that the deviance residuals are better than Pearson residuals in all four cases; no matter the value of λ, the normal quantile-quantile plot of deviance residuals is closer to a straight line of slope 1. We thus prefer the deviance residual in all cases, and certainly when the expected number of deaths is below 100.

However, Figure 6.3 also shows that deviance residuals are no panacea; if the number of expected deaths is below 20, the distribution still does not approach a continuous random variable, as shown by the step function in place of a

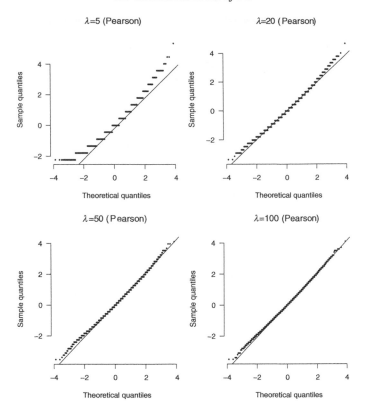

Figure 6.2 Normal quantile-quantile plots for Pearson residuals for 10,000 simulated Poisson counts with varying expected values ($\lambda \in \{5, 20, 50, 100\}$).

straight line. While the deviance residual is a more reliable definition to use for the χ^2 test statistic, it is still important to avoid cells with fewer than five expected deaths. This needs to be borne in mind when testing the quality of fit using the Poisson assumption, and it is also a reason to be careful of Poisson mortality models for grouped counts.

6.6.2 The Standardised-Deviations Test

An alternative to the χ^2 test of Section 6.6.1 is to divide the residuals into m groups, where m is the integer part of the square root of the number of residuals, to ensure that the number of residual groups is set suitably. Clearly, this test cannot be used unless $m \geq 2$.

As the N(0,1) distribution is open on both the left and right, we calculate $m - 1$ breakpoints of the inverse cumulative distribution function, Φ^{-1},

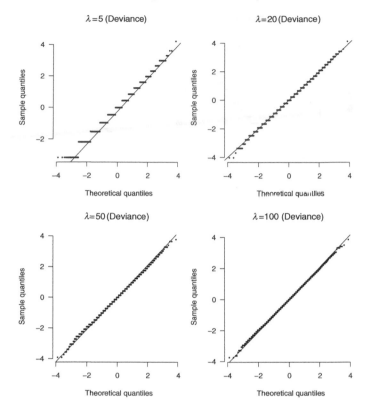

Figure 6.3 Normal quantile-quantile plots for deviance residuals for 10,000 simulated Poisson counts with varying expected values ($\lambda \in \{5, 20, 50, 100\}$).

i.e. the breakpoints are $\{\Phi^{-1}(\frac{k}{m}), k = 1, 2, \ldots, m - 1\}$. The intervals are therefore $\left(-\infty, \Phi^{-1}(\frac{1}{m})\right), \left(\Phi^{-1}(\frac{1}{m}), \Phi^{-1}(\frac{2}{m})\right), \ldots, \left(\Phi^{-1}(\frac{m-1}{m}), +\infty\right)$. This is illustrated in Figure 6.4.

On average, there should be $f = n/m$ residuals in each interval, and this forms the basis of the test statistic:

$$Y = \sum_{j=1}^{m} \frac{(c_j - f)^2}{f}, \qquad (6.17)$$

where c_j is the count of residuals falling into the jth non-overlapping interval. If the deviance residuals are i.i.d. N(0,1), then Y will have a χ^2 distribution with $m - 1$ degrees of freedom. An example of this calculation is given in Table 6.3.

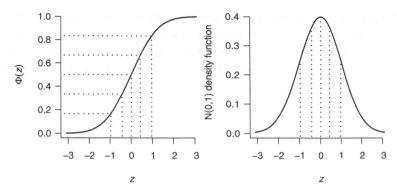

Figure 6.4 Setting breakpoints for standardised-deviations test with $m = 6$. The left panel shows the cumulative distribution function for N(0,1), showing that the five breakpoints divide the range of z into $m = 6$ ranges of equal probability. The right panel shows these same breakpoints on the N(0,1) density function.

Table 6.3 *Standardised-deviations test for residuals in Table 6.2.*

Interval	f	c_j	$\dfrac{(c_j - f)^2}{f}$
$(-\infty,-0.967)$	7.333	12	2.970
$(-0.967,-0.431)$	7.333	5	0.742
$(-0.431,0.000)$	7.333	4	1.515
$(0.000,0.431)$	7.333	6	0.242
$(0.431,0.967)$	7.333	9	0.379
$(0.967,+\infty)$	7.333	8	0.061
Sum	44	44	5.909

The test statistic is $Y = 5.909$ with $m - 1 = 5$ degrees of freedom, which gives us a p-value of 0.315 and the test is passed.

6.6.3 The Bias Test

We count the number of residuals which are strictly greater than zero, calling this n_1, and the number which are strictly less than zero, calling this n_2. For completeness we also count the number of residuals which are exactly zero, calling this n_3, although in practice n_3 is almost always zero. The total number of residuals is $n_1 + n_2 + n_3$. Our hypothesis is that a residual is negative or non-negative each with probability 1/2 independent of all the others. Note that this is slightly weaker than assuming the residuals to be i.i.d. N(0,1), but not by much.

Without loss of generality we can find the probability that there should have been $n_1 + n_3$ non-negative residuals under a binomial$(n_1 + n_2 + n_3, 1/2)$ distribution. We work out $P[N \leq n_1 + n_3]$ under this model and compare the probability with the desired test level, in practice commonly 5%. Among the residuals in Table 6.2 we have 23 non-negative values out of $n = 44$, giving a p-value of 0.6742, and so the residuals pass the test (as would be expected from a simple visual inspection of Figure 6.1). It is rare for a properly specified survival model to fail this test unless it has a wholly inappropriate shape.

The bias test is traditionally also known as the signs test.

6.6.4 The Runs Test

For ordinal variables like age and pension size we have the option of the runs test. We count the number of changes in sign of the residuals in $\{r_i\}$, which is then one less than the number of runs of residuals of the same sign; call this statistic u.

Using the same hypothesis as in the bias test in Section 6.6.3, we work out the probability that the number of runs U, a random variable, could be as small as u. Without loss of generality, we focus on the probability of u runs or fewer arising among $n_1 + n_3$ non-negative residuals and n_2 negative residuals. From Kendall and Stuart (1973, Exercise 30.8, p.480), the probability of u runs is:

$$
P[U = u] =
\begin{cases}
\dfrac{2\dbinom{n_1 + n_2 - 1}{k - 1}\dbinom{n_3 - 1}{k - 1}}{\dbinom{n_1 + n_2 + n_3}{n_1 + n_2}}, & u = 2k \\[3em]
\dfrac{\dbinom{n_1 + n_2 - 1}{k}\dbinom{n_3 - 1}{k - 1} + \dbinom{n_1 + n_2 - 1}{k - 1}\dbinom{n_3 - 1}{k}}{\dbinom{n_1 + n_2 + n_3}{n_1 + n_2}}, & u = 2k + 1.
\end{cases}
$$

$$(6.18)$$

For our test we want the probability that the number of runs should have been less than or equal to u, $P[U \leq u]$. In our example in Table 6.2 we have $u = 24$ runs over $n = 44$ residuals, which has a p-value of 0.6825 and the model fit passes the runs test (as is obvious from a visual inspection of Figure 6.1).

6.6.5 The lag-1 Autocorrelation Test

The usual test statistic is that from Durbin and Watson (1971). However, a widely used test for CMI graduations in the UK is described by Forfar et al.

(1988, p.46). As with the runs test in Section 6.6.4, if the residuals are for an ordinal variable like age or pension size, we can calculate the correlation coefficient between successive residuals. We define $\bar{z}_1$ and $\bar{z}_2$ as the mean of the first $n-1$ and last $n-1$ residuals in $\{r_j, j = 1, 2, \ldots n\}$, respectively, i.e.:

$$\bar{z}_1 = \frac{1}{n-1} \sum_{i=1}^{n-1} r_j \qquad (6.19)$$

$$\bar{z}_2 = \frac{1}{n-1} \sum_{i=2}^{n} r_j. \qquad (6.20)$$

We then define the lag-1 sample autocorrelation coefficient, c_1, for the residuals as follows:

$$c_1 = \frac{\sum_{j=1}^{n-1}(r_j - \bar{z}_1)(r_{j+1} - \bar{z}_2)}{\sqrt{\left(\sum_{j=1}^{n-1}(r_j - \bar{z}_1)^2\right)\left(\sum_{j=2}^{n}(r_j - \bar{z}_2)^2\right)}}. \qquad (6.21)$$

Adapting Forfar et al. (1988), the test statistic $Z = c_1 \sqrt{n-1}$ should have an approximately $N(0,1)$ distribution. For the data in Table 6.2 we calculate:

$$\bar{z}_1 = -0.134 \qquad (6.22)$$
$$\bar{z}_2 = -0.095 \qquad (6.23)$$
$$c_1 = -0.010 \qquad (6.24)$$
$$Z = -0.064. \qquad (6.25)$$

We can see that $Z = -0.064$ is nowhere near the tail of the $N(0,1)$ distribution and so the residuals in Table 6.2 pass the lag-1 autocorrelation test. This is clear from a visual inspection of Figure 6.1.

6.7 Financial Tests of Fit

Thus far we have considered standard statistical procedures and tests for examining models. For a model to be useful for actuarial purposes we also require that all financially significant risk factors are included. All the tests and procedures discussed in this chapter so far are statistical. In particular, each life had equal weight. However, not all lives are equal in their financial impact. All else being equal, a pensioner receiving £10,000 per annum has ten times the financial importance of a pensioner receiving £1,000 per annum. Thus, one area where actuaries' needs differ from those of statisticians is testing the *financial* suitability of a model, in addition to its statistical acceptability. In this section we will describe a general procedure and illustrate it for a pension scheme by using the annual pension. However, it is up to the analyst to decide what the most appropriate financial measure would be for the risk concerned and the purpose. For a term-insurance portfolio one would replace pension amount with the sum assured. Alternatively, the amount used might be the policy reserve or some measure of profitability.

This can be done through a process of repeated sampling to test whether the model predicts variation not only in the number of deaths, but also in the amounts-weighted number of deaths. We use the process of bootstrapping described in Richards (2008), as follows:

(i) We randomly sample $b = 1,000$ records with replacement.
(ii) We use the fitted model to predict the number of deaths and the pension amounts ceasing due to death in the random sample.
(iii) We calculate the ratios of the actual number of deaths and pension amounts ceasing compared to the model's predicted number and amounts. The ratios A/E (of actual mortality experience against expected deaths) to calculate for a given sample are:

$$\text{Bootstrap A/E}_{\text{lives}} \quad = \sum_{i=1}^{b} d_i \bigg/ \sum_{i=1}^{b} \Lambda_{x_i,t_i} \qquad (6.26)$$

$$\text{Bootstrap A/E}_{\text{amounts}} = \sum_{i=1}^{b} w_i d_i \bigg/ \sum_{i=1}^{b} w_i \Lambda_{x_i,t_i}, \qquad (6.27)$$

where w_i is the benefit size, such as annual pension, as in Section 8.2, and Λ_{x_i,t_i} is the integrated hazard function for the ith individual from equation (3.23). Here we are making use of equation (6.13) again: the number of deaths in each sub-group has an approximate Poisson distribution with the parameter equal to the sum of the integrated hazard functions. R source code to perform this procedure is in the `bootstrap()` function described in Appendix H.2.

(iv) We repeat the three steps above a large number of times; we have used 10,000. Suitable summary statistics of this sample of A/E ratios (we use the medians) should be close to 100% if the fitted model is generating samples consistent with the observations.

The results of this are shown for three models in Table 6.4. These are models of the 2012 experience in the Case Study, in which first gender and then

Table 6.4 *Median bootstrap ratios for Gompertz model for the Case Study, 2012 mortality experience; 10,000 samples of 1,000 lives with replacement.*

Model	$100\% \times A/E_{lives}$	$100\% \times A/E_{amounts}$
Age	100.4	91.7
Age+Gender	99.8	87.2
Age+Gender+Size	99.8	96.1

pension amount are included as covariates (see Chapter 7). The ratios in the lives column are close to 100%. This means that all three models have a good record in predicting the number of deaths, as one would expect for models fitted by the method of maximum likelihood. However, we can see that the first two models perform poorly when the ratio is weighted by pension size. This is because those with larger pensions have lower mortality, thus leading the first two models to overstate pension-weighted mortality. The first two models in Table 6.4 are therefore unacceptable for financial purposes. The third model is closer to being acceptable, but would require more development to get the bootstrapped A/E ratios closer to 100% for both lives- and amounts-weighted measures. A further option, discussed in Richards (2008), would be to weight the log-likelihood by pension size, but this should be viewed very much as a last resort when all other approaches fail. The reason is that weighting the log-likelihood abandons any underlying theory to allow the analyst to obtain statistical properties of the estimates.

7

Modelling Features of the Portfolio

7.1 Categorical and Continuous Variables

In most portfolios of interest to actuaries, the members have differing attributes, other than age, that we believe might affect the risk of dying. Some of these are discrete attributes, called *categorical variables*, that partition the portfolio into distinct subsets. Examples are:

- gender
- smoking status
- occupation
- socio-economic class
- nationality or domicile.

Even a straightforward categorical variable may introduce some delicate points of definition, requiring judgements to be made. For example, to what gender should we assign a person who has undergone transgender surgery? Or, is someone who gave up smoking cigarettes ten years ago a smoker or a non-smoker? Is someone who has switched from cigarettes to e-cigarettes a smoker or a non-smoker? We may decide to exclude such difficult cases from the analysis, or we may devise a rule that will at least ensure consistency.

In other cases, attributes may take any value within a reasonable range. These we may call *continuous attributes* or *variables*. The amount of pension or assurance benefit is an example; it is a proxy for affluence, and there is abundant evidence that affluence has a major influence on mortality and morbidity (see, for example, Madrigal et al., 2011). Figure 7.1 shows empirical estimates of survival functions from age 60 for the highest and lowest pension size bands in a large multi-employer German pension scheme described in Richards et al. (2013), for males only. (These are Kaplan–Meier non-parametric estimators, which we will meet in Chapter 8.)

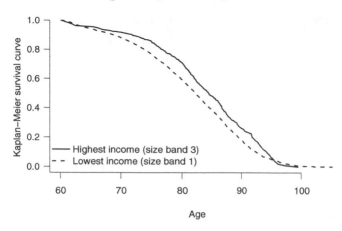

Figure 7.1 Empirical estimates of survival functions from age 60 for the highest and lowest pension size bands in a large multi-employer German pension scheme described in Richards et al. (2013), males only.

The division between discrete and continuous attributes is not hard and fast. For example, year of birth is known to affect mortality in a systematic way in some populations – the so-called *cohort effect* (see Willets, 1999; Willets, 2004 or Richards et al., 2006) . We may choose to regard this as a discrete attribute with a very large number of values, or approximately as a continuous attribute, whichever is more convenient.

Continuous attributes can always be turned into discrete attributes by dividing the range of values into a discrete number of intervals and grouping the data. For example, pension amounts might be grouped into the four quartiles of the distribution of its values.

Legislation may limit the attributes that may be used for certain actuarial purposes. For example, in the EU since 2012 it has been illegal to use gender to price an insurance contract for individuals, but it is not illegal to take it into account in pricing transactions between corporate entities, such as bulk buy-outs of pension schemes. For internal risk management and reserving, gender is almost always used.

To represent attributes, we associate a *covariate vector* z_i with the ith individual in the experience. For example, suppose we have two discrete covariates:

- gender, labelled 0 = male and 1 = female
- current smoking status, labelled 0 = does not smoke cigarettes and 1 = smokes cigarettes,

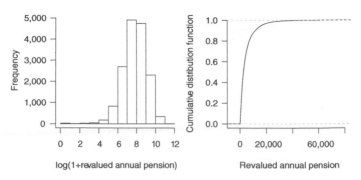

Figure 7.2 Distribution of pension size (British £) for the scheme in the Case Study, all ages, 2007–2012 and with pensions to deceased cases revalued by 2.5% per annum from the date of death to the end of 2012. Note that log(1 + pension size) is used in case pension size may legitimately be recorded as zero.

and one continuous covariate:

• annual amount of pension.

Figure 7.2 shows the distribution of pension amount for the scheme in the Case Study, all ages, 2007–2012 and with pensions to deceased persons revalued by 2.5% per annum from the date of death to the end of 2012. Notice that the histogram on the left shows log(1 + pension size): we take logarithms, as the distribution is so extremely skewed, and then in case there may be legitimate records with pension size zero, we use (1 + pension size), which has a negligible effect on the analysis. Alternatively, such cases could be excluded if we are confident they are anomalous.

Then if the covariate vector represents these attributes in the order given above, $z_i = (z_{i1}, z_{i2}, z_{i3}) = (1, 0, 15704)$ (for example) means that the ith individual is a female who does not smoke cigarettes and who has a pension amount of £15,704 per year. Note that both categorical variables here are labels representing a qualitative attribute, but some categorical variables can be ordinal, for example year of birth or pension size band.

We suppose that the hazard rate may now be a function of the covariates as well as age:

$$\text{hazard rate for } i\text{th individual} = \mu(x, z_i). \tag{7.1}$$

Age is listed separately because it is usually of primary interest in actuarial investigations. In medical statistics, this may not be the case, and age might be modelled as just one covariate among many others, perhaps quite crudely, for example, age 60 and over, or age less than 60.

The actuary's task, now, has two additional parts, namely:

- determining which, if any, of the available covariates has a large enough and clear enough influence on mortality to be worth allowing for
- in such cases, fitting an adequate model as in equation (7.1).

7.2 Stratifying the Experience

The UK pension scheme used in the Case Study has both male and female members, and each record includes the benefit amount (annual amount of pension). It is commonly observed that males and females differ in their mortality, and that socio-economic status or affluence, for which benefit amount may be a proxy, also affects mortality. Instead of analysing the combined experience, as in Chapter 5, we might explore the possible effects of these covariates. Note that we are able to do so only because their values are included in each member's record. Other covariates of possible interest, for example smoking status, are not recorded so no analysis is possible. Mathematically, we have defined a discrete set of covariate values $\mathcal{Z}$ say, and for each $z \in \mathcal{Z}$ we fit a separate model of the hazard rate:

$$\text{hazard rate for covariate } z = \mu_z(x). \tag{7.2}$$

7.2.1 The Case Study: Stratifying by Gender

To stratify by gender, we simply carry out the fitting procedure in Chapter 5 separately for males and females, thus producing two separate models. For simplicity, we will again assume that a Gompertz model will be adequate, postponing any discussion of model adequacy.

Figure 7.3 shows the R outputs resulting from fitting a Gompertz model to the 2007–2012 data for male lives only, in the pension scheme in the Case Study, and Figure 7.4 shows the Poisson deviance residuals (see Section 6.3.1) by single ages. Figures 7.5 and 7.6 show the corresponding results for the female lives in the same scheme. These can be compared with Figures 5.1 and 6.1, where a Gompertz model was fitted to the combined data, and also with Figure 4.3. Note that comparison of the AICs is meaningless (because we are not comparing models fitted to exactly the same data; see Section 6.4).

```
nlm() return code 1.
Data cover 5626 lives and 878 deaths.
Total exposure time is 25196.82 years.

Iterations=16
Log-likelihood=-3376.48
AIC=6756.95

Parameter    Estimate Std. error
------------ -------- ----------
Intercept    -12.2271  0.315336
Age           0.116823 0.0039043
```

Figure 7.3 Output from fitting Gompertz model to 2007–2012 data from the pension scheme in the Case Study, males only.

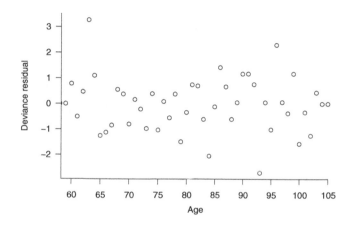

Figure 7.4 Deviance residuals from fitting Gompertz model to 2007–2012 data from the pension scheme in the Case Study, males only.

Figure 7.7 shows the fitted (log) hazard rates. As can be seen from the R outputs, the Gompertz functions have slightly different slopes.

7.2.2 The Case Study: Stratifying by Benefit Amount

As shown in Figure 7.2, pension size is so skewed that it is appropriate to transform it for inclusion as a covariate. Using $z_{i2} = \log(1 + \text{pension size})$ is an obvious choice (see the discussion of Figure 7.2).

Strictly speaking, benefit amount is a categorical variable, since it is measured in minimum units that may be pounds or pence (in the example from

```
nlm() return code 1.
Data cover 9147 lives and 1150 deaths.
Total exposure time is 40885.59 years.

Iterations=17
Log-likelihood=-4497.46
AIC=8998.92

Parameter     Estimate Std. error
------------  -------- ----------
Intercept     -12.8487  0.263683
Age            0.120288 0.0031716
```

Figure 7.5 Output from fitting Gompertz model to 2007–2012 data from the pension scheme in the Case Study, females only.

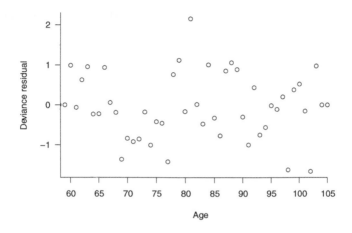

Figure 7.6 Deviance residuals from fitting Gompertz model to 2007–2012 data from the pension scheme in the Case Study, females only.

the UK). However, the number of categories needed to accommodate all benefit amounts would be so great that in practice it is easier to adopt one of two approaches:

- regard benefit amount as essentially continuous, and model it as a covariate directly (see Section 7.4)
- stratify the analysis by benefit amount, by defining a reasonable number of relatively homogeneous bands.

Stratifying an essentially continuous covariate can be done in any convenient way. One approach that has the advantage of not prejudging the outcome is to

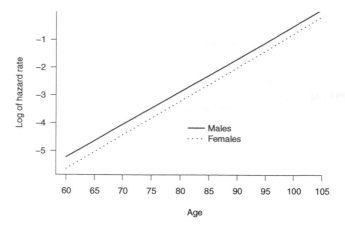

Figure 7.7 Log hazard rates from fitting Gompertz model to 2007–2012 data from
the pension scheme in the Case Study, separately to males and females.

```
nlm() return code 1.
Data cover 12995 lives and 1831 deaths.
Total exposure time is 57988.09 years.

Iterations=17
Log-likelihood=-7120.01
AIC=14244.01

Parameter     Estimate Std. error
------------  -------- ----------
Intercept     -12.2998  0.210117
Age           0.115322 0.0025588
```

Figure 7.8 Output from fitting Gompertz model to 2007–2012 data from the pension scheme in the Case Study, pension < £10,000 only.

divide the benefit amounts into percentiles (for example, quartiles or deciles).
Here, for a simple illustration, we sub-divide pension size into two groups,
below £10,000 and £10,000 or above.

Figure 7.8 shows the R outputs resulting from fitting a Gompertz model to
the 2007–2012 data for persons with pensions less than £10,000 only, in the
pension scheme in the Case Study, and Figure 7.9 shows the deviance residuals
by single ages. Figures 7.10 and 7.11 show the corresponding results for persons with pensions of £10,000 or more in the same scheme. Figure 7.11 shows
that nobody over age 100 has a pension of £10,000 or more. These figures can
also be compared with Figures 5.1 and 6.1.

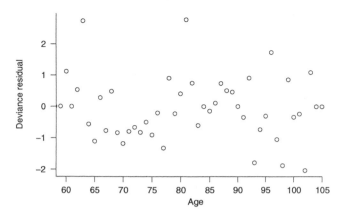

Figure 7.9 Deviance residuals from fitting Gompertz model to 2007–2012 data from the pension scheme in the Case Study, pension < £10,000 only.

```
nlm() return code 1.
Data cover 1778 lives and 197 deaths.
Total exposure time is 8094.32 years.

Iterations=17
Log-likelihood=-779.70
AIC=1563.40

Parameter    Estimate Std. error
------------ -------- ----------
Intercept    -13.5823  0.630846
Age           0.130057 0.0076727
```

Figure 7.10 Output from fitting Gompertz model to 2007–2012 data from the pension scheme in the Case Study, pension ≥ £10,000 only.

Figure 7.12 shows the (log) hazard rates obtained for the data split by pension size. This time, we see that the fitted Gompertz functions cross over. At lower ages persons with the smaller pensions have markedly higher mortality, but at higher ages this is reversed. One possibility is that the Gompertz model (which is very simple) is a poor fit at higher ages, and this could be investigated. The actuary has to decide whether this feature is acceptable for the purpose at hand.

Note that by stratifying the benefit amount in this way we obtain a subset with only 197 deaths of persons with large pensions (as we defined them), whereas in the complete data set there were 2,028 deaths. This shows that as

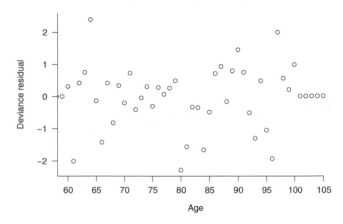

Figure 7.11 Deviance residuals from fitting Gompertz model to 2007–2012 data from the pension scheme in the Case Study, pension ≥ £10,000 only.

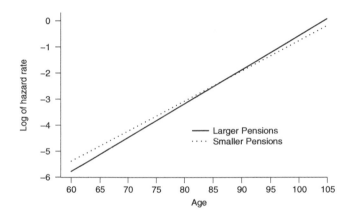

Figure 7.12 Log hazard rates from fitting Gompertz model to 2007–2012 data from the pension scheme in the Case Study, separately to persons with pension size < £10,000 and ≥ £10,000.

we sub-divide the experience, we will generate subsets with fewer and fewer deaths. We discuss this in more detail next.

7.3 Consequences of Stratifying the Data

Stratifying the experience, as above, has the advantage of simplicity and of requiring no more statistical sophistication to fit the individual models than

Table 7.1 *Deaths categorised by risk-factor combination for a large portfolio of German pension schemes. Source: Richards et al. (2013).*

Member of largest scheme	Region	Scheme type	Pension size-band	Normal retirees: Females	Normal retirees: Males	Ill-health retirees: Females	Ill-health retirees: Males	Widow(er)s: Females	Widow(er)s: Males
No	B	1	1	5,142	5,313	525	738	4,434	618
			2	824	725	39	98	36	0
			3	282	413	14	33	24	1
		2	1	2,200	1,323	308	183	628	222
			2	305	275	20	39	18	0
			3	140	206	15	18	15	1
	P	1	1	695	811	51	99	798	89
			2	138	122	7	22	9	0
			3	59	72	1	5	3	1
		2	1	174	274	26	33	166	23
			2	26	56	3	4	4	0
			3	8	41	5	2	5	0
Yes	B	1	1	480	338	41	45	224	47
			2	108	65	12	3	4	0
			3	60	45	1	3	4	0
Totals				10,641	10,079	1,068	1,325	6,372	1,002

to fit the combined model. For each model we used the same R code as in Chapter 5, reading an appropriate input file. In many circumstances, this will be a satisfactory approach. However, it raises questions, mentioned in Section 1.7, and begins to encounter problems, especially as the number of separate experiences to be fitted becomes large:

• By dividing the whole experience into smaller parts, each on its own is a smaller set of data and so statistical procedures lose power when applied to each separately. For example, Table 7.1 shows the number of deaths in a large portfolio of German pension schemes, stratified by gender (two categories), type of pensioner (normal, ill-health or widow(er), three categories), pension size (three categories), scheme type (public sector or private sector) region of residence (two categories) and membership, or not, of the largest scheme (two categories), a total of $2 \times 3 \times 3 \times 2 \times 2 \times 2 = 144$ cells. Many of these cells have too few deaths to support any analysis of mortality by age.

- We have not really avoided the need for greater statistical sophistication. We did not need a more sophisticated fitting procedure, but we now face questions about how to compare the individual fitted models. For example, if we compare the models fitted separately for males and females and find that they are *not* significantly different, we may conclude that we would be better to return to the combined model of Chapter 5. We considered methods of comparing different experiences in Chapter 6.

- Stratifying the data emphasises differences between parts of the experience, at the expense of losing any advantage to be gained from remaining similarities between parts of the experience. For example, if it is the case that hazard rates for males and females have a similar shape but different levels, then for the sake of allowing for the different levels we have lost the ability to use the combined data to fit the common shape.

An alternative to stratification, therefore, is to include the covariates in the model of the hazard rate, as in equation (7.1), so that they contribute directly to the likelihood. This does require a suitable form of equation (7.1) to be found, which must pass tests of being a reasonable representation of the data. Sometimes this may not be possible; for example it may be found that males and females have such differently shaped hazard rates that it is difficult to bring them into a simple functional relationship. However, if it can be done, it may overcome the drawbacks of stratifying the data listed above. In particular, looking back to Chapter 6, because the underlying data remain unchanged, the AIC and other information criteria may be used to assess the significance of covariates, singly or in combination.

7.4 Example: a Proportional Hazards Model

As a simple example of modelling instead of stratifying the data, we formulate a *proportional hazards* model with gender and benefit amount as covariates and apply it to the data in the Case Study. Define the two covariates for the ith scheme member:

$$z_{i1} = 0 \text{ if the } i\text{th member is male, 1 if female} \qquad (7.3)$$

$$z_{i2} = \text{benefit amount of the } i\text{th member}, \qquad (7.4)$$

and for brevity let $z_i = (z_{i1}, z_{i2})$ be the covariate vector of the ith member. Then specify the model of the hazard rate as follows:

$$\mu(x, z_i) = \mu(x)\, e^{\zeta_1 z_{i1} + \zeta_2 z_{i2}}. \tag{7.5}$$

In practice, the distribution of pension sizes is so skewed that we would usually define the covariate z_{i2} as some suitable transform of the pension size, as we will see. We shall examine in detail some important features of this model:

- The hazard rate has two factors. The first is $\mu(x)$, which is a function of age alone and is the same for all individuals. This determines the general shape of the hazard rates and is usually called the *baseline hazard*. The second is the term $\exp(\zeta_1 z_{i1} + \zeta_2 z_{i2})$, which depends on the ith member's covariates but not on age.

- The exponentiated term takes the form of a linear regression model on two covariates, with regression coefficients ζ_1 and ζ_2 modelling the effect of gender and benefit amount, respectively. These additional parameters must be fitted in addition to any parameters of the baseline hazard.

- We will assume that the baseline hazard is a Gompertz function as in equation (5.1).

- The name "proportional hazards" derives from the fact that the hazard rates of two individuals, say the ith and jth members, are in the same proportion at all ages x:

$$\frac{\mu(x, z_i)}{\mu(x, z_j)} = \frac{\mu(x)\, e^{\zeta_1 z_{i1} + \zeta_2 z_{i2}}}{\mu(x)\, e^{\zeta_1 z_{j1} + \zeta_2 z_{j2}}} = \frac{e^{\zeta_1 z_{i1} + \zeta_2 z_{i2}}}{e^{\zeta_1 z_{j1} + \zeta_2 z_{j2}}}. \tag{7.6}$$

- The likelihood (5.15) and log-likelihood (5.16) now become:

$$L(\alpha, \beta, \zeta_1, \zeta_2) \propto \prod_{i=1}^{n} \exp\left(-\int_0^{t_i} \exp(\alpha + \beta(x_i + t) + \zeta_1 z_{i1} + \zeta_2 z_{i2})\, dt\right)$$

$$\times \exp(\alpha + \beta(x_i + t_i) + \zeta_1 z_{i1} + \zeta_2 z_{i2})^{d_i} \tag{7.7}$$

$$\ell(\alpha, \beta, \zeta_1, \zeta_2) = -\sum_{i=1}^{n} \int_0^{t_i} \exp(\alpha + \beta(x_i + t) + \zeta_1 z_{i1} + \zeta_2 z_{i2})\, dt$$

$$+ \sum_{d_i=1} (\alpha + \beta(x_i + t_i) + \zeta_1 z_{i1} + \zeta_2 z_{i2}). \tag{7.8}$$

Although more complicated in appearance than equations (5.15) and (5.16), the only modifications to the fitting procedure used in Chapter 5 are as follows:

- The analyst must enlarge the function NegLogL() to accommodate the additional regression terms of the model. The Gompertz model is based on a function linear in age. At the heart of the proportional hazards model is a function linear in age and the covariate values, the simplest possible extension of the Gompertz model.

- The R function nlm() must be given a vector of length 4 of starting parameter values and a vector of length 4 of parameter sizes.

• The additional coding work for the analyst is very slight.

7.5 The Cox Model

The proportional hazards model is most closely associated with the famous Cox model used in medical statistics (Cox, 1972). However, actuaries are perhaps not very likely to use the proportional hazards model as it is used in the Cox model.

The Cox model assumes a proportional hazards formulation, exactly as in equation (7.5) above, but then assumes that the baseline hazard $\mu(x)$ is not important and need not be estimated. This can often be justified in a clinical trials setting. If the research question of interest is purely about the effect of the covariates, as measured by the regression coefficients ζ, the baseline hazard is a nuisance. Actuaries, however, usually need to estimate the entire hazard function.

Cox's contribution was to observe that if all factors involving periods between observed events were dropped from the likelihood, yielding a *partial likelihood*, then the estimates of the regression coefficients obtained by maximising this partial likelihood have an extremely simple form not involving the baseline hazard at all. Moreover, he showed by heuristic arguments (later justified formally) that these estimates should possess all the attractive features of true maximum likelihood estimates (MLEs). Cox (1972) went on to become one of the most cited statistics papers ever written. Any text on survival analysis, such as Collett (2003), will give details of the Cox model and the partial likelihood estimates.

The fact that the Cox model avoids the need to estimate the baseline hazard is its great strength for medical statisticians but, for an actuary, also its greatest weakness. Since it is not difficult, with modern software, to fit the full likelihood of the proportional hazards model, it should now be a useful addition to the actuary's toolkit, even if the Cox model itself sometimes may not be.

7.6 Analysis of the Case Study Data

With two covariates, gender and amount of pension, we have four basic choices
of proportional hazards models to fit: the Gompertz model with no covariates;
a model with gender as the only covariate; a model with pension amount as
the only covariate; and a model with both covariates. In fact, this does not ex-
haust our choices, because we can examine interactions between covariates,
but we shall omit that here for brevity. We regard a model in which all possible
covariates are included as a "full" model, and models in which one or more
covariates are omitted as "sub-models". Sub-models may be thought of as the
full model with some of the regression parameters set to zero.

7.6.1 The Case Study: Gender as a Covariate

The model is that presented in equation (7.5), with $\zeta_2 = 0$. Figure 7.13 shows
the R outputs resulting from fitting a Gompertz model to the same data as in
Section 7.2, but as a single model with gender as a covariate, and for the years
2007–2012. Figure 7.14 shows the deviance residuals against single ages.

The AIC is not comparable with those in Section 7.2, but it is comparable
with that in Figure 5.2, because both models were fitted to the same data. The
reduction in the AIC from 15,808.13 to 15,754.34 appears relatively small, but
the absolute value of the AIC depends mostly on the volume of data, and it is
the absolute reduction in the AIC that matters (see Section 6.4 and Pawitan,
2001). Section 6.4.1 suggested that an absolute reduction of about 4 AIC units
would usually be regarded as significant, so an absolute reduction of over 50
is in fact large, and is strong evidence that including gender as a covariate is
worthwhile.

However, care is needed in interpreting the AIC. The reduction tells us that
using gender as a covariate makes better use of the information in the data. It
does not say that either model is a satisfactory fit for actuarial use. Thus the
AIC is usually supplemented with a selection of tests of goodness-of-fit, checking that particular features of the fitted model are satisfactory (see Chapter 6).
These may regarded as formalising the visual impression given by Figure 5.3
and the like.

7.6.2 The Case Study: Pension Size as a Covariate

If we consider the deviance residuals against deciles of pension amount, shown
in Figure 7.15, there appears to be a non-random relationship. This suggests
that there is a trend, such that the fitted model is understating mortality for

```
nlm() return code 1.
Data cover 14773 lives and 2028 deaths.
Total exposure time is 66082.41 years.

Iterations=30
Log-likelihood=-7874.17
AIC=15754.34

Parameter    Estimate Std. error
------------ -------- ----------
Intercept    -12.7353    0.205048
Age          0.118914 0.00245967
Gender.M      0.34008   0.0450958
```

Figure 7.13 Output from fitting Gompertz model to 2007–2012 data from the pension scheme in the Case Study, with gender as a discrete covariate.

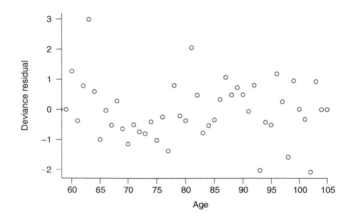

Figure 7.14 Deviance residuals by single ages after fitting a Gompertz model to 2007–2012 data from the pension scheme in the Case Study, with gender as a discrete covariate.

small benefits and overstating mortality for large benefits, which is clearly of financial importance. Therefore it is worth exploring pension size as a further covariate. First we look at pension size on its own.

Figure 7.16 shows the result of fitting a model with pension size as a continuous covariate. The AIC has increased from 15,808.13 to 15,809.69, which is unacceptable. The likely reason is that the proportional hazards assumption is not appropriate for this covariate; even transformed by taking logarithms, larger pensions are having larger effects. While we would reject this particular

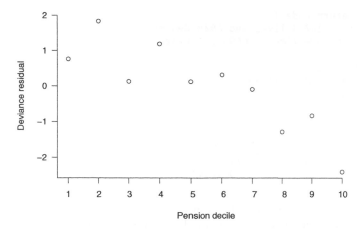

Figure 7.15 Deviance residuals by pension size (deciles) after fitting a Gompertz model to 2007–2012 data from the pension scheme in the Case Study, with gender as a discrete covariate.

```
nlm() return code 2.
Warning! nlm() did not return cleanly.
Data cover 14773 lives and 2028 deaths.
Total exposure time is 66082.41 years.

Iterations=17
Log-likelihood=-7901.84
AIC=15809.69

Parameter     Estimate Std. error
-----------   -------- ----------
Intercept     -12.3588    0.254476
Age            0.116893 0.00242597
log(Pension)  -0.0100209  0.0194228
```

Figure 7.16 Output from fitting Gompertz model to 2007–2012 data from the pension scheme in the Case Study, with log(1 + pension size) as a continuous covariate. (Return code 2 was OK in this example.)

model, we would not therefore ignore pension size, as Figure 7.15 is evidence that it is influential. We may proceed in one of two ways:

- We could search for a transformation of pension size such that proportional hazards are restored.
- We could sub-divide pension size into a small number of ranges and fit a separate covariate for each, as in Section 7.2.2.

```
nlm() return code 1.
Data cover 14773 lives and 2028 deaths.
Total exposure time is 66082.41 years.

Iterations=31
Log-likelihood=-7901.39
AIC=15808.77

Parameter     Estimate Std. error
------------  -------- ----------
Intercept     -12.4228   0.199703
Age           0.116828 0.00242661
Pension.L     -0.0864888  0.0750378
```

Figure 7.17 Output from fitting Gompertz model to 2007–2012 data from the pension scheme in the Case Study, with two categories of pension size; below £10,000 and £10,000 or more.

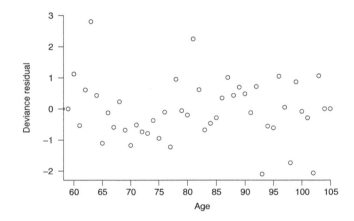

Figure 7.18 Deviance residuals by single ages after fitting a Gompertz model to 2007–2012 data from the pension scheme in the Case Study, with two categories of pension size; below £10,000 and £10,000 or more.

Searching for a suitable transformation of pension size could be time consuming and ultimately frustrating, since typically pension size has a small to modest effect except for the very largest pensions, and then it has a large effect. For simplicity, we sub-divide pension size as we did in Section 7.2.2, namely pensions below £10,000 and pensions of £10,000 or more. Figure 7.17 shows the result, which not very different from Figure 7.16, and Figure 7.18 shows the corresponding deviance residuals. We conclude as follows:

```
nlm() return code 1.
Data cover 14773 lives and 2028 deaths.
Total exposure time is 66082.41 years.

Iterations=44
Log-likelihood=-7869.13
AIC=15746.26

Parameter    Estimate Std. error
------------ -------- ----------
Intercept     -12.717   0.205403
Age          0.118808 0.00246363
Gender.M     0.377154  0.0463981
Pension.L    -0.238433  0.0772122
```

Figure 7.19 Output from fitting Gompertz model to 2007–2012 data from the pension scheme in the Case Study, with gender and pension size as discrete covariates.

- Including gender as a covariate improves the model fit significantly.
- Including gender as a covariate leaves unexplained variability associated with pension size.
- Pension size on its own does not improve the model fit.

This suggests fitting both gender and pension size as covariates, as the next step.

7.6.3 The Case Study: Gender and Pension Size as Covariates

Figure 7.19 shows the model of equation (7.5) with both gender and pension size fitted as discrete covariates, pension size having two levels as before. Figure 7.20 shows the corresponding deviance residuals.

Compared with the model which had gender as the only covariate, the AIC has reduced by about 8 units, from 15,754.34 to 15,746.26. This indicates a significant concentration of risk on larger annuities, with lower mortality, and suggests that we should retain pension size as a covariate.

7.7 Consequences of Modelling the Data

We will compare these examples of modelling the data using covariates, with the alternatives of: (a) fitting crude hazards at single years of age and then graduating the results; and (b) stratifying the data and fitting separate models.

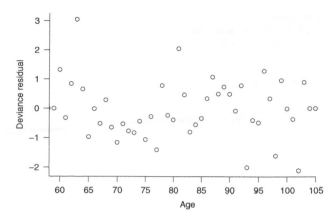

Figure 7.20 Deviance residuals by single ages after fitting a Gompertz model to 2007–2012 data from the pension scheme in the Case Study, with gender and pension size as discrete covariates.

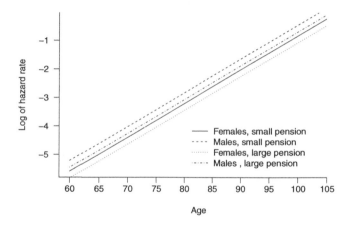

Figure 7.21 Log hazard rates from fitting Gompertz model to 2007–2012 data from the pension scheme in the Case Study, with gender and pension size as discrete covariates.

We chose to fit a mid-sized experience for these examples, with just 2,028 deaths over more than 40 years of age. Stratifying the data by gender alone resulted in two even smaller experiences, with 878 deaths among males and 1,150 among females. Stratifying the data by large pension amount ($\geq$ £10,000) yielded an even smaller experience, with just 197 deaths, yet one that we may suppose to be financially significant.

Our analyses above of the stratified data proceeded independently and, as Table 7.1 showed, were inherently limited in their scope. Any useful knowledge to be gained about male mortality from that of females, and *vice versa*, was lost, and similarly for pension size. Indeed, we would now have to check that the separately fitted models did not violate sensible *a priori* conditions, such as males having higher mortality than females. In situations where a great number of experiences have been graduated separately, as in some large-scale CMI analyses, arbitrary adjustments have sometimes been needed to restore the "common-sense" orderings of the graduated hazards, in particular at the highest ages. We saw a possible example of this in Figure 7.12, when separate graduations of persons with smaller and larger pensions crossed over. The chances of such violations of "common sense" increase if the separate graduations are each unconstrained in respect of the parametric function chosen, for example different members of the Gompertz–Makeham family (equations (5.34) or (5.35)).

The advantages gained by modelling the data are mainly the following:

• We avoid the fragmentation of the experience that stratification causes. This means that we retain the statistical power of modelling all of the data, and, if it is useful to do so, we can incorporate more covariates than stratification would allow.

• By choosing a single parametric model we make it less likely that "common sense" constraints will be violated, although this is not guaranteed. Figure 7.21 shows, in our simple example, that all the expected constraints are met.

• We have, through the AIC or other information criteria, an objective measure of the effect of covariates, an important tool in model selection.

To illustrate the last of these, consider the inclusion of pension size as a covariate in the example above. Figure 7.13 had already justified including gender as a covariate, reducing the AIC from 15,808.13 to 15,754.34. Figure 7.16 suggested that pension size by itself did not improve the fit, as the AIC increased, but Figure 7.15 showed that pension size had an appreciable effect in the presence of gender as a covariate. Finally, Figure 7.19 showed that including both gender and pension size, even though the latter was represented very crudely, improved the fit further, decreasing the AIC to 15,746.26. In other words, we find that one covariate significantly enhances the model in the presence of another covariate, but not on its own. None of this insight could be obtained by stratifying the data.

The process of model-fitting illustrated in this chapter is not the end of the story. Goodness-of-fit as described by the AIC does not guarantee that a model will be suitable for any actuarial application. In Chapter 6 we described a battery of other tests that check particular features of the model fit in more detail.

8

Non-parametric Methods

8.1 Introduction

In this chapter we look at non-parametric methods:

- comparison of mortality experience against a reference table

- non-parametric estimators of the survival function.

Actuaries will be familiar with the idea of comparing the mortality of a portfolio against a reference table. The idea is to express portfolio mortality as a proportion of the reference-table mortality rates, where this proportion is estimated empirically and may vary by age. Less commonly, a rating to age can be used.

A particularly useful non-parametric procedure is to use one of the available estimators of the survival function. We define the two main approaches, Kaplan–Meier and Fleming–Harrington estimators, although there is little practical difference between them for the sizes of data sets typically used by actuaries. Although they have limitations for multi-factor analysis, non-parametric methods have specific and useful roles to play in communication, data validation and high-level model checking, as described in Section 8.7. The Kaplan–Meier and Fleming–Harrington estimators have their origins in medical statistics. Actuarial applications have different requirements, most obviously the need to define the origin relative to a starting age, rather than time zero in a medical trial. Thus, the definitions for the Kaplan–Meier and Fleming–Harrington estimators given here are slightly different to those usually given elsewhere, i.e. the definitions given here are tailored for actuarial use.

8.2 Comparison against a Reference Table

It can be useful to check the mortality rates experienced by a portfolio with reference to an externally available mortality table. For example, with pensioner mortality a comparison could be made with the relevant population mortality table. We can calculate the expected number of deaths using the reference table and compare this to the observed deaths. Where the exposure times are relatively short for each individual, say no more than a year, we can calculate the aggregate actual-to-expected (A/E) ratio as follows:

$$A/E_{\text{lives}} = \frac{\sum_x d_x}{\sum_x E_x^c \mu_{x+1/2}}, \tag{8.1}$$

where summation is over the single years of age x, d_x is the number of deaths aged x last birthday and E_x^c is the central exposure time between age x and $x + 1$. The reference table hazard rate $\mu_{x+1/2}$ is supposed to represent mortality between ages x and $x + 1$. The "lives" subscript shows that this is a lives-weighted calculation, as opposed to a money-weighted calculation.

The structure of equation (8.1) is dictated by the structure of the reference table; such tables are most commonly available only for integral single years of age. Population tables in the UK provide both m_x and q_x values, in which case $\mu_{x+1/2} \approx m_x$ can be used. Some actuarial tables are only available in q_x form, in which case we use the approximation $\mu_{x+1/2} \approx -\log(1 - q_x)$. It is preferable to use equation (8.1) based on $\mu_{x+1/2}$, rather than any alternative based on q_x, as it is simpler to construct central exposures and more of the available data can be used; see Section 2.9.1. The R function `splitExperienceByAge()` in the online resources will calculate d_x and E_x^c for a file of individual exposure records.

A traditional actuarial approach is to weight such calculations by the benefit size, w_i. This gives rise to an alternative amounts-weighted aggregate A/E ratio:

$$A/E_{\text{amounts}} = \frac{\sum_i w_i d_i}{\sum_x \mu_{x+1/2} \sum_i w_i t_{i,x}}, \tag{8.2}$$

where $t_{i,x}$ is the time spent by the ith individual under observation between ages x and $x + 1$. An alternative is to use equation (8.1) and substitute the amounts-weighted deaths, $d_x^a = d_x w_x$, for d_x and the amounts-weighted exposures, $E_x^{ca} = E_x^c w_x$, for E_x^c, where w_x is the total pension payable to lives aged x. The R

Table 8.1 *Case Study, 2012 mortality experience, compared against UK interim life tables for 2011–2013.*

Weighting	Males	Females
Lives	86.5%	89.1%
Amounts	72.7%	79.3%

function `splitExperienceByAge()` in the online resources will calculate d_x^a and E_x^{ca} for a file of individual records (see also Appendix H.1).

Since population mortality tables are typically available for both males and females, we would calculate separate A/E ratios for each gender. Table 8.1 shows the ratios for the data behind the model in Figure 5.1. Since only a single calendar year's experience is used, we compare it against the corresponding UK population mortality rates. The lives-weighted A/E ratios in Table 8.1 suggest mortality which is lighter than the general population. However, the amounts-weighted A/E ratios are even lower, showing the impact of lower mortality rates for those with larger pensions.

There are several drawbacks with A/E calculations like those in Table 8.1:

- They are rough point estimates without any measure of uncertainty.
- A single percentage can conceal wide variation by age. The mortality rates of a portfolio tend to converge on the rates of the reference table with increasing age, converging from either below or above. This can lead to distortions, particularly in valuing benefits where the A/E percentage is far from 100% (say below 70% or above 130%). It will often be necessary to change the reference table such that the A/E percentage is as close as possible to 100% across a wide range of valuation ages.
- The lower mortality of wealthier pensioners is only indirectly and crudely reflected in the aggregate ratio, whereas a proper model of mortality will consider pension size and other risk factors simultaneously.

Nevertheless, an A/E calculation such as in Table 8.1 can be a useful summary for communication.

8.3 The Kaplan–Meier Estimator

Kaplan and Meier (1958) introduced a non-parametric estimator for the survival function. It is the equivalent of the ordinary empirical distribution function allowing for both left-truncation (Section 4.3) and right-censoring (Section 4.2). The main features of the Kaplan–Meier approach are as follows:

- It is a non-parametric approach – the Kaplan–Meier estimator of the survival function requires no parameters to be estimated.
- It is based around q-type probabilities, but where the interval over which each q applies is not determined *a priori* by the analyst, for example by choosing single years of age, but varies along the curve and is determined by the actual data *a posteriori*.
- Being non-parametric, it can be used to model the mortality of sub-groups only by stratifying the data and fitting separate functions; see Sections 1.7 and 7.3 for issues relating to stratification.
- Although the Kaplan–Meier survival function is non-parametric, it is still a statistical estimator of the survival function and so confidence intervals can be derived.

An example of the Kaplan–Meier estimator was given in Figure 2.8, highlighting its usefulness during data validation. One wrinkle for actuaries is that the standard Kaplan–Meier estimator is typically defined with reference to the time since a medical study commenced. In actuarial work it makes more sense to define the Kaplan–Meier estimator with respect to age, which we will do here.

Calculation of the basic Kaplan–Meier data can be time-consuming, as it involves traversing the entire data set as many times as there are deaths. We therefore illustrate the creation of the Kaplan–Meier estimator by using a small subset of data. Table 8.2 shows the mortality experience of centenarian females in the 2007–2012 experience of the Case Study. We are interested in calculating the survival function from age 100, so we extract the subset of lives who have any time lived after this age.

Figure 8.1 presents the data in Table 8.2 graphically; cases are drawn in ascending order of exit ages with crosses representing the deaths and a dotted line to the horizontal axis. This makes it easier to see which lives are alive immediately prior to a death age; in general we label the number of lives alive immediately before age $x + t$ as l_{x+t^-}. For example, at the first death age of 100.117 we can see that there were 12 lives. At the second death age of 100.533 we also have 12 lives immediately beforehand, since the third life from the top entered the investigation at an age above the first death age, but before the second death age. This process continues until the final death age of 103.203, where there were just two cases alive beforehand. The results of this process are shown in Table 8.3 and Figure 8.2.

Table 8.2 *Example data for calculation of Kaplan–Meier estimate ordered by exit age. Source: Case Study, 2007–2012 experience, females with exposure time above age 100 only. Status "0" at age $x_i + t_i$ represents a right-censored observation.*

x_i	$x_i + t_i$	Status at age $x_i + t_i$
98.184	100.117	DEATH
96.161	100.533	DEATH
100.476	100.648	DEATH
95.290	100.684	DEATH
96.993	100.873	DEATH
94.948	100.947	0
97.954	100.996	DEATH
97.062	101.270	DEATH
95.342	101.341	0
99.496	101.645	DEATH
96.353	102.351	0
98.099	103.203	DEATH
97.289	103.288	0

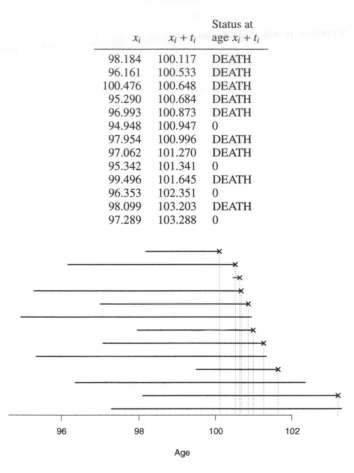

Figure 8.1 Exposure times for centenarian females in the Case Study. Data from Table 8.2 with crosses representing observed deaths. This is an alternative to the Lexis diagram in Figure 4.2, as calendar time is unimportant here.

Note that a right-censored observation does nothing at the time of censoring, but it does reduce the number exposed-to-risk at the next death age. Similarly, a new individual joining the portfolio does nothing at the time of entry, but it does increase the number exposed to the risk at the next death age. Examples of this can be seen in Table 8.4.

Table 8.3 *Kaplan–Meier estimate of the survival function $S_{100}(t)$ of centenarian females. Source: Case Study, 2007–2012 experience, females reaching age 100 or more only.*

$x + t_i$	$l_{x+t_i^-}$	d_{x+t_i}	$\hat{S}_{100}(t_i)$
100			1
100.117	12	1	0.91667
100.533	12	1	0.84028
100.648	11	1	0.76389
100.684	10	1	0.68750
100.873	9	1	0.61111
100.996	7	1	0.52381
101.270	6	1	0.43651
101.645	4	1	0.32738
103.203	2	1	0.16369

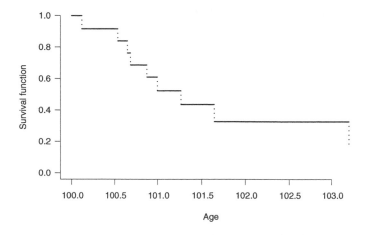

Figure 8.2 Kaplan–Meier survival function calculated in Table 8.3.

Users familiar with the *survival* package in R might wonder why we did not simply use the fit() function to generate the table in Table 8.3. The answer lies with the third record in Table 8.2 and Figure 8.1: this life was not observed from the start of the investigation, i.e. from age 100. This is a specific example of the atypical nature of survival data available to actuaries, i.e. there is usually continuous recruitment in every portfolio in the form of new policies and benefits being set up.

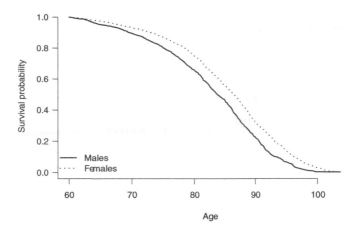

Figure 8.3 Kaplan–Meier survival function as per equation (8.3) for the Case Study (2007–2012 experience data).

Under certain weak conditions, Kaplan and Meier (1958) derived a maximum-likelihood estimator of the survival function as follows:

$$\hat{S}_x(t) = \prod_{t_i \le t} \left(1 - \frac{d_{x+t_i}}{l_{x+t_i^-}}\right), \qquad (8.3)$$

where x is the outset age for the survival function, $\{x + t_i\}$ is the set of n distinct ages at death, $l_{x+t_i^-}$ is the number of lives alive immediately before age $x + t_i$ and d_{x+t_i} is the number of deaths occurring at age $x + t_i$. The product in equation (8.3) is over all death times up to and including time t. Informally the term $\left(1 - d_{x+t_i}/l_{x+t_i^-}\right)$ is the probability that a life survives the interval of time between death ages.

Implicit in equation (8.3) is that the data are ordered by death ages, and such an approach is possible only with individual-level data, not aggregated data. The form of equation (8.3) explains the alternative description of the Kaplan–Meier estimator as the *product-limit estimator* of the survival function.

The Kaplan–Meier estimator straddles the concepts of q_x and μ_x. The definition in equation (8.3) is clearly based around q_x, but where the discretisation of the age range is decided by the data *a posteriori*, rather than by the analyst *a priori*. As the number of events in life-office portfolios is typically rather large, the discretisation steps can be quite small and the results quickly look like μ_x because of the relationship in equation (3.16). For example, the median gap between death ages for the male lives in Figure 8.3 is just ten days. However, the data set used in Richards (2012) had a median age gap of one day, which is the smallest interval possible when using dates to measure survival times. For this reason a purist might argue that in this case the Kaplan–Meier estimates are not for μ_x but actually for q_x with a daily interval, i.e. $_h q_x$ with $h = 1/365$.

An example calculation of a Kaplan–Meier estimate for the survival function is given in Table 8.4. Note that in theory age is measured continuously and that therefore each age at which a death occurs should have only one death. In practice, age at death is usually only recorded at a granularity of one day, and so it is possible for large portfolios to have more than one death at each age, as is the case for age 63.878 in Table 8.4.

Table 8.4 *Example calculation of Kaplan–Meier estimate $\hat{S}_{60}(t)$ with approximate standard errors. The real-valued ages are determined by the exact age at death, which leads to a non-constant spacing. Source: Case Study, 2012 experience only, females only.*

$x + t_i$	$l_{x+t_i^-}$	d_{x+t_i}	$\hat{S}_{60}(t_i)$	s.e.$\left(\hat{S}_{60}(t_i)\right)$
60	384	0	1	0
60.872	473	1	0.99789	0.00211
61.118	498	1	0.99588	0.00291
61.214	535	1	0.99402	0.00345
61.274	540	1	0.99218	0.00390
61.775	592	1	0.99050	0.00424
61.874	586	1	0.98881	0.00456
61.901	582	1	0.98711	0.00485
62.457	622	1	0.98553	0.00510
62.498	613	1	0.98392	0.00534
62.591	596	1	0.98227	0.00558
62.602	594	1	0.98062	0.00581
63.013	620	1	0.97903	0.00601
63.090	607	1	0.97742	0.00621
63.180	612	1	0.97582	0.00641
63.665	648	1	0.97432	0.00657
63.878	665	2	0.97139	0.00687
63.938	663	1	0.96992	0.00701
⋮	⋮	⋮	⋮	⋮

As a statistical estimate of the survival function, it is possible to compute an approximate standard error for $\hat{S}_x(t_j)$ using the formula from Greenwood (1926):

$$\text{s.e}\left(\hat{S}_x(t)\right) \approx \hat{S}_x(t) \times \sqrt{\sum_{t_i \le t} \frac{d_{x+t_i}}{l_{x+t_i^-}\left(l_{x+t_i^-} - d_{x+t_i}\right)}}. \qquad (8.4)$$

The standard errors from equation (8.4) are shown in Table 8.4. Note that this approximation must be applied intelligently – for small numbers of deaths the approximate standard error can equal or exceed the estimate $\hat{S}_x(t_j)$, as shown in Table 8.5, which is a continuation of Table 8.4 to the highest ages. To avoid producing a confidence interval including negative values for $\hat{S}_x(t_j)$ (or values above 1), the analyst should apply sensible bounds.

Table 8.5 *Example calculation of Kaplan–Meier estimate $\hat{S}_{60}(t)$ with approximate standard errors. Source: data from Case Study, 2012 experience only, females only.*

$x + t_i$	$l_{x+t_i^-}$	d_{x+t_i}	$\hat{S}_{60}(t_i)$	s.e.$\left(\hat{S}_{60}(t_i)\right)$
⋮	⋮	⋮	⋮	⋮
96.035	25	1	0.07386	0.01532
96.068	23	1	0.07065	0.01499
96.139	19	1	0.06693	0.01465
96.259	18	1	0.06322	0.01430
96.465	17	1	0.05950	0.01394
96.624	15	1	0.05553	0.01356
97.916	11	1	0.05048	0.01323
99.236	7	1	0.04327	0.01316
99.274	7	1	0.03709	0.01265
99.408	6	1	0.03091	0.01196
99.476	5	1	0.02473	0.01105
99.627	3	1	0.01648	0.00998
100.684	3	1	0.01099	0.00802
103.203	3	1	0.00733	0.00613

8.4 The Nelson–Aalen Estimator

Nelson (1958) defined what was then called the Nelson estimator of the integrated hazard function, $\Lambda_{x,t}$, as defined in equation (3.23). A version of this suitable for actuarial work on closed and open portfolios is given by:

$$\hat{\Lambda}_{x,t} = \sum_{t_j \leq t} \frac{d_{x+t_i}}{l_{x+t_i^-}}, \tag{8.5}$$

where the definitions of x, $l_{x+t_i^-}$ and d_{x+t_i} are the same as for the Kaplan–Meier estimator in equation (8.3). Summation in equation (8.5) is over all death times up to and including time t. Aalen's name was added after 1978, when he introduced counting processes to survival analysis and showed them to be of fundamental importance (Aalen, 1978). We will see in Chapter 17 that the Nelson–Aalen estimator is the simplest non-parametric estimator and the Kaplan–Meier estimator is derived from it.

The Nelson–Aalen estimator is seldom used on its own in actuarial work, but it is an intermediate step in the Fleming–Harrington estimator of the survival curve.

8.5 The Fleming–Harrington Estimator

Fleming and Harrington (1991) derived a non-parametric estimator for the survival function by using the Nelson–Aalen estimator for the integrated hazard:

$$\hat{S}_x(t) = \exp(-\hat{\Lambda}_{x,t}), \tag{8.6}$$

where $\hat{\Lambda}_{x,t_j}$ is defined in equation (8.5). The Fleming–Harrington estimator produces a survivor function higher than the Kaplan–Meier estimator, although the difference is negligible for the size of data sets actuaries typically work with; these two points are demonstrated in Table 8.6. Note that what Collett (2003, pp.20–22) refers to as the Nelson–Aalen estimator for the survivor function is in fact the Fleming–Harrington estimator.

8.6 Extensions to the Kaplan–Meier Estimator

Table 8.1 suggests a difference between lives- and amounts-weighted measures, so a logical extension of equation (8.3) for actuaries would be to weight by benefit amount:

$$\hat{S}_x(t) = \prod_{t_i \leq t} \left(1 - \frac{d_{x+t_i}^{\text{amounts}}}{l_{x+t_i^-}^{\text{amounts}}} \right), \tag{8.7}$$

where $d_{x+t_i}^{\text{amounts}}$ and $l_{x+t_i^-}^{\text{amounts}}$ are directly analogous to the lives-based definitions in equation (8.3) but where each life contributes its benefit amount, w_i, instead of a count of 1, to the counts of deaths and lives. The results of doing this are shown in Figure 8.4, which shows the higher amounts-weighted survivor function (which is more prominent for males). However, amounts-weighted calculations are volatile and can be heavily influenced by the death of an individual with a large benefit amount. Also, amounts-weighted measures do not have simple approaches for generating confidence intervals and there is no simple equivalent to equation (8.4) for the standard error of the amounts-weighted Kaplan–Meier survivor function in equation (8.7).

Table 8.6 *Comparison of Kaplan–Meier and Fleming–Harrington estimates for* $S_{60}(t)$. *Source: data from Case Study, 2012 experience only.*

	$\hat{S}_{60}(t_i)$ according to:	
$x + t_i$	Kaplan–Meier	Fleming–Harrington
60	1	1
60.872	0.99789	0.99789
61.118	0.99588	0.99589
61.214	0.99402	0.99403
61.274	0.99218	0.99219
61.775	0.99050	0.99051
61.874	0.98881	0.98882
61.901	0.98711	0.98713
62.457	0.98553	0.98554
62.498	0.98392	0.98393
62.591	0.98227	0.98228
62.602	0.98062	0.98063
63.013	0.97903	0.97905
63.090	0.97742	0.97744
63.180	0.97582	0.97584
63.665	0.97432	0.97434
63.878	0.97139	0.97141
63.938	0.96992	0.96995
⋮	⋮	⋮

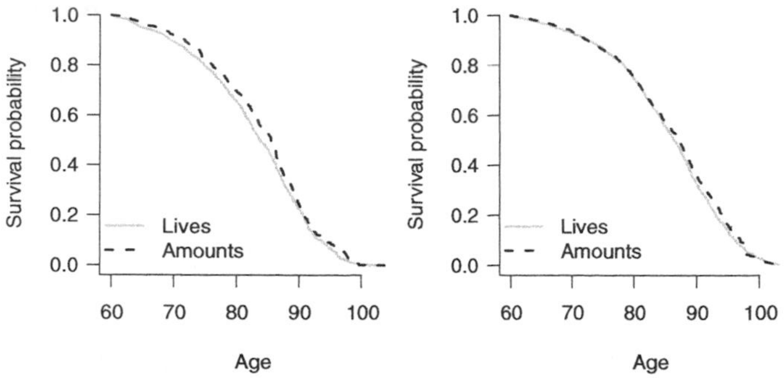

Figure 8.4 Kaplan–Meier estimators of the survivor function weighted by amounts; males (left panel) and females (right panel). Source: Case Study, 2007–2012 experience.

8.7 Limitations and Applications

A major limitation of the non-parametric methods described in this chapter is that they can handle the mortality of sub-groups only by stratification. That is, a separate Kaplan–Meier or Fleming–Harrington estimator has to be calculated for each sub-group. As we have seen in Sections 1.7 and 7.3, this severely limits the number of risk factors which can be used. This quickly becomes a problem even for very large data sets, as some combinations will be much rarer than others. Also, since there is no modelled relationship between the sub-groups we cannot say much that is quantitative about differences between the estimators.

The comparison against a reference table offers the ability indirectly to allow for financially significant risk factors by weighting the A/E ratios by pension size (or some other financial measurement). However, amounts-weighted Kaplan–Meier or Fleming–Harrington curves may not be practical because of the volatility of amounts-weighted experience.

A key aim of modelling is the reduction of data into a handful of key parameters which summarise the essence of the process or risk. A further limitation of the Kaplan–Meier and Nelson–Aalen estimators is that no such summarisation takes place: we require all the data to create the survival curve. A non-parametric method is therefore not a model in the sense we typically seek, i.e. a concise summary of the main features of the risk with statistical properties.

Nevertheless, despite these limitations there are a number of important actuarial applications for non-parametric methods:

- *Communication.* As shown in Figure 8.3, non-parametric survival analysis lends itself well to visual presentation of information. Non-parametric methods can be particularly useful when communicating with an audience which is unfamiliar with (or even hostile towards) statistical models.
- *Checking data quality.* As shown in Figure 2.8, non-parametric methods are a useful visual check on the validity of the basic data. Every new analysis should start with a Kaplan–Meier plot before model-fitting commences.
- *Checking model fit.* Figure 8.5 shows how the fitted survival curve can be compared with the Kaplan–Meier curve as a high-level check of model suitability. This is a practical way to demonstrate how a parametric model is essentially a simpler, concise and smoothed version of the actual data. Further detailed checks of model fit were discussed in Chapter 6.

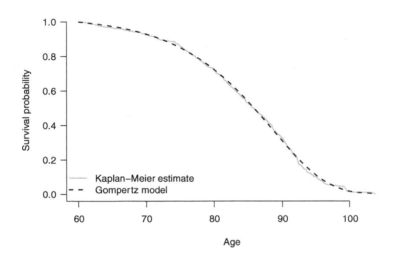

Figure 8.5 Kaplan–Meier estimator of the survival function for data from Tables 8.4 and 8.5, together with fitted Gompertz model with $\hat{\alpha} = -12.972$ and $\hat{\beta} = 0.122872$.

9

Regulation

9.1 Introduction

In the preface we said that this book was concerned with practical problems. One of the ineluctable aspects of actuarial work is dealing with regulations, especially regulations on how to value liabilities, how to reserve for them and what solvency capital to hold. This chapter departs from the focus hitherto on mortality modelling and looks at the real-world application of the results of analysis in setting insurer reserves. It will not be of interest to every mortality modeller or statistician, and so this chapter can be skipped by those not interested in insurance-company reserving.

9.2 Background

Historically, insurance-company regulations have been deterministic rules of thumb for calculating a static balance sheet. These deterministic rules typically specified such things as the margins to use in the assumptions for calculating the statutory liability, together with additional capital margins based on the size and nature of the liability valuations. However, the past decade has seen a shift in regulation towards a more probabilistic view of insurance-company balance sheets. In particular, regulators are keen that each insurer holds enough capital and reserves to cover a high proportion of adverse scenarios which might arise over the short term. One specification is that each insurer has to hold enough capital and reserves to cover at least 99.5% of scenarios which might arise over the coming year. Alternatively, each insurer has to hold enough capital and reserves such that the probability of it being insolvent at the end of the year is less than 0.005, i.e. the insurer can withstand adverse events with a probability of "one in two hundred over one year". Such a specification lies behind the

Solvency II regulations for insurers in the European Union. Solvency II is a *value-at-risk* (VaR) methodology, and we can write VaR_p, where p denotes the probability level; Solvency II would therefore be a $\mathrm{VaR}_{0.995}$ regulatory standard. A related concept is the *conditional tail expectation* (CTE), i.e. where an insurer has to hold reserves large enough to cover the expected cost of an event occurring in the tail of a distribution. The Swiss Solvency Test (SST) is an example of this. We denote by CTE_p the reserve required to meet the average liability in the tail of events with probability $1 - p$. There is a relationship between VaR_p and CTE_p (see Section 9.3) and both can be estimated from the same set of simulated losses or reserves.

Regulatory regimes of this kind are inherently probabilistic in nature. A model has to be specified for each risk, and these models have to be calibrated to data wherever possible. Nevertheless, insurance-company regulations of this sort typically also allow for actuarial judgement to be used in cases where data are unavailable. The usual procedure is to calculate the capital required for each risk individually, and then to aggregate allowing for possible diversification. For example, consider an insurer that is exposed to just two risks, and where the regulatory capital margin required for each of the two risks in isolation is r_A and r_B, respectively; we assume that "regulatory capital margin" here means the relative excess on top of the best estimate, i.e. $r = Q_{0.995}/Q_{0.5} - 1$ in the case of a $\mathrm{VaR}_{0.995}$ regulatory environment ($Q_{0.5}$ is the median liability and $Q_{0.995}$ is the 99.5th quantile). Then, the total regulatory capital margin, r, for the insurer is at most $r_A + r_B$, and this can only happen if the two risks are perfectly positively correlated. If the two risks are uncorrelated, or only weakly positively correlated, then $r < r_A + r_B$. This situation would be referred to as a *benefit of diversification*, and it is one reason why insurers do not benefit from focusing exclusively on one particular type of risk or product line, as is often recommended for businesses in non-insurance industries. Theoretically, if the two risks were negatively correlated, it would even be possible to have $r < \max(r_A, r_B)$. This latter result would be of considerable interest to insurers, because it would mean that the total regulatory capital would be less than the capital for the larger of the two risks in isolation. However, in practice few regulators would permit this. In the discussion of Figure 9.1 we will see an example of why regulators are reluctant to allow too much credit for diversification, and are very sceptical indeed of claims of negative correlation.

If the two risks came from separate business lines, and the risks were less than perfectly correlated, this would mean a bigger overall business could be sustained for the same available regulatory capital. An increasing number of insurers are restructuring their business with these capital results in mind – the goal is to minimise regulatory capital, and combinations of risks are sought to minimise this overall capital requirement. This goal increasingly drives the sale and purchase of subsidiaries and portfolios (or the conclusion of reinsurance treaties which can have a similar effect).

To illustrate these points, we will consider three risks in the calculation of regulatory capital for a portfolio of annuities. To make it simple we will only consider three demographic risks: the risk of mis-estimation of current mortality rates, the risk posed by the long-term trend in mortality and the risk posed by random variation. This is clearly an artificially simple example: there are other aspects to demographic risk in annuity business, some of which are listed in Richards et al. (2014) and Richards (2016). Furthermore, there are many other non-demographic risks, such as interest rates and default risk on the assets. Nevertheless, the basic concepts can be illustrated with an artificially simple, three-risk view.

9.3 Approaches to Probabilistic Reserving

We assume that a liability is a continuous random variable, X, for which we want to set a reserve amount, R. There are two common probabilistic approaches to setting R:

- *The value-at-risk (VaR) measure.* Here the reserve is set such that it will cover a high proportion of scenarios, i.e. R is set such that $\Pr(X \leq R) = p$. Therefore R is equal to Q_p, the p-quantile of the distribution of X. This is the style of reserving under the Solvency II regime in the EU, and was the style of reserving under the UK's former Individual Capital Assessment (ICA) regime; in both cases $p = 0.995$ was used, and a time horizon of one year.
- *Conditional tail expectation (CTE).* Here the reserve is the expected value of the liability conditional on an event of a given extremity having occurred, i.e. $R = E[X|X > Q_p]$. This style of reserving is common in North America and for the Swiss Solvency Test (SST). For a given value of p the CTE approach naturally produces a higher reserve than the VaR approach; the SST uses $p = 0.99$.

We can see that the VaR and CTE methods are linked: $\mathrm{CTE}_p = E[X|X > Q_p]$. One immediate result from this is $\mathrm{CTE}_p \geq Q_p$, so a regulatory or reporting environment based around the CTE may use a lower value for p than a VaR-based regime with $p = 0.995$. However, this does not mean that the CTE-based regulatory environment is weaker. Indeed, according to Hardy (2006) the CTE approach has the advantage that "[a]s a mean it is more robust with respect to sampling error than the quantile".

A second advantage of the CTE is that it is *coherent*, while a quantile approach is not (or not always). For two risks X and Y, for example, it is always the case that $\text{CTE}_p(X + Y) \leq \text{CTE}_p(X) + \text{CTE}_p(Y)$, a property known as *subadditivity* – diversification cannot make the total risk greater, but it might make the overall risk smaller if the risks are not perfectly correlated. For quantile methods, however, it is possible to construct examples whereby $Q_p(X + Y) > Q_p(X) + Q_p(Y)$, which violates an intuitive principle: it should not be possible to reduce the capital requirement for a risk by splitting it into constituent parts. The above feature of quantiles is one reason why regulators in quantile-regulated territories require capital to be identified for each risk separately before being aggregated into a company-wide capital requirement. This way, the regulator can ensure that the failure of quantiles to always be subadditive is not being used to game the capital requirements.

A more detailed comparison between the quantile and CTE approaches is given in Hardy (2006). A key element in both of the above approaches to reserving is the time horizon, which is usually over a single year. This presents new challenges for actuaries, since many risks – such as longevity trend risk – exist primarily in run-off, not a single year. This requires rethinking how some models are used, an example of which we will consider in Section 9.7. Before we come to the annuity example, however, we will look first at quantile estimation, then consider an example of a risk which *does* lend itself well to the idea of a catastrophe occurring in a single year: mortality risk.

9.4 Quantile Estimation

Quantiles are points taken at regular intervals from the cumulative distribution function of a random variable. They are generally described as q-quantiles, where q specifies the number of intervals which are separated by $q - 1$ points. For example, the 2-quantile is the median, i.e. the point where values of a random variable are equally likely to be above or below this point.

A percentile is the name given to a 100-quantile. In Solvency II work we most commonly look for the 99.5th quantile, i.e. the point at which the probability that a random event exceeds this value is 0.005. The simplest approach to estimating the 99.5th quantile might be to simulate 1,000 times and take the 995th or 996th largest value. However, there are several alternative ways of estimating a quantile or percentile, as documented by Hyndman and Fan (1996). One of the commonest approaches is the definition used by the PERCENTILE() function in Microsoft Excel, and which is also option type=7 in the R function quantile(). In general, we seek a quantile level p which lies in the interval

(0, 1). If $x[i]$ denotes the ith largest value in a data set, then the quantile sought by Excel is $x[(n-1)p + 1]$.

To illustrate the calculation of sample quantiles, consider the following R commands to generate a sample of 1,000 losses from the N(0,1) distribution and display the largest values:

```
# Generate some pseudo-random N(0,1) variates
set.seed(1)
temp = rnorm(1000)
sort(temp)[995:1000]
```

When these commands are run you should see the six largest values as follows:

```
2.446531, 2.497662, 2.649167, 2.675741, 3.055742, 3.810277
```

In this example, $n = 1000$ and $p = 0.995$, so $(n-1)p + 1 = 995.005$. This is not an integer, so we must interpolate between the 995th and 996th largest values. The final answer is then $0.995 \times 2.446531 + 0.005 \times 2.497662 = 2.447$, which is perhaps further from the theoretical 99.5th quantile of the N(0,1) distribution (2.576) than might be expected.

So we now have our estimate of the 99.5th quantile. What is sometimes overlooked is the fact that the sample quantile is an estimate, i.e. there is uncertainty over what the true underlying value is for the 99.5th quantile. In fact, the quantile is part of a branch of probability theory called order statistics, and it turns out that the sample quantile above is not the most efficient estimator. There are many other estimators, of which one is due to Harrell and Davis (1982). One reason the Harrell–Davis estimator is more efficient is because it uses all of the data, rather than just the order statistics. In the example above, the Harrell–Davis estimate of the 99.5th quantile can be found with the following extra R commands:

```
# Calculate the 99.5th quantile and its standard error:
library(Hmisc)
hdquantile(temp, 0.995, se=TRUE)
```

This yields an estimate of the 99.5th quantile of 2.534, and we can see that the Harrell–Davis estimator is closer to the known quantile of the N(0,1) distribution (2.576). Perhaps even more useful is the fact that the Harrell–Davis estimator comes with a standard error, which here is 0.136. Table 9.1 shows the sample size required to get within a given level of closeness to the true underlying 99.5th quantile for a N(0,1) distribution.

Table 9.1 *Harrell–Davis estimates and standard errors of estimated 99.5th quantile of N(0,1) variates.*

Number of simulations	99.5th quantile	Standard error	Coefficient of variation
1,000	2.534	0.136	5.4%
10,000	2.517	0.047	1.9%
25,000	2.564	0.027	1.1%
50,000	2.577	0.020	0.8%
100,000	2.564	0.014	0.5%

9.5 Mortality Risk

In a portfolio of term insurances an important risk is that there could be some kind of catastrophe which leads to many more deaths occurring in a short period than expected. There is a potentially long list of hard-to-quantify potential causes, including wars and epidemics. The question is therefore one of judgement, rather than modelling *per se*. A major problem in setting the regulatory capital for mortality risk is the lack of data, which leads to the stress scenario being set somewhat subjectively. One starting point for such a mortality shock would be the so-called Spanish Influenza pandemic of 1918–19, the effect of which in Sweden is depicted in Figure 9.1.

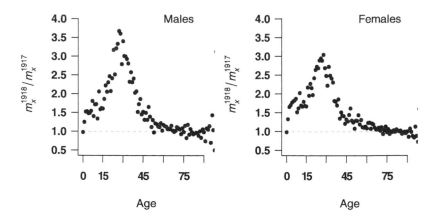

Figure 9.1 Excess mortality in Sweden in 1918 compared with 1917. Source: own calculations using data from Human Mortality Database.

Figure 9.1 demonstrates a number of tricky features for actuaries looking to calibrate a mortality shock:

- Is the mortality shock demonstrated in Figure 9.1 a "one in 200 year" event, as required for Solvency II? Since it occurred not quite a century ago at the time of writing, it is arguably only a "one in 100 year" event.
- Excess mortality peaks for lives in their late twenties, whereas excess mortality is negligible above age 65 (say); a mortality shock may not affect all ages equally. In the case of the 1918 pandemic, an insurer with a portfolio of term insurances and annuities would have seen losses on the term insurances without any compensating profits on the annuities.
- Peak male excess mortality is greater than peak female excess mortality; a mortality shock may not affect the genders equally.

There are other candidate epidemics besides influenza, of course. The Black Death in Europe was an even more lethal killer, but the relevance of a six-hundred-year-old bacterial plague might be questioned in the age of antibiotics. The likes of Ebola and other haemorrhagic fevers would be very modern viral alternatives, but any mortality-shock scenario would require transplantation from the handful of known outbreaks in countries with weak or non-existent healthcare systems to the sorts of countries with substantial life insurance liabilities. This all serves to emphasise that the calibration of a mortality shock for term-insurance liabilities is a question of judgement, rather than mathematical modelling.

A mortality shock would normally be expected to lead to the release of reserves in an annuity portfolio, but an insurer reserving under a VaR- or CTE-style regulatory regime would not be allowed to take credit for this. More specifically, Figure 9.1 shows why regulators would be unwilling to permit insurers to reduce capital requirements by having a negative correlation between mortality shocks on their term-assurance portfolios and longevity risks on their annuity portfolios.

9.6 Mis-estimation Risk

Mis-estimation risk is sometimes also called *level risk*. It is the risk that the insurer's current estimate of mortality (or any other demographic assumption) is incorrect. Since an insurer will only have a finite amount of relevant experience data for a given portfolio, there is a risk that even an unbiased estimate based on these data will be an incorrect assessment of the underlying rate; this is mis-estimation risk.

Richards (2016) presents a way of assessing mis-estimation risk using a portfolio's own mortality experience. The basic idea is to take the best-estimate

Table 9.2 *Mis-estimation capital requirements for two portfolios: percentage addition on top of the best-estimate reserve. Source: Richards (2016).*

Portfolio	95% interval for $VaR_{0.995}$ mis-estimation risk capital
Case Study	4.40–4.68%
Richards et al. (2013)	1.10-1.19%

model, together with the estimated covariance matrix for the model parameters, and use this to consider plausible alternative parameter vectors which are internally consistent. The portfolio is then valued repeatedly using these alternative parameter vectors, giving rise to a set of portfolio valuations which can be used to calculate, say, $VaR_{0.995}$ or $CTE_{0.99}$. Key elements of this procedure are:

- The model must be rich enough in risk factors that it is applicable for financial purposes. A check for this can be done using the bootstrapping procedure described in Section 6.7.
- If the risk is varying over time, then the time trend must be included in the model. However, the impact of the time trend need only be considered up to the point of valuation, since future mortality rates are the subject of a separate assessment in Section 9.7.

Further details of mis-estimation can be found in Richards (2016), which gives the $VaR_{0.995}$ mis-estimation capital requirements for two portfolios shown in Table 9.2. The smaller portfolio in our Case Study has less data and so has higher mis-estimation risk (and thus a higher mis-estimation capital requirement). Since we estimate the 99.5th quantile using a finite data set, there is uncertainty even over this, so we use the estimation procedure from Harrell and Davis (1982) to establish a 95% confidence interval for the true 99.5th quantile. In practice an insurer would err on the side of prudence and pick the upper limits of each 95% confidence interval.

One point to note about the methodology in Richards (2016) and the figures in Table 9.2 is that there is no mention of a time horizon. These figures are therefore not strictly in accordance with the "one in 200 over one year" principle behind Solvency II and the SST. Nevertheless, we will use the figures in Table 9.2 in our example solvency calculation for illustration purposes.

9.7 Trend Risk

Trend risk is the risk that the development of future mortality rates will deviate from the best estimate. Longevity trend risk, however, does not naturally fit the "one in 200 over one year" approach required for modern reserving. As *The Economist* (2012) puts it, "whereas a catastrophe can occur in an instant, longevity risk takes decades to unfold". Richards et al. (2014) proposed a framework for putting trend risk into the one-year horizon demanded by modern insurance-company regulation, such as Solvency II. The basic procedure is as follows:

(i) Take a data set, usually of population data, and fit a stochastic model capable of projecting mortality rates in future years. Examples of such models are presented in Chapter 12.

(ii) Use the stochastic model to project the following year's mortality rates, and simulate the deaths which would occur.

(iii) Using the actual data and the extra year's pseudo-data from step (ii), refit the stochastic model and calculate the central projection. Use this projection to calculate either a portfolio valuation or sample annuity factors.

(iv) Repeat steps (ii) and (iii) a suitably large number of times to obtain a set of financial values which reflect the variability in projections stemming from an extra year's data.

An illustration of the first ten simulations using this procedure is shown in Figure 9.2.

Two further aspects of trend risk need to be considered:

- *Model risk.* There is no way of knowing which of the models illustrated is the correct one (if any). In Figure 9.2 we used the Lee–Carter model (see Section 12.2), but there is no way of knowing whether this model is appropriate. We need to make an allowance for this lack of knowledge.

- *Basis risk.* The figures in Table 9.3 have been calculated using models calibrated to population data, since life-insurer and pension-scheme portfolios rarely have long enough time series of mortality data. However, the members of an annuity portfolio or pension scheme are a select sub-group of the national population, so it is possible that their mortality rates could change faster than the population-based forecast. We need to make an allowance for this risk, too.

The allowances for model risk and basis risk are subjective and require actuarial judgement. An insurer would separately itemise these risks in its

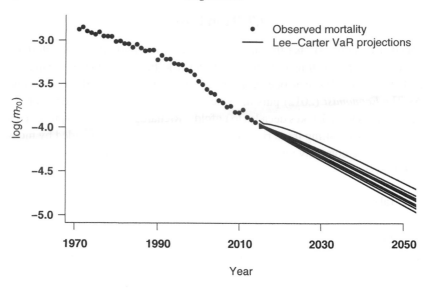

Figure 9.2 Observed mortality rate at age 70, together with the first ten central forecasts from a Lee–Carter VaR simulation. Source: own calculations using ONS data for males in England and Wales, ages 50–104 over the period 1971–2014.

Table 9.3 *Value-at-risk capital requirements at age 70 under various stochastic models. 10,000 simulations according to steps* (ii) *and* (iii) *in Section 9.7. Annuity factors calculated using male mortality 1961–2014, payments made continuously and discounted at 1% per annum. Approach to ARIMA parameter risk from Kleinow and Richards (2016).*

Model	Best-estimate reserve age 70	99.5th quantile	95% interval for $VaR_{0.995}$ capital
Lee–Carter (Section 12.2)	15.17	15.84	4.21–4.59%
Lee–Carter (Delwarde et al., 2007)	15.18	15.83	4.12–4.48%
Lee–Carter (smoothed α_x & β_x)	15.18	15.83	4.13–4.50%
Age-Period-Cohort	15.55	16.09	3.40–3.55%
Age-Period-Cohort (smoothed α_x)	15.55	16.09	3.40–3.55%
Cairns–Blake–Dowd (Section 12.3)	14.37	15.05	4.53–4.90%

documentation to demonstrate to the regulator that they had been considered, but in practice they would be implicitly allowed for by picking a value for longevity trend risk towards the upper end of the range in Table 9.3. For the sake of argument let us assume that the insurer chooses a value for $VaR_{0.995}$ for trend-risk of 5% of the best-estimate reserve.

9.8 Number of Simulations

Table 9.1 shows that a very large number of simulations can be required to achieve a close level of precision in estimating the 99.5th quantile, even for a well-behaved distribution like N(0,1). One practical challenge lies in the run-times: even if a single projection model is fitted in ten seconds (say), a thousand VaR simulations would take nearly three hours. A larger number of simulations would be generally preferred for estimating a VaR or CTE reserve, and Richards et al. (2014) recommended using parallel processing, since each pair of steps (ii) and (iii) in Section 9.7 is independent of all the others.

An alternative approach would be to see if the simulated reserves looked like they approximately followed a known distribution. For example, the 10,000 values from the Cairns–Blake–Dowd (CBD) simulation for Table 9.3 have a sample skewness at age 70 of 0.05 and a sample kurtosis of 3.00. Since the normal distribution has a theoretical value of 0 for skewness and 3 for kurtosis, one approach to estimating the VaR or CTE values with fewer simulations would be to use the equivalent $N(\mu, \sigma^2)$ distribution with μ and σ^2 estimated from the sample. Estimation of these two moments is usually robust enough with a sample of a few hundred values, so 1,000 simulations would suffice. However, this approach will not always work, as some VaR output samples are bimodal. In the case of the Lee–Carter model in Table 9.3, the simulated annuity factors at age 70 are unimodal, but the skewness is -1.00 and the kurtosis is 6.37. Under such circumstances more simulations than 1,000 seems unavoidable.

9.9 Idiosyncratic Risk

Idiosyncratic risk also goes by the name stochastic risk. It refers to the fact that, even if the underlying mortality rates are known, and even if the future trend path is known, individual lifetimes are still random variables whose future values are unknown. Idiosyncratic risk is best dealt with by simulating each individual lifetime in the portfolio. One important reason for this is *concentration risk*, i.e. the tendency for liabilities to be unequally distributed across all lives. Pension schemes and annuity portfolios tend to be far more unequal than wider society, with large pension amounts concentrated in a relatively small number of lives, as shown in Table 9.4. There are a number of ways in which this concentration of financial risk can be illustrated. One way is to sort the membership by pension size and calculate the proportion of pension benefits paid to each decile (say), or to calculate what proportion of the membership receives

Table 9.4 *Concentration of pension benefits by membership decile. Source: Richards (2008).*

Membership decile	Percentage of total pension amount:	
	Life office	Pension schemes
1	54.3%	46.3%
2	15.2%	17.8%
3	9.4%	11.4%
4	6.6%	8.0%
5	4.9%	5.8%
6	3.6%	4.1%
7	2.7%	2.9%
8	1.8%	2.0%
9	1.1%	1.2%
10	0.4%	0.5%
Total	100.0%	100.0%
Gini coefficient	66.0%	60.9%

half of all the benefits. Another way is to calculate the Gini coefficient (Gini, 1921), which is used widely in social statistics to measure income inequality. The Gini coefficient takes the value of 0% when everyone has the same income (equality), and 100% when one individual has everything (perfect inequality). The Gini coefficient for the UK as a whole was 36.8% in 2005 according to the CIA World Factbook, and we find that most pension schemes and annuity portfolios are generally much more unequal than society as a whole.

Table 9.4 shows that we need to simulate each lifetime when assessing indiosyncratic risk. A crucial aspect of this simulation is that deduplication (Section 2.5) has to have been carried out, as idiosyncratic risk will be under-estimated if policies to the same lives are erroneously simulated as if they are independent. Simulation of future lifetime is made rather neat with survival models because of the relationship (equation (3.21)) between the survival probability, $_tp_x$, and the integrated hazard function, $\Lambda_{x,t}$:

$$_tp_x = e^{-\Lambda_{x,t}} \tag{9.1}$$

$$\Rightarrow \Lambda_{x,t} = -\log {_tp_x}. \tag{9.2}$$

Since $_tp_x \in (0,1)$, we can use equation (9.2) to simulate the future lifetime with $\Lambda_{x,t} = -\log U$, where U is a random number uniformly distributed over the interval $(0,1)$. Depending on the nature of $\Lambda_{x,t}$ we can either evaluate an explicit formula for the future time lived, t, or else iterate a solution. Richards (2012) gives a table of formulae for a number of mortality laws, together with an algorithm for laws without closed-form expressions. Using the simulated

Table 9.5 *Idiosyncratic risk capital requirements for two portfolios; 95% confidence intervals for 99.5th quantile calculated using methodology of Harrell and Davis (1982).*

Portfolio	95% interval for $\text{VaR}_{0.995}$ idiosyncratic risk capital
Case Study	0.28–0.30%
Richards et al. (2013)	0.07–0.08%

lifetime we can work out the present value of benefits paid, and thus the total present value for the whole portfolio. Repeating this 1,000 times gives us a set of run-off values from which we can calculate the financial impact of idiosyncratic risk; the resulting values for our Case Study and another, larger portfolio are shown in Table 9.5. Note that, as with our calculation of mis-estimation risk in Section 9.6, our methodology does not strictly qualify because there is no one-year time horizon. Nevertheless, the values in Table 9.5 will suffice to illustrate our longevity risk-capital calculation.

9.10 Aggregation

In Sections 9.6–9.9 we have three values for $\text{VaR}_{0.995}$ for each of mis-estimation risk, trend risk and idiosyncratic risk separately. Under regimes like Solvency II these values are itemised separately for the regulator to assess their reasonableness in isolation. However, to arrive at an overall capital requirement these individual values have to be aggregated. Simply adding the margins together would be over-cautious, so aggregation is done using a correlation matrix, C. If r is the vector of individual margins, then the overall capital requirement is given by $\sqrt{r'Cr}$, where $'$ denotes the transpose of a vector (or matrix).

9.10.1 The Correlation Matrix

The correlation matrix is usually set using expert judgement. The entries on the leading diagonal all have to be 1 by definition, and the matrix obviously needs to be symmetric. The insurer has to document the reason behind its choice of each off-diagonal correlation, including justifying any choices of zero correlation. Negative correlations reduce the overall capital requirement, and for this reason regulators will usually push back strongly on them. Positive correlations increase the overall capital requirement, and even reducing a positive correlation can lead to regulator scrutiny. A matrix of zero correlations, i.e. $C = I$, would in practice be unlikely to pass muster with a regulator.

To be a valid correlation matrix, C obviously has to be symmetric about the leading diagonal. However, to be a true correlation matrix it also needs to be positive semi-definite, i.e. $r'Cr$ is non-negative for every non-zero column vector r. In other words, with a set of positive individual risk margins in r, it cannot be possible to have a negative overall capital requirement. A symmetric matrix can be quickly checked as to whether it is positive semi-definite in R with a call to the chol() function; if the matrix is not positive semi-definite, the call will fail with an error. Note that R's chol() function does not check if the supplied matrix is symmetric, which the user needs to check beforehand.

For simplicity we will assume a 20% correlation between mis-estimation risk and trend risk, and zero correlation between the other pairings, i.e.:

$$C = \begin{pmatrix} 1 & 0.2 & 0 \\ 0.2 & 1 & 0 \\ 0 & 0 & 1 \end{pmatrix}.$$

We first check that our (symmetric) correlation matrix is positive semi-definite using R's chol() function:

```
C <- matrix(c(1, 0.2, 0, 0.2, 1, 0, 0, 0, 1), nrow=3)
chol(C)
```

which produces the following output:

```
> C <- matrix(c(1, 0.2, 0, 0.2, 1, 0, 0, 0, 1), nrow=3)
> chol(C)
      [,1]      [,2] [,3]
[1,]    1 0.2000000    0
[2,]    0 0.9797959    0
[3,]    0 0.0000000    1
```

showing that C is positive semi-definite because the chol() function has succeeded.

9.10.2 Example Aggregation: Case Study

We assume that the vector of capital requirements covers mis-estimation risk, trend risk and idiosyncratic risk as follows:

$$r_1 = \begin{pmatrix} 0.0468 \\ 0.05 \\ 0.0030 \end{pmatrix}, \qquad (9.3)$$

where 0.0468 is the capital requirement for mis-estimation risk from Table 9.2, 0.05 is the capital requirement for trend risk from Section 9.7 and 0.0030 is the capital requirement for idiosyncratic risk from Table 9.5. The overall capital requirement for the longevity-risk sub-module for the Case Study is therefore $\sqrt{r_1' C r_1} = 0.07506824$, or 7.51% of best-estimate reserves.

9.10.3 Example Aggregation: Richards et al. (2013)

We assume that the vector of capital requirements covers mis-estimation risk, trend risk and idiosyncratic risk as follows:

$$r_2 = \begin{pmatrix} 0.0119 \\ 0.05 \\ 0.0008 \end{pmatrix}, \qquad (9.4)$$

where 0.0119 is the capital requirement for mis-estimation risk from Table 9.2, 0.05 is the capital requirement for trend risk from Section 9.7 and 0.0008 is the capital requirement for idiosyncratic risk from Table 9.5. The overall capital requirement for the longevity-risk sub-module for the portfolio in Richards et al. (2013) is therefore $\sqrt{r_2' C r_2} = 0.05366796$, or 5.37% of best-estimate reserves. Although the portfolio in Richards et al. (2013) has markedly lower mis-estimation risk and idiosyncratic risk courtesy of its scale, the overall longevity risk capital is only modestly lower than for the portfolio in the Case Study because both are exposed to the same trend risk, model risk and basis risk.

   In practice an insurer or pension scheme would have other longevity-related items beyond the three illustrated here. Examples include the proportion married and the age difference between spouses. Richards et al. (2014) gives a table of some of the elements that might appear in a longevity-risk sub-module.

9.10.4 Further Levels of Aggregation

In Sections 9.10.2 and 9.10.3 we calculated a single capital requirement for a longevity-risk sub-module by aggregating the various components. In practice an insurer would face many more risks, such as interest-rate risk, credit default risk, inflation risk, expense risk and operational risk, to name just a few. Each of these risks would appear in other sub-modules, which would have their own breakdown of component risks and their own correlation matrices. In each case a similar aggregation calculation to those of Sections 9.10.2 and 9.10.3 would be carried out, and then a further aggregation step at the topmost level would be peformed to arrive at the insurer's overall capital requirement.

PART TWO

REGRESSION AND PROJECTION MODELS

10

Methods of Graduation I: Regression Models

10.1 Introduction

The mathematical study of mortality began with Benjamin Gompertz's landmark paper (Gompertz, 1825). To quote Gompertz: "[A]t the age x the intensity of man's mortality might be denoted by aq^x." We can illustrate Gompertz's argument and this law with data from the Human Mortality Database on UK males. Table 10.1 shows deaths, d_x, and central exposures, E_x^c, on UK males in 2000 for ages 40 through 90 in steps of 10 years. Column 4 gives the logarithm of "the intensity of (his) mortality" (this is the logarithm of the mortality ratio $\hat{r}_x$ in equation (4.5)). Column 5 gives the first differences of column 4. Gompertz observed that these differences were approximately constant and his law follows. We could estimate this constant as the mean of the differences, i.e. 0.982. This constant is a measure of the rate of ageing and is a characteristic of UK males in 2000. Here we have $q \approx \exp(0.982) = 2.670$.

Gompertz was working in the early nineteenth century. He was restricted by the limited data available and his paper contains not a single graph. Figure 10.1 shows $\log(d_x/E_x^c)$ for ages $x = 40, \ldots, 90$; the data are evidently well approxi-

Table 10.1 *Extract from the Human Mortality Database on UK males in 2000. The estimate $\hat{r}_x$ is the mortality ratio in equation (4.5).*

Age x	Deaths, d_x	Exposure, E_x^c	$\log \hat{r}_x = \log(d_x/E_x^c)$	$\Delta(\log \hat{r}_x)$
40	706	423,059	−6.395652	−
50	1,652	379,712	−5.437427	0.958
60	3,334	288,066	−4.459017	0.978
70	7,115	230,010	−3.475919	0.983
80	11,737	128,203	−2.390872	1.085
90	4,950	21,877	−1.486068	0.905

Figure 10.1 Observed $\log(d_x/E_x^c)$ and fitted $\log(\mu_{x+1/2})$. Data: UK males in 2000 from Human Mortality Database.

mated by a straight line and thus in approximate arithmetical progression, just as Gompertz observed.

In Gompertz's case it is doubtful if the mortality ratios $\hat{r}_x$ can be said to estimate the hazard at any single age, since the hazard is not even approximately constant over a ten-year age range. However, that just demonstrates the power of a simple parametric model. Gompertz used his observed common differences to estimate his parameters a and q, and used these values to interpolate the missing entries in his life table. He then observed that this "affords a very convenient mode of calculating values connected with life contingencies".

In this chapter we will describe methods of graduating or smoothing data similar to that in Table 10.1. We will start with simple least squares methods and build up to generalised linear models with polynomial functions. We will discuss general smooth models of mortality in one and two dimensions in Chapters 11 and 12, respectively; forecasting of mortality is described in Chapter 13.

We will be using data from the Human Mortality Database (HMD) throughout this chapter, so first we show how to read data from the HMD into R.

10.2 Reading Data from the Human Mortality Database into R

Many of our examples use data from the Human Mortality Database (HMD). Here we show how to obtain deaths and exposures for UK males in 2000 for single ages 40 to 90 (those used to produce Figure 10.1) from data available on the HMD website.

We assume that the reader has already downloaded the UK data from the HMD (these are the Deaths and Exposure-to-risk (the central exposed-to-risk) at single ages and single calendar years) and that the data are stored in the files HMD_UK_Deaths.txt and HMD_UK_Exposures.txt. HMD copyright does not allow us to make these files directly available but the reader may download the files subject to the conditions of use on the HMD website. The data can now be read into R with the function Read.HMD() in the file Read_HMD.r. This function is available on the website; see Appendix H.3. We outline how this function is used. First, we use the R function source() to read the file Read_HMD.r from disk:

```
#   Load function Read.HMD from file Read_HMD.r
source("Read_HMD.r")
#   Read data and check what data are available
Death.Data = Read.HMD("HMD_UK_Deaths.txt")
names(Death.Data)
Exposure.Data = Read.HMD("HMD_UK_Exposures.txt")
names(Exposure.Data)
#   Retrieve and check ages and years
Age = Death.Data$Age; Age
Year = Death.Data$Year; Year
#   Select male data for ages 40-90 for year 2000
Dth = Death.Data$Male.Matrix[(39 < Age) & (Age < 91),
          Year == 2000]
Exp = Exposure.Data$Male.Matrix[(39 < Age) & (Age < 91),
          Year == 2000]
AGE = 40:90
#   Display the data in column form
cbind(Dth, Exp)
```

We shall explain some features of these R commands. In R we specify subsets of data by using logical commands as in (39 < Age); the colon (:) is used to specify a range of values as in AGE = 40:90, and the ampersand (&) is used for logical AND. The object `Death.Data` itself contains a list of objects. The names of these objects are given by the `names()` function, as in:

```
names(Death.Data)
[1] "Age"      "Year"     "Female.Matrix" "Male.Matrix"
```

The $ symbol is used to access the objects in a list, as in `Death.Data$Age`. The ; symbol allows more than one command to be issued on a single line; we will use this sparingly. In this case, it is wise to check the contents of `Age` and `Year` before trying to extract any data for particular ages and years. Finally, the `cbind()` function displays two or more variates in column form, as in:

```
cbind(Dth, Exp)
      Dth        Exp
40    706 423059.05
41    785 413821.46
42    748 405747.70
 .     .       .

 .     .       .
```

The R variables `Dth` and `Exp` now contain the deaths and exposures for UK males for ages 40 to 90 in the year 2000.

10.3 Fitting the Gompertz Model with Least Squares

The method of least squares makes minimal assumptions and yields estimates of the Gompertz parameters that best fit the data, where "best" is interpreted in a least squares sense. Formally, let $y_x = \log(\hat{r}_x) = \log(d_x/E_x^c)$, $x = 40, \ldots, 90$, where $\hat{r}_x$ is as in Chapter 4, and then assume:

$$y_x = \alpha + \beta x + \epsilon_x, \quad E[\epsilon_x] = 0, \quad \text{Var}[\epsilon_x] = \sigma^2, \quad \text{Cov}[\epsilon_{x_1}, \epsilon_{x_2}] = 0, \quad x_1 \neq x_2.$$
(10.1)

We note that this specifies the simplest form of linear regression model. This is another example of formulating a probabilistic model capable of generating the data. In this case, it leads to an estimation procedure that produces smoothed estimates of the y_x, so when we obtain our estimates $\hat{y}_x$ (in vector form) in equation (10.11), they will already be smooth functions of age, and no further graduation is needed. See the discussion in Section 1.11 on the "hat" notation for estimates.

10.3.1 R Code to Fit a Gompertz Model with Least Squares

We fit a linear model like (10.1) in R with the `lm()` function:

```
Fit.Gompertz = lm(log(Dth/Exp) ~ AGE)
summary(Fit.Gompertz)
```

We do not need to specify an intercept term in the code since R inserts one by default. From the `summary()` function we learn that $\hat{\alpha} = -10.50$ and $\hat{\beta} = 0.1002$ and so $\log(\hat{\mu}_x) = \hat{\alpha} + \hat{\beta}x = -10.50 + 0.1002x$ or, in Gompertz's original formulation, $\hat{\mu}_x = 0.0000275 \times 1.105^x$. This fitted Gompertz line is shown in Figure 10.1.

The `summary()` function returns rather few significant figures, but R calculates $\hat{\alpha}$ and $\hat{\beta}$ to much greater accuracy. This may matter if we were to use the Gompertz law in actuarial work. We use the `names()` function to list the objects in `Fit.Gompertz` and display the fitted coefficients with:

```
names(Fit.Gompertz)
Fit.Gompertz$coef
```

We learn that $\hat{\alpha} = -10.4951552$ and $\hat{\beta} = 0.1001714$, from which we find $\hat{\mu}_x = 0.00002767 \times 1.1054^x$.

We also require an estimate of σ^2. The `summary()` function tells us that the `Residual standard error` is 0.04347. Thus:

$$\hat{\sigma} = 0.04347 \Rightarrow \hat{\sigma}^2 = 0.00189. \tag{10.2}$$

As an aside, we note that the difference method used by Gompertz gives $\hat{\beta} = 0.098$ (compared to the least squares answer of 0.100).

The `summary()` function displays the fitted regression and other useful information, but sometimes we might want to access this information directly. For example, `Fit.Gompertz` does not contain (as a displayable object) the estimate of σ, but the object returned by the `summary()` function does. We use the `names()` function on the `summary()` object:

```
names(summary(Fit.Gompertz))
summary(Fit.Gompertz)$sigma^2
```

and learn that $\sigma^2 = 0.001889942$, in agreement with equation (10.2).

10.3.2 Least Squares Formulae for the Gompertz Model

R makes it easy to fit the Gompertz model but we also need to derive the underlying mathematical formulae, for two reasons. First, these formulae explain how R fits the model, and second, they enable us to compute confidence and prediction intervals for our estimates.

It is convenient to formulate the least squares problem in matrix/vector form. We use boldface symbols to denote matrix or vector quantities. We write $y = (y_1, \ldots, y_n)'$, where now the suffix indicates the age index and not the ages themselves, and n is the number of ages. Let Y be the random variable corresponding to the observed y. Further, let $\mathbf{1}_n = (1, \ldots, 1)'$ with length n, $x = (x_1, \ldots, x_n)'$, $\theta = (\alpha, \beta)'$ and $X = [\mathbf{1}, x]$; we will set $\mathbf{1}_n = \mathbf{1}$ when there is no confusion, as here. Least squares requires the minimisation of:

$$\sum_{j-1}^{n}(y_j - \alpha - \beta x_j)^2 \tag{10.3}$$

$$= (y - X\theta)'(y - X\theta) \tag{10.4}$$

$$= y'y - 2y'X\theta + \theta'X'X\theta. \tag{10.5}$$

We differentiate (10.5) with respect to θ with the help of Appendix F. We find that the least squares estimate of θ satisfies:

$$-2X'y + 2X'X\hat{\theta} = \mathbf{0} \tag{10.6}$$

$$\Rightarrow \hat{\theta} = (X'X)^{-1}X'y, \tag{10.7}$$

with variance-covariance matrix given by:

$$\text{Var}\left[\hat{\theta}\right] = \text{Var}\left[(X'X)^{-1}X'Y\right] \tag{10.8}$$

$$= (X'X)^{-1}X'\text{Var}[Y]X(X'X)^{-1} \quad \text{by Appendix E} \tag{10.9}$$

$$= \sigma^2(X'X)^{-1} \quad \text{since Var}(Y) = \sigma^2 I_n, \tag{10.10}$$

where I_n is the identity matrix of size n; we will often write $I_n = I$. The fitted log hazard rates are given by:

$$\hat{y} = X\hat{\theta} = X(X'X)^{-1}X'y = Hy, \tag{10.11}$$

where $H = X(X'X)^{-1}X'$ is known as the *hat-matrix* (because it "puts the hat on y").

We can now compute the variance-covariance matrix of our fitted values $\hat{y}$, again with the help of Appendix E. We find:

$$\text{Var}[\hat{Y}] = \text{Var}[HY] \quad \text{by (10.11)} \tag{10.12}$$

$$= H\text{Var}[Y]H' \tag{10.13}$$

$$= \sigma^2 HH' \quad \text{since Var}[Y] = \sigma^2 I_n \tag{10.14}$$

$$= \sigma^2 X(X'X)^{-1}X'X(X'X)^{-1}X' \tag{10.15}$$

$$= \sigma^2 X(X'X)^{-1}X'. \tag{10.16}$$

The standard errors of the fitted values are given by the square roots of the diagonal elements of (10.16); we have:

$$\text{StErr}[\hat{y}] = \sqrt{\text{diag}\left\{\sigma^2 X\,(X'X)^{-1}\,X'\right\}} = \sigma\,\sqrt{\text{diag}\left\{X\,(X'X)^{-1}\,X'\right\}}. \quad (10.17)$$

It remains to estimate σ^2. We have, from standard regression theory:

$$\hat{\sigma}^2 = \frac{\text{Residual sum of squares}}{n-2} = \frac{(y - X\hat{\theta})'(y - X\hat{\theta})}{n-2}. \quad (10.18)$$

The divisor of the residual sum of squares is $n - 2$ since the parameter vector θ has length 2; see Dobson (2002). The 95% confidence intervals computed point-wise at each age are:

$$\hat{y} \pm t_{n-2,\,0.975}\,\text{StErr}[\hat{y}], \quad (10.19)$$

where $t_{n-2,\,0.975}$ is the 97.5% point on the t-distribution with $n - 2$ degrees of freedom.

The `predict()` function makes it simple to compute fitted values and the associated confidence intervals. On page 167 the object `Fit.Gompertz` contains the output from the regression fitted with the `lm()` function. We write:

```
#   Fitted values and confidence intervals
Fit.and.CI = predict(Fit.Gompertz, interval = "confidence")
Fit.and.CI[1:3, ]
          fit        lwr       upr
40 -6.488297 -6.512408 -6.464186
41 -6.388126 -6.411525 -6.364727
42 -6.287954 -6.310649 -6.265260
 .         .          .         .
 .         .          .         .
```

and `Fit.and.CI` is a matrix whose columns contain the fitted values, `fit`, and the lower and upper confidence limits, `lwr` and `upr`, respectively.

A confidence interval tells us about the accuracy of our estimate of the mean, but sometimes we are more interested in the accuracy of a predicted value. The stochastic error is an additional source of error in $\text{Var}[\hat{y}]$. We add $\sigma^2 I$ to (10.16), which becomes:

$$\sigma^2\left[I + X\,(X'X)^{-1}\,X'\right] \quad (10.20)$$

and the square roots of the diagonal elements give our prediction intervals . In R we modify the call of `predict()` to:

```
#   Fitted values and prediction intervals
Fit.and.PI = predict(Fit.Gompertz, interval = "prediction")
Fit.and.PI[1:3, ]
          fit        lwr        upr
40 -6.488297 -6.578927 -6.397668
41 -6.388126 -6.478568 -6.297683
42 -6.287954 -6.378217 -6.197692
    .        .        .         .
    .        .        .         .
```

and `Fit.and.PI` is a matrix whose columns contain the fitted values, `fit`, and the lower and upper prediction limits, `lwr` and `upr`, respectively. We note that the width of the 95% confidence interval at age 40 is 0.048 whereas the width of the 95% prediction interval is 0.181, almost four times as great. These results emphasise that the confidence interval allows for parameter error while the prediction interval allows for both parameter and stochastic error. We shall have much more to say on this topic, particularly when we come to forecasting in Chapter 13.

The confidence and prediction intervals obtained with the `predict()` function agree with those obtained with our formulae (10.16) and (10.20), as demonstrated by the program `Gompertz.r` described in Appendix H.3.

A good introduction to regression models with matrix notation is Dobson (2002).

10.3.3 Weighted Least Squares

We saw in the second example in Section 4.8 that a plausible model for the distribution of the number of deaths in the year of life x to $x + 1$ was $D_x \sim$ Poisson$(E_x^c \mu_{x+1/2})$. It follows that $\hat{r}_x = d_x/E_x^c$ is the maximum likelihood estimate of $\mu_{x+1/2}$; thus, if $\tilde{\mu}_{x+1/2} = D_x/E_x^c$ is the estimator corresponding to $\hat{r}_x$, then:

$$E[\tilde{\mu}_{x+1/2}] = \mu_{x+1/2}, \quad \text{Var}[\tilde{\mu}_{x+1/2}] = \mu_{x+1/2}/E_x^c. \qquad (10.21)$$

The least squares set-up at the beginning of this section assumed all observations are made with the same precision, but the second equation in (10.21) tells us that the variance of the observation $\log(d_x/E_x^c)$ depends on the exposure. How should we reflect this in our regression of $\log(d_x/E_x^c)$ on x? We need to compute $\text{Var}[\log(D_x/E_x^c)]$. We derive here a general expression for $\text{Var}[f(X)]$ for some function $f(\cdot)$ of a random variable X. This very useful result is sometimes known as the *delta method* or Δ-method.

Proposition 10.1   *Let X be a random variable with mean μ, and let $f(\cdot)$ be some differentiable function. Then:*

$$\mathrm{Var}[f(X)] \approx [f'(\mu)]^2 \mathrm{Var}[X]. \tag{10.22}$$

Proof: We use Taylor's theorem and expand $f(X)$ about μ:

$$\mathrm{Var}[f(X)] = \mathrm{Var}[f(\mu) + (X - \mu)f'(\mu) + \cdots] \tag{10.23}$$
$$\approx \mathrm{Var}[(X - \mu)f'(\mu)] \tag{10.24}$$

and the result follows.   □

Using equations (10.21) and the Δ-method, we approximate $\mathrm{Var}[\log(D_x/E_x^c)]$. We simplify the notation and set $\mu_{x+1/2} = \mu$, and find:

$$\mathrm{Var}[\log(D_x/E_x^c)] \approx \left(\frac{1}{\mu}\right)^2 \mathrm{Var}[D_x/E_x^c] \quad \text{by the } \Delta\text{-method} \tag{10.25}$$

$$= \left(\frac{1}{\mu}\right)^2 \frac{\mu}{E_x^c} \quad \text{by (10.21)} \tag{10.26}$$

$$= \frac{1}{E_x^c \mu} \tag{10.27}$$

$$\approx \frac{1}{d_x}. \tag{10.28}$$

We can reflect this by modifying (10.1) as follows:

$$y_x = \alpha + \beta x + \epsilon_x, \quad \mathrm{E}[\epsilon_x] = 0, \quad \mathrm{Var}[\epsilon_x] = \sigma^2/d_x, \quad \mathrm{Cov}[\epsilon_{x_1}, \epsilon_{x_2}] = 0 \ (x_1 \neq x_2). \tag{10.29}$$

The R code to fit this model is:

```
Fit.Gompertz.Wt = lm(log(Dth/Exp) ~ AGE, weights = Dth)
```

Note that the weight is inversely proportional to the variance of $y_x = \log(d_x/E_x^c)$; that is, the greater the number of deaths, the greater the weight. We find $\hat{\alpha} = -10.508$ and $\hat{\beta} = 0.1004$, very close to our previous estimates.

We can derive formulae for weighted least squares similar to those given in Section 10.3.2 for unweighted least squares. We first write the weighted model (10.29) in matrix/vector form:

$$\boldsymbol{y} = \boldsymbol{X}\boldsymbol{\theta} + \boldsymbol{\epsilon}, \ \mathrm{Var}[\boldsymbol{\epsilon}] = \sigma^2 \boldsymbol{W}^{-1}, \ \boldsymbol{W} = \mathrm{diag}\{d_x\}. \tag{10.30}$$

Multiplying through by $W^{1/2}$ gives an unweighted model:

$$y^* = X^*\theta + \epsilon^*, \; \text{Var}[\epsilon^*] = \sigma^2 I, \tag{10.31}$$

where $y^* = W^{1/2}y$, $X^* = W^{1/2}X$ and $\epsilon^* = W^{1/2}\epsilon$. It follows immediately from equation (10.7) that:

$$\hat{\theta} = (X'WX)^{-1} X'Wy, \tag{10.32}$$

from equation (10.10) that:

$$\text{Var}\left[\hat{\theta}\right] = \sigma^2 (X'WX)^{-1} \tag{10.33}$$

and from equation (10.11) that:

$$\hat{y} = X\hat{\theta} = X(X'WX)^{-1} X'Wy = Hy, \tag{10.34}$$

where $H = X(X'WX)^{-1} X'W$ is the hat-matrix for weighted least squares.

The program `Gompertz.r` described in Appendix H.3 can be used to produce Figure 10.1, fit the Gompertz model and check many of the formulae derived in this section.

10.4 Poisson Regression Model

In the previous section we used the Poisson distribution for the number of deaths to improve our least squares fit of the Gompertz model. Can we do better than this and use the Poisson distribution directly in the regression model? Let D_x be the random variable with $D_x \sim \text{Poisson}(E_x^c \mu_{x+1/2})$. The probability that D_x takes the observed value d_x is:

$$P[D_x = d_x] = \exp(-E_x^c \mu_{x+1/2})(E_x^c \mu_{x+1/2})^{d_x}/d_x!. \tag{10.35}$$

The likelihood, $L(\mu_{x+1/2})$, is proportional to $P[D_x = d_x]$:

$$L(\mu_{x+1/2}) \propto \exp(-E_x^c \mu_{x+1/2}) \mu_{x+1/2}^{d_x} \tag{10.36}$$

and the log-likelihood follows as:

$$\ell(\mu_{x+1/2}) = d_x \log(\mu_{x+1/2}) - E_x^c \mu_{x+1/2}. \tag{10.37}$$

Note that, as in equation (5.24), the likelihood in equation (10.36) is defined up to a constant of proportionality, so the log-likelihood in equation (10.37) is defined up to an additive constant. We will adopt the convention that this additive constant is omitted.

The assumption that the number of deaths at each age x follows a Poisson distribution with mean $E_x^c \mu_{x+1/2}$ is found widely in the mortality literature; see

Brouhns et al. (2002), Cairns et al. (2009) or Renshaw and Haberman (2006), for example. Within this framework we can make the Gompertz assumption that $\log(\mu_{x+1/2}) = \alpha^* + \beta x$ (where, for consistency with equation (5.1), $\alpha^* = \alpha + \beta/2$) and we can estimate α^* and β by maximum likelihood. Let $\psi = (\alpha^*, \beta)'$; then from (10.37) we have:

$$\ell(\psi) = \sum_x [d_x(\alpha^* + \beta x) - E_x^c \exp(\alpha^* + \beta x)] \tag{10.38}$$

on summing over the ages x. It is evident that this likelihood leads to a pair of non-linear equations in α^* and β. Fortunately, there is a large class of models, of which our Gompertz model is an example, which can be fitted simply in R, as we shall see in Section 10.8.

Although the Poisson and binomial assumptions are found widely in the actuarial literature, they rarely hold. Mortality data generally show more variation than is allowed for under these models. This is the phenomenon of *overdispersion*. We devote Sections 11.9 and 11.10 to this important and oft overlooked topic.

10.5 Binomial Regression Model

In the previous section we used the Poisson distribution to derive the log-likelihood for the Gompertz model for $\mu_{x+1/2}$. An alternative approach is to use the binomial distribution and q_x to derive the log-likelihood for the Gompertz model. In the first example in Section 4.8 we saw that if we observed n lives, all age x at the start of the year and not subject to right-censoring during the year, then the number of deaths D_x in the year of life x to $x + 1$ has a binomial(n, q_x) distribution.

These are rather restrictive conditions. An individual may come under observation part way through the year of life x to $x + 1$, for example upon buying a life insurance policy or immigrating to a country. There usually will be right-censoring during the year of age. We can deal with this only approximately. We accept E_x^c as a measure of the exposure incorporating any left-truncation and right-censoring, but note that individuals who die do not contribute to E_x^c beyond their death. A better estimate of the number of lives at the start of the year is therefore $E_x^c + \frac{1}{2}d_x$, where we assume that deaths occur, on average, mid-way through the year of age. Actuaries call $E_x^c + \frac{1}{2}d_x$ the *initial exposed-to-risk* and denote it by E_x. Further, d_x is the number of deaths observed in the year, so we have approximately:

$$D_x \sim \text{binomial}(E_x, q_x). \tag{10.39}$$

Thus, following the argument in the previous section, we have:

$$P[D_x = d_x] = \binom{E_x}{d_x} q_x^{d_x}(1 - q_x)^{E_x - d_x} \qquad (10.40)$$

and so the likelihood for q_x is:

$$L(q_x) \propto q_x^{d_x}(1 - q_x)^{E_x - d_x}, \qquad (10.41)$$

from which it follows that the maximum likelihood estimate of q_x is:

$$\hat{q}_x = \frac{d_x}{E_x} = \frac{d_x}{E_x^c + \frac{1}{2}d_x}. \qquad (10.42)$$

This formula is very familiar to actuaries and is usually known as the *actuarial estimate*. Note that, like the estimates $\hat{r}_x$, the $\hat{q}_x$ are not smoothed by age.

Some readers may be concerned that E_x in (10.39) is not an integer so strictly D_x cannot have a binomial distribution. However, we are concerned with inference, and the likelihood in equation (10.41) is perfectly well defined for non-integer E_x.

One of the aims of this book is to promote estimation of the hazard rate μ_x in preference to the annual rate of mortality q_x. In equation (3.22) we derived the approximate relationship between $\mu_{x+1/2}$ and q_x:

$$q_x \approx 1 - \exp(-\mu_{x+1/2}). \qquad (10.43)$$

We obtain a second estimate of q_x if we substitute $\hat{r}_x = d_x/E_x^c$ in (10.43), namely:

$$\dot{q}_x = 1 - \exp(-d_x/E_x^c). \qquad (10.44)$$

It is then of interest to compare the actuarial estimate $\hat{q}_x$ in equation (10.42) with this second estimate, $\dot{q}_x$. We simplify notation and set $\hat{r}_x = d_x/E_x^c = \hat{\mu}$. We expand each estimate in powers of $\hat{\mu}$ and retain terms as far as $\hat{\mu}^3$. We find:

$$\hat{q}_x - \dot{q}_x = \frac{d_x}{E_x^c + \frac{1}{2}d_x} - [1 - \exp(-d_x/E_x^c)] \qquad (10.45)$$

$$= \hat{\mu}\left(1 + \tfrac{1}{2}\hat{\mu}\right)^{-1} - [1 - \exp(-\hat{\mu})] \qquad (10.46)$$

$$\approx \hat{\mu}\left(1 - \tfrac{1}{2}\hat{\mu} + \tfrac{1}{4}\hat{\mu}^2\right) - \left(\hat{\mu} - \tfrac{1}{2}\hat{\mu}^2 + \tfrac{1}{6}\hat{\mu}^3\right) \qquad (10.47)$$

$$= \tfrac{1}{12}\hat{\mu}^3. \qquad (10.48)$$

This is a very small number indeed; for example, if $\mu = 0.02$ (roughly UK male mortality at age 70 in 2009 from the HMD), then $\frac{1}{12}\hat{\mu}^3 \approx 7 \times 10^{-7}$. See Currie (2016) for further discussion of this result.

It is convenient to express the likelihood (10.41) in terms of $y_x = d_x/E_x$. We find:

$$L(q_x) \propto q_x^{E_x y_x}(1 - q_x)^{E_x - E_x y_x} \tag{10.49}$$

$$\Rightarrow \ell(q_x) = E_x y_x \log q_x + E_x(1 - y_x) \log(1 - q_x) \tag{10.50}$$

$$= E_x\{y_x \log[q_x/(1 - q_x)] + \log(1 - q_x)\}. \tag{10.51}$$

The function:

$$\text{logit}(q) = \log[q/(1 - q)], \ 0 < q < 1 \tag{10.52}$$

is known as the *logit* function and occurs widely in statistics in the analysis of proportions. An important property of the logit function is that, while q has range $[0, 1]$, $\text{logit}(q)$ has range $[-\infty, +\infty]$, so fitting models to $\text{logit}(q)$ is often easier than fitting models to q. Furthermore, we have, from (10.43) (dropping the suffix $x + 1/2$ for the moment):

$$\text{logit}(q) \approx \log\left(\frac{1 - \exp(-\mu)}{\exp(-\mu)}\right) \tag{10.53}$$

$$= \log(\exp(\mu) - 1) \tag{10.54}$$

$$\approx \log(1 + \mu - 1) \text{ expanding the exponential} \tag{10.55}$$

$$= \log(\mu). \tag{10.56}$$

Thus, we have $\text{logit}(q_x) \approx \log(\mu_{x+1/2})$ and so the quantity $\text{logit}(q_x)$ may be regarded as the analogy in the binomial model of $\log(\mu_{x+1/2})$ in the Poisson model. We define a Gompertz-style model in the binomial setting by:

$$\text{logit}(q_x) = \alpha + \beta x. \tag{10.57}$$

(There is no need here to define α^* rather than α since this is not the same model as in equation (5.1).)

Let $\hat{q}_x = d_x/E_x$ be the maximum likelihood estimate of q_x from equation (10.42). Figure 10.2 shows the observed $\text{logit}(\hat{q}_x)$ plotted against age, x. As expected, the plot looks very similar to that of $\log(\hat{r}_x)$ against age in Figure 10.1. A slight modification to the code in Section 10.3.1 gives the least squares fit of $\text{logit}(\hat{q}_x)$ to x:

```
#   Fit a straight line by least squares
Logit = function(x) log(x/(1 - x))
Exp.Init = Exp + Dth/2
Fit.Gompertz.Bin = lm(Logit(Dth/Exp.Init) ~ AGE)
summary(Fit.Gompertz.Bin)
```

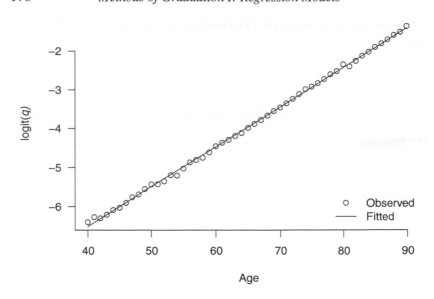

Figure 10.2 Observed logit (d_x/E_x) and fitted logit q_x. Data: UK males in 2000 from Human Mortality Database.

The logit function is not supplied in R so we have used R's `function()` function to define our own logit function, which can then be used in any R expression. We find that $\hat{\alpha} = -10.585$ and $\hat{\beta} = 0.1019$, again similar to our previous results. The resulting estimates of logit(q_x) are smooth over age, as were the $\hat{y}_x$ in Section 10.3.2; these estimates have been added to Figure 10.2.

We leave it as an exercise to the reader to use the Δ-method to show that

$$\text{Var}\left[\text{logit}\left(\hat{q}_x\right)\right] \approx \frac{1}{E_x q_x (1 - q_x)}, \tag{10.58}$$

from which we can derive a weighted version of the immediately preceding code.

Finally, we use the Gompertz assumption to write the binomial log-likelihood (10.51) in a similar form to the Poisson log-likelihood in (10.38). From equation (10.57):

$$q_x = \frac{\exp(\alpha + \beta x)}{1 + \exp(\alpha + \beta x)} \tag{10.59}$$

and with $\psi = (\alpha, \beta)'$ as before, equation (10.51) becomes:

$$\ell(\psi) = \sum_x E_x \{ y_x (\alpha + \beta x) - \log[1 + \exp(\alpha + \beta x)] \} \tag{10.60}$$

on summing over all ages. The similarity between the Poisson and binomial log-likelihoods (10.38) and (10.60) is striking and leads to general definitions of the exponential family and generalised linear models.

10.6 Exponential Family

We suppose a random variable Y has a probability density function of the form:

$$f_Y(y; \theta, \phi) = \exp\{[y\theta - b(\theta)]/a(\phi) + c(y, \phi)\} \tag{10.61}$$

for some functions $a(\cdot)$, $b(\cdot)$ and $c(\cdot, \cdot)$. If ϕ is known, then Y is a member of the *exponential family* with *canonical parameter* θ. We will be performing regression and it is worth pointing out at this early stage that it is the θ parameter that will correspond to the regression coefficients; the ϕ parameter will correspond to such things as the error variance, σ^2, in normal regression.

Suppose y is an observation from the distribution of Y. Then the log-likelihood is:

$$\ell(\theta, \phi; y) = [y\theta - b(\theta)]/a(\phi) + c(y, \phi). \tag{10.62}$$

Comparing (10.62) and (10.37), we see that our Poisson example, $D_x \sim$ Poisson$(E_x^c \mu_{x+1/2})$, is in the exponential family with $y = d_x$, $\theta = \log(\mu_{x+1/2})$, $b(\theta) = E_x^c \exp(\theta)$ and $a(\phi) = 1$ (we may ignore the term $c(y, \phi)$ since it is free of θ and so has no effect on its estimation). In a similar fashion, we compare equations (10.62) and (10.51) and deduce that Y_x is also in the exponential family where Y_x is the random variable corresponding to the observed mortality rate d_x/E_x with $D_x \sim$ binomial(E_x, q_x), $\theta = \text{logit}(q_x)$, $b(\theta) = \log[1 + \exp(\theta)]$ and $a(\phi) = 1/E_x$. Again, we can ignore the term $c(y, \phi)$ since it is free of θ.

The normal, gamma and inverse Gaussian distributions are other well-known members of the exponential family, but we do not use them in this book. Mc-Cullagh and Nelder (1989, Chapter 2) give a table of $a(\cdot)$, $b(\cdot)$ and $c(\cdot, \cdot)$ for all these distributions.

We can gain some insight into the general nature of a random variable with the density in (10.61) since the score function, the partial derivative of the log-likelihood with respect to the parameter θ, is:

$$U(\theta) = \frac{\partial \ell}{\partial \theta} = \frac{y - b'(\theta)}{a(\phi)}. \tag{10.63}$$

Now $E[U(\theta)] = 0$ by Proposition B.3 (Appendix B), so:

$$E[Y] = b'(\theta) \tag{10.64}$$

for a random variable with the density in equation (10.61). Proposition B.3

also tells us that:

$$\text{Var}\left[\frac{Y - b'(\theta)}{a(\phi)}\right] = \text{Var}[U(\theta)] = -\text{E}\left[\frac{\partial^2 \ell}{\partial \theta^2}\right] = \frac{b''(\theta)}{a(\phi)}, \tag{10.65}$$

which gives:

$$\text{Var}[Y] = a(\phi)b''(\theta). \tag{10.66}$$

For example, in the Poisson case, $b(\theta) = E_x^c \exp(\theta)$ and so $b'(\theta) = E_x^c \exp(\theta) = E_x^c \mu_{x+1/2} = \text{E}[D_x]$ and $a(\phi)b''(\theta) = E_x^c \mu_{x+1/2} = \text{Var}[D_x]$ as required. We leave it as an exercise to the reader to verify that if $Y_x = D_x/E_x$ then equations (10.64) and (10.66) reduce to $\text{E}[Y_x] = q_x$ and $\text{Var}[Y_x] = q_x(1 - q_x)/E_x$ in the binomial case.

10.7 Generalised Linear Models

The year 1972 was an important year for statistics, with the publication of Nelder and Wedderburn's paper, Generalized linear models (Nelder and Wedderburn, 1972). Before this paper appeared, multiple linear regression with normal errors was well developed, but when other error distributions arose (such as the Poisson or binomial distributions) methods were *ad hoc* with little or no indication that a common framework existed. Nelder and Wedderburn's paper provided this framework, and software which exploited the theory soon followed. Nowadays, the R language is just one of many statistical languages in which fitting multiple regression models with normal, Poisson or binomial errors is both simple and routine.

We first review ordinary linear regression with normal errors. We identify three components:

1. a *random component* $Y = (Y_1, \ldots, Y_n)'$ with mean $\text{E}(Y) = \mu = (\mu_1, \ldots, \mu_n)'$
2. a *systematic component* $\eta = (\eta_1, \ldots, \eta_n)'$, known as the *linear predictor*, where:

$$\eta = \sum_{1}^{p} \beta_j z_j = Z\beta, \tag{10.67}$$

where $z_1, \ldots, z_p$ is a set of covariates with $z_j = (z_{1,j}, \ldots, z_{n,j})'$, $Z = [z_1, \ldots, z_p]$ and $\beta = (\beta_1, \ldots, \beta_p)'$
3. a *link* between μ, the mean of the random component, and η, the linear predictor of the systematic component.

The function that provides the link between μ and η is called the *link function*;

it is usually denoted $g(\mu_i) = \eta_i$. In the case of ordinary regression with normal errors the link function is the identity function since $\mu = \eta$.

We have given a general description of ordinary regression with normal errors. In a generalised linear model or GLM we allow two extensions, in which we may replace:

1. the normal distribution with a member of the exponential family
2. the identity link function with any monotonic differentiable function (such as the log or logit functions).

We now give two important examples of GLMs: the Gompertz model with (a) Poisson errors and (b) binomial errors.

10.8 Gompertz Model with Poisson Errors

In Section 10.4 we argued that the number of deaths followed the Poisson distribution, $D_x \sim \text{Poisson}(E_x^c \mu_{x+1/2})$, where D_x is the random variable corresponding to the number of deaths, d_x, in the year of age x to $x+1$ and E_x^c is the central exposure. We now show that the Gompertz assumption $\log(\mu_{x+1/2}) = \alpha^* + \beta x$ (see Section 10.4) together with Poisson errors gives rise to a GLM. We have, for any age x:

$$D_x \sim \text{Poisson}(E_x^c \mu_{x+1/2}) \tag{10.68}$$

$$\Rightarrow \text{E}(D_x) = E_x^c \mu_{x+1/2} \tag{10.69}$$

$$\Rightarrow \log[\text{E}(D_x)] = \log E_x^c + \log(\mu_{x+1/2}) \tag{10.70}$$

$$= \log E_x^c + \alpha^* + \beta x \tag{10.71}$$

by the Gompertz assumption. Suppose we have data $[(d_{x_1}, E_{x_1}^c), \ldots, (d_{x_n}, E_{x_n}^c)]$. We know from Section 10.6 that D_x is a member of the exponential family, so, following the definition given on page 178, we have a GLM with

1. *random component* $\boldsymbol{D} = (D_{x_1}, \ldots, D_{x_n})'$ with

$$\boldsymbol{\mu} = \text{E}[\boldsymbol{D}] = (E_{x_1}^c \mu_{x_1+1/2}, \ldots, E_{x_n}^c \mu_{x_n+1/2})' \tag{10.72}$$

2. *systematic component* $\boldsymbol{\eta}$ where

$$\boldsymbol{\eta} = \log[\text{E}(\boldsymbol{D})] \tag{10.73}$$

with linear predictor

$$\eta_x = \log E_x^c + \alpha^* + \beta x \tag{10.74}$$

or

$$\eta = \log E^c + X\theta \qquad (10.75)$$

in matrix/vector notation where $\eta = (\eta_{x_1}, \ldots, \eta_{x_n})'$, $E^c = (E^c_{x_1}, \ldots, E^c_{x_n})'$, $X = [1, x]$ and $\theta = (\alpha^*, \beta)'$ along the same lines as before

3. *link function* equal to the log function.

An important point in this example is that the linear predictor contains not only the usual regression coefficients α^* and β which we need to estimate, but also the term $\log E^c_x$; we can think of $\log E^c_x$ as a regression variable with known coefficient (equal to 1). Such a term in a linear predictor is known as an *offset*. We fit a GLM in R with the glm() function. The form of the glm() function is very similar to the lm() function but we note the following three points: first, the Poisson distribution is specified with the family option; second, there is no mention of the log link since R assumes this by default with Poisson errors; third, the offset is handled with the offset() function in the specification of the linear predictor.

```
#   Fit the Gompertz model with Poisson errors
Gompertz.P = glm(Dth ~ AGE + offset(log(Exp)),
                  family = poisson)
summary(Gompertz.P)
```

and the fitted Gompertz line is $\log(\hat{\mu}_{x+1/2}) = -10.45975 + 0.1005x$, very close to the estimated line with our simple least squares fit in Section 10.3.1, where we found $\log(\hat{\mu}_{x+1/2}) = -10.50 + 0.1002x$.

We can add the call of the predict() function to the above code; it is important to note that in a GLM the predict() function operates on the scale of the linear predictor and not on the scale of the response variable. Thus, the predict() function returns the fitted linear predictors and their standard errors. It is illuminating to compare the fitted values and the standard errors from the weighted least squares model and from the GLM with Poisson errors. We obtain these with the predict() function as follows:

```
#   Normal and Poisson regression prediction
Normal.Reg = predict(Fit.Gompertz.Wt, se.fit = TRUE)
Poisson.GLM = predict(Gompertz.P, se.fit = TRUE)
```

For example, at age 40 we have Normal.Reg$fit[1] = -6.4905 while the fitted linear predictor is Poisson.GLM$fit[1] = 6.4636, which leads to a fitted log hazard of $6.4636 - \log E^c_{x_1} = -6.4917$ from equation (10.75); the two values are in very close agreement.

However, we find something very different when we look at the standard errors. Again, at age 40 we have `Normal.Reg$se.fit[1] = 0.0171` while `Poisson.GLM$se.fit[1] = 0.00622`, and the normal model returns a standard error about 2.75 times larger than that of the Poisson model.

It is not difficult to see how this problem arises. In the normal model there is an independent variance parameter, σ^2, which reflects the variation about the Gompertz line. The variance-covariance matrix of the estimated regression coefficients is given by equation (10.33) as $\sigma^2(X'WX)^{-1}$; the `summary()` function tells us that $\hat{\sigma} = 2.73$, very close to the ratio of the standard errors. There is no such parameter in the Poisson model since the variance of the number of deaths is constrained to equal the mean. This is a strong assumption and there is no *a priori* reason why it should be true. We will return to this problem and how to deal with it in Section 11.9.

10.9 Gompertz Model with Binomial Errors

In the binomial model of Section 10.5 we had $D_x \sim$ binomial(E_x, q_x), where E_x is the initial exposed-to-risk and we expressed the Gompertz law as logit $(q_x) = \alpha + \beta x$. We now show that these two assumptions also lead to a GLM. We follow the derivation on page 179 closely. We have, for any age x:

$$D_x \sim \text{binomial}(E_x, q_x) \tag{10.76}$$

$$\Rightarrow \text{E}(D_x) = E_x q_x \tag{10.77}$$

$$\Rightarrow \text{E}(Y_x) = q_x \text{ where } Y_x = D_x/E_x \tag{10.78}$$

$$\Rightarrow \text{logit}[\text{E}(Y_x)] = \text{logit}(q_x) \tag{10.79}$$

$$= \alpha + \beta x \tag{10.80}$$

by the Gompertz assumption.

We saw in Section 10.6 that Y_x is in the exponential family, and we again follow the definition on page 178. We suppose that we have data $(y_{x_1}, \ldots, y_{x_n})$; then we have a GLM with:

1. *random component* $Y = (Y_{x_1}, \ldots, Y_{x_n})'$ with $\mu = \text{E}(Y) = (q_{x_1}, \ldots, q_{x_n})'$
2. *systematic component* η, where

$$\eta = \text{logit}\,[\text{E}(Y)] \tag{10.81}$$

with linear predictor

$$\eta_x = \alpha + \beta x \tag{10.82}$$

3. *link function* equal to the logit function.

In addition, we have $a(\phi) = 1/E_x$ from Section 10.6. The R code to fit the model also uses the `glm()` function. As in the Poisson case, we make three remarks: first, the binomial distribution is specified with the `family` option; second, the logit link is the default with binomial errors, so there is no need to specify it explicitly; third, the weight function $a(\phi) = 1/E_x$ is specified with the `weights` option.

```
#   Fit the Gompertz model with binomial errors
Y = Dth/Exp.Init
Gompertz.B = glm(Y ~ AGE, family = binomial,
                 weights = Exp.Init)
summary(Gompertz.B)
```

and we find $\text{logit}(q_x) = -10.65 + 0.1028x$, very similar to our findings with least squares in Section 10.5.

Some readers may be familiar with an alternative call of the `glm()` function to fit a binomial model. This approach uses the `cbind()` function on the left-hand side of the $\sim$ sign. The two methods are exactly equivalent.

10.10 Polynomial Models

In the previous sections we saw how to fit the Gompertz model with GLMs with Poisson and binomial errors. The model structure was very simple: linear in age. In this section we consider how mortality varies with time. This is an important topic for actuaries since rapidly improving rates of mortality have major implications for the pricing of annuities. The simple linear structure of the Gompertz model is no longer sufficient. However, the `lm()` and `glm()` functions are very versatile and in this section we show how we can fit models with a polynomial structure.

We assume that the USA data from the Human Mortality Database have been downloaded and read into R with the function `Read.HMD()` as in Section 10.2, and that male and female data are available. Figure 10.3 shows the raw log mortality ratios $\hat{r}_{70}$ at age 70 for calendar years 1961 to 2009: female mortality is substantially lighter than male, as is widely known.

We show how to model the male/female mortality in terms of a quadratic/cubic function in time. We assume that the number of deaths, D_t, in year t has the Poisson distribution $\text{Poisson}(E_t^c \mu_{t+1/2})$ and (for the cubic model):

$$\log[\text{E}(D_t)] = \log E_t^c + \alpha_0 + \alpha_1 t + \alpha_2 t^2 + \alpha_3 t^3, \qquad (10.83)$$

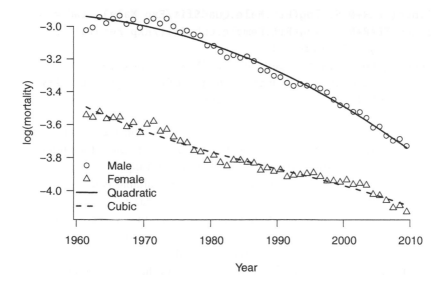

Figure 10.3 Observed $\log(d_t/E_t^c)$ and fitted $\log(\mu_{t+1/2})$. Data: USA males and females age 70 from 1961 to 2009 from Human Mortality Database.

and fit on a transformed timescale to avoid possible numerical problems:

```
YEAR = 1960:2009
x = YEAR - mean(YEAR)
x2 = x^2
x3 = x^3
Fit.Male.Quad = glm(Dth.Male ~ x + x2 + offset(log(Exp.Male)),
                     family = poisson)
names(Fit.Male.Quad)
Fit.Female.Cubic = glm(Dth.Female ~ x + x2 + x3 +
                       offset(log(Exp.Female)), family = poisson)
```

Figure 10.3 can be produced with R code along the following lines:

```
Obs.Male = log(Dth.Male/Exp.Male)
Obs.Female = log(Dth.Female/Exp.Female)
Y.lim = range(Obs.Male, Obs.Female)
plot(YEAR+0.5, Obs.Male, ylim = Y.lim, xlab = "Year",
     ylab = "log(mortality)")
points(YEAR+0.5, Obs.Female, pch = 2)
```

```
lines(YEAR+0.5, log(Fit.Male.Quad$fit/Exp.Male), lwd = 2)
lines(YEAR+0.5, log(Fit.Female.Cube$fit/Exp.Female),
      lty = 2, lwd = 2)
legend("bottomleft", legend = c("Male", "Female",
      "Quadratic", "Cubic"), pch = c(1, 2, -1, -1), lty =
      c(-1, -1, 1, 2), lwd = c(-1, -1, 2, 2), bty="n")
```

The `legend()` function has various options for controlling the plotting symbol pch, the line type `lty` and the line width `lwd`. See `help(par)` for a full list of graphical parameters and `par()` to see the current settings.

We note that in Poisson models the dependent variable is d_t, so the fitted values are the fitted deaths; these are stored in `Fit.Male.Quad$fit`. We can use essentially the same code to model the mortality rate with a logit link and binomial errors, but note that the fitted values are the fitted mortality rates since the dependent variable is $\hat{r}_t$, the observed mortality ratios.

It is perhaps worth sounding a word of caution here. Polynomial models often fit data very well but their forecasting properties are poor. Forecasting with polynomial models is not recommended.

The program to fit these time dependent models is `Time_Models.r`, as described in Appendix H.3.

11

Methods of Graduation II: Smooth Models

11.1 Introduction

Figure 10.3 shows a quadratic model fitted to the male USA data and a cubic model fitted to the female data. The curves follow the general trends in the data but there are sections of each curve that are systematically above or below the data values. We could try increasing the degree of the fitted polynomial but this seems rather clumsy; the curve is fitted *globally* to the data. Can we turn this round and make the curve fit the data *locally*; i.e. can we adjust the amount of smoothing to the local behaviour of the data? This is what smoothing methods attempt to do.

Actuaries have had a very long association with smoothing. The Gompertz model (Gompertz, 1825) could be considered an early smoothing method, albeit a very simple one. Makeham's extension (Makeham, 1860) of the Gompertz model improved the fit to mortality tables, but neither model was sufficiently flexible to be applicable outside a limited age range, say 40 to 90. There were many other efforts, all grouped under the general heading of *mathematical formulae*, of which Perks (1932) is perhaps the best known. The basic idea is that adding a parameter will improve the fit, so we have the Gompertz model with two parameters, Makeham's with three and the two Perks formulae with four and five parameters. English Life Tables No. 11 and No. 12 were graduated using a mathematical formula with seven parameters. Forfar et al. (1988) gave a general family of mathematical models of mortality and used extensive statistical testing to choose the best fitting model. See Section 5.9 for a summary of graduation by mathematical formulae, and Benjamin and Pollard (1980, Chapter 14) for a full discussion of this approach.

Moving averages are another general approach. These all suffer from an "edge effect" problem, in that smooth values for initial and final ages have to be estimated by separate and generally arbitrary means. Spencer's famous

21-point formula is among the best known of these methods (Spencer, 1904). Again, see Benjamin and Pollard (1980, Chapter 13) for a discussion of moving average methods.

Moving average methods are examples of *local methods*, by which we mean that the smoothed value at age x, say, depends largely on the observed values in the vicinity of x. There are many ways of approaching the smoothing problem: smoothing splines, locally weighted regression and kernel smoothing are three of the most widely used. Hastie and Tibshirani (1990), Green and Silverman (1994) and Wood (2006) are three excellent general references on smoothing. Actuaries have used smoothing spline methods, most notably in the construction of English Life Tables No. 13 (McCutcheon and Eilbeck, 1975), No. 14 (McCutcheon, 1985), No. 15 and No. 16 (Kaishev et al., 2009).

In this book we will use a fourth widely used method: penalised *B*-spline regression, generally known as *P*-splines (Eilers and Marx, 1996). To some extent, choice of smoothing method is a matter of taste, but we prefer *P*-splines for the following four reasons: first, the user can include a number of regression variables, some of which can be smooth; second, the method extends naturally and flexibly to smoothing in more than one dimension; third, the fitting algorithms are essentially least squares and look just like the algorithms for ordinary regression; fourth, highly efficient algorithms exist for smoothing in more than one dimension where the data are arranged in an array (for example, mortality data). Currie et al. (2004) applied *P*-splines to the smoothing and forecasting of mortality data. We will use *P*-splines to smooth one-dimensional mortality data in this chapter, to smooth two-dimensional mortality data in the next chapter and to forecast both one-dimensional and two-dimensional mortality data in Chapter 13. Richards et al. (2006) introduced actuaries to *P*-spline methods.

There is a further reason which the new user should find attractive. Smoothing methods are often described as non-parametric. Locally weighted regression is a good example of a non-parametric smoother; there are no regression coefficients and the smooth curve emerges from the algorithm. In contrast, *P*-splines is not really a non-parametric method: there are regression coefficients which are readily interpreted and the method is more accurately described as over-parametrised rather than non-parametric; the remarks on Figure 11.5 and Figure 11.7 support this idea.

We begin by discussing a smoothing method that will be familiar to many actuaries: Whittaker smoothing (Whittaker, 1923). Readers may be surprised to know that Whittaker smoothing is, in fact, a simple example of the method of *P*-splines. Whittaker's argument is simple and persuasive, and provides a nice introduction to our more general approach. After describing Whittaker

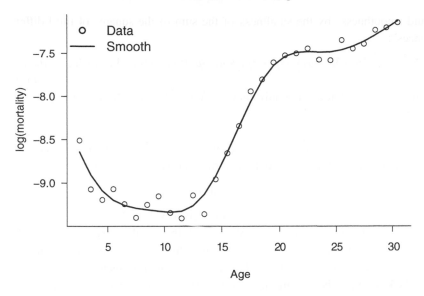

Figure 11.1 Observed $\log(d_x/E_x^c)$ and fitted $\log(\mu_{x+1/2})$ with Whittaker smoothing. Data: UK males in 2011 from Human Mortality Database.

smoothing we build up to the full method of *P*-splines by describing in turn (a) a single *B*-spline, (b) *B*-spline bases, (c) *B*-spline regression and (d) *P*-splines. We will give skeleton R code as we go.

11.2 Whittaker Smoothing

Figure 11.1 is a typical plot of log mortality of the kind of data we might want to smooth; here we use UK male data for 2011 for ages 2 to 30 taken from the Human Mortality Database. We have included younger ages and it is clear that a low-order polynomial will not provide an adequate fit. Let $y_i = \log(d_i/e_i), i = 1, \ldots, n$ be our data, where the d_i are the observed deaths at age i and the e_i are the corresponding central exposures, and let $\mu_i, i = 1, \ldots, n$ be the candidate smooth values.

Whittaker (1923) realised that smoothing data is achieved by striking a balance between *fidelity to the data* and *smoothness of the fitted curve*. He measured fidelity by the sum of squares (SS) of deviations:

$$\mathrm{SS} = \sum_{i=1}^{n}(y_i - \mu_i)^2, \tag{11.1}$$

and smoothness "by the smallness of the sum of the squares of third differences":

$$M_3 = (\mu_4 - 3\mu_3 + 3\mu_2 - \mu_1)^2 + (\mu_5 - 3\mu_4 + 3\mu_3 - \mu_2)^2 + \ldots + (\mu_n - 3\mu_{n-1} + 3\mu_{n-2} - \mu_{n-3})^2. \tag{11.2}$$

This second definition is equivalent to saying that the chosen smooth curve will be approximately locally quadratic, and has long been a popular criterion for smoothness among actuaries; see Benjamin and Pollard (1980, p.242). Whittaker's breakthrough was to find a way of balancing the competing criteria of fidelity and smoothness. He defined the relative balance function:

$$SS + \lambda M_3, \tag{11.3}$$

where λ is a constant to be chosen; for a given value of λ our smooth curve is obtained by minimising (11.3). We remark that small values of λ will yield a curve that follows the data more closely, while large values will result in a smoother curve. We will refer to λ as the *smoothing parameter*.

Back in 1923, the minimisation of (11.3) for given λ required some pretty daunting arithmetic, certainly by today's standards. Nowadays we can easily write down an expression for an explicit solution and then use the power of a modern computer to evaluate it. We write (11.3) in matrix/vector form. First, we define $\mathbf{y} = (y_1, \ldots, y_n)'$ and $\boldsymbol{\mu} = (\mu_1, \ldots, \mu_n)'$; then (11.1) and (11.2) become:

$$SS = (\mathbf{y} - \boldsymbol{\mu})'(\mathbf{y} - \boldsymbol{\mu}) \tag{11.4}$$

$$M_3 = \boldsymbol{\mu}' \mathbf{D}' \mathbf{D} \boldsymbol{\mu}, \tag{11.5}$$

where:

$$\mathbf{D} = \begin{bmatrix} 1 & -3 & 3 & -1 & 0 & 0 & \cdots \\ 0 & 1 & -3 & 3 & -1 & 0 & \cdots \\ 0 & 0 & 1 & -3 & 3 & -1 & \cdots \\ \vdots & \vdots & \vdots & \vdots & \vdots & \vdots & \ddots \end{bmatrix}, \ (n-3) \times n \tag{11.6}$$

is the *third-order difference matrix*. With this notation (11.3) becomes:

$$(\mathbf{y} - \boldsymbol{\mu})'(\mathbf{y} - \boldsymbol{\mu}) + \lambda \boldsymbol{\mu}' \mathbf{D}' \mathbf{D} \boldsymbol{\mu} \tag{11.7}$$

$$= \mathbf{y}'\mathbf{y} - 2\mathbf{y}'\boldsymbol{\mu} + \boldsymbol{\mu}'\boldsymbol{\mu} + \lambda \boldsymbol{\mu}' \mathbf{D}' \mathbf{D} \boldsymbol{\mu}. \tag{11.8}$$

We differentiate with respect to $\boldsymbol{\mu}$ (see Appendix F for rules on differentiating a scalar function of a vector $\mathbf{x}$ with respect to $\mathbf{x}$); the minimum of (11.3) occurs when:

$$-2\mathbf{y} + 2\boldsymbol{\mu} + 2\lambda \mathbf{D}' \mathbf{D} \boldsymbol{\mu} = \mathbf{0} \tag{11.9}$$

$$\Rightarrow (\mathbf{I} + \lambda \mathbf{D}' \mathbf{D})\boldsymbol{\mu} = \mathbf{y}, \tag{11.10}$$

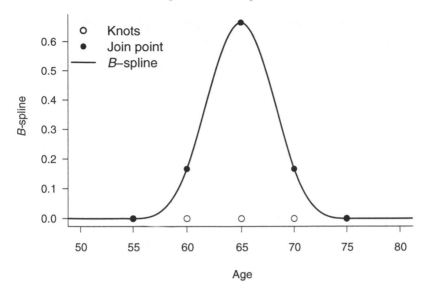

Figure 11.2 Single cubic B-spline with knots at 55, 60, 65, 70 and 75.

where I is the identity matrix of size n. Solving (11.10) we find the Whittaker smoother for given λ can be written:

$$\hat{\mu} = (I + \lambda D'D)^{-1}y. \qquad (11.11)$$

Whittaker took a very practical approach to the choice of λ; he suggested that "we try two or three different values and see which gives the most satisfactory result". The R function WhittakerSmooth() in the file Whittaker_Smooth.r is described in Appendix H.4 and enables us to perform this task with a minimum of effort. The function call looks like:

```
Whittaker_Smooth(Age, Dth, Exp, Lambda)
```

where Age, Dth and Exp are our ages, deaths and central exposures, and Lambda is the proposed value of λ. The program Whittaker.r loads data, calls the function and produces graphical output. We can quickly decide that $\lambda = 10$ is a suitable value; Figure 11.1 results. In Section 11.8 we will suggest a less subjective way of selecting λ.

11.3 B-Splines and B-Spline Bases

A B-spline is a piecewise polynomial of degree d; commonly $d = 0, 1, 2$ or 3, which correspond to constant, linear, quadratic and cubic B-splines.

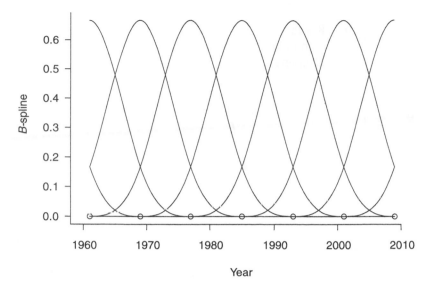

Figure 11.3 A cubic *B*-spline basis with 9 *B*-splines in the basis and 5 internal knots.

Figure 11.2 shows a single cubic *B*-spline. A *B*-spline is defined with reference to a set of *knots*; in Figure 11.2 these knots are located at ages 55, 60, 65, 70 and 75. *B*-splines are *local* functions, i.e. they are zero everywhere outside the range of their knots, here below 55 and above 75. Between 55 and 75 the *B*-spline consists of four cubic polynomial pieces smoothly bolted together at the knots; for a cubic *B*-spline "smoothly bolted" means the spline is continuous and has continuous first and second derivatives at each of its knots. In a similar fashion, a linear *B*-spline, for example, is defined with reference to three knots and consists of two linear pieces continuous but not differentiable at each of its knots. We will work exclusively with cubic *B*-splines.

Suppose we want a smooth model for the mortality rates for years 1961 to 2009 in Figure 10.3 in Section 10.10. We use a set of *B*-splines which covers the required years; such a set is called a *basis*. Figure 11.3 shows a basis of nine *B*-splines which span the required years. Note that the *B*-splines at either end of the basis are only partially visible. We write the set of functions which comprise the basis as $\mathcal{B} = \{B_1, \ldots, B_9\}$. We use these basis functions in exactly the same way as we use $\{1, x\}$ or $\{1, x, x^2\}$ as basis functions for linear or quadratic regression.

Appendix H.4 describes an R program, `Basis.r`, which produces a figure similar to Figure 11.3.

11.4 *B*-Spline Regression

In Section 10.8 on the Gompertz model with Poisson errors, we wrote the systematic part of the model as:

$$\log[E(D_x)] = \log E_x^c + \alpha + \beta x, \tag{11.12}$$

which we wrote compactly as:

$$\log[E(\boldsymbol{D})] = \log \boldsymbol{E}^c + \boldsymbol{X}\boldsymbol{\theta}, \tag{11.13}$$

where $\boldsymbol{D}$ is the random vector for deaths, $\boldsymbol{E}^c$ is the vector of central exposures, $\boldsymbol{\theta} = (\alpha, \beta)'$ is the vector of regression coefficients and $\boldsymbol{X}$ is the regression matrix where, as before:

$$\boldsymbol{X} = [\boldsymbol{1}, \boldsymbol{x}], \tag{11.14}$$

$\boldsymbol{1} = (1, \ldots, 1)'$ with length n and $\boldsymbol{x} = (x_1, \ldots, x_n)'$.

For a smooth model we want to replace the linear assumption in (11.12) with a smooth function defined in terms of our *B*-spline basis, i.e.:

$$\log[E(D_x)] = \log E_x^c + \sum_{j=1}^{c} \theta_j B_j(x), \tag{11.15}$$

where c is the number of splines in the basis. Just as in the linear case above, we write this compactly as:

$$\log[E(\boldsymbol{D})] = \log \boldsymbol{E}^c + \boldsymbol{B}\boldsymbol{\theta}, \tag{11.16}$$

where $\boldsymbol{\theta} = (\theta_1, \ldots, \theta_c)'$ is the vector of regression coefficients and $\boldsymbol{B}$ is the regression matrix, where:

$$\boldsymbol{B} = \begin{bmatrix} \boldsymbol{b}_1' \\ \vdots \\ \boldsymbol{b}_n' \end{bmatrix}, \quad \boldsymbol{b}_i' = (B_1(x_i), \ldots, B_c(x_i)). \tag{11.17}$$

Thus $\boldsymbol{b}_i'$ is the ith row of $\boldsymbol{B}$, i.e. the elements of $\boldsymbol{b}_i$ consist of the basis functions evaluated at x_i, exactly as in the linear regression above.

We now obtain the regression matrix $\boldsymbol{B}$ for our years $1961, \ldots, 2009$ by evaluating our basis functions at each year in turn; the R function `bspline()` does exactly this. See Appendix H.4 for access to the code for `bspline()`, and note that R's splines library is used. The arguments `ndx` and `bdeg` are used to set the number of knots and the degree of the *B*-splines, respectively. The function `bspline()` places a knot at the minimum and the maximum years and uses `ndx-1` equally spaced internal knots:

```
source("Bspline.r") # Function to compute regression matrix
library(splines)     # Load spline library
ndx = 6; bdeg = 3    # 5 internal knots, cubic B-splines
Setup = bspline(YEAR, min(YEAR), max(YEAR), ndx, bdeg)
names(Setup)
B = Setup$B          # B is the regression matrix
dim(B)               # There are 9 B-splines in the basis
Knots = Setup$Knots; Knots        # The knots
```

The function $\mathtt{dim()}$ returns the dimension of a matrix and $\mathtt{bspline()}$ returns the regression matrix $\boldsymbol{B}$ and the set of knots $\mathtt{Knots}$. Here $\boldsymbol{B}$ has dimension 49×9, the number of years by the number of B-splines in the basis. We can use $\boldsymbol{B}$ just like any other regression matrix. Here we fit the GLM for mortality by year with Poisson errors and log link. A subtle point is that since $\boldsymbol{B1} = \boldsymbol{1}$ we do not need to fit an intercept when we regress on $\boldsymbol{B}$. This is the opposite of our code in Section 10.1 for the basic Gompertz model, where we did not specify an intercept. In the present case we remove R's intercept and fit the model as follows:

```
# B-spline regression with Poisson errors
Fit.Bspline = glm(Dth ~ -1 + B + offset(log(Exp)),
                    family = poisson)
```

It is worth emphasising that the above code fits an ordinary GLM in exactly the same way as we would for a quadratic or cubic regression, the only difference being that this code uses the regression matrix $\boldsymbol{B}$ instead of specifying the individual regressors with $\mathtt{x1 + x2}$, as on page 183. The alternative to this earlier method is to specify the regression matrix as follows:

```
X = cbind(x1, x2)
Fit.Male.Quad = glm(Dth.Male ~ X + offset(log(Exp.Male)),
                    family = poisson)
```

This parallels our B-spline regression code and is precisely equivalent to the earlier code on page 183.

Figure 11.4 shows the output from our B-spline regression, and repays careful study. In the lower panel we see the original basis functions; in the middle panel each basis function has been scaled by its regression coefficient, as in (11.15); the upper panel shows the fitted regression obtained by summing the contribution of each B-spline at each x.

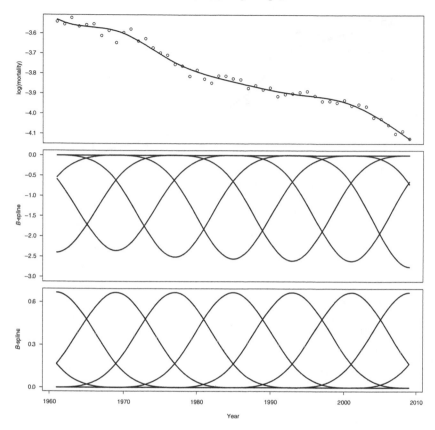

Figure 11.4 Lower panel: *B*-spline basis; middle panel: scaled basis; upper panel: fitted regression.

Figure 11.5 shows that the fitted regression tracks the data very well and does so in a flexible, smooth way. We also see the fitted regression coefficients and these too track the data; this bears out our remark on page 186 that in *B*-spline regression the coefficients do have an interpretation.

11.5 The Method of *P*-Splines

In the previous section we used nine *B*-splines in our basis; the resulting fit was clearly satisfactory. We say that the basis has *dimension* 9. One obvious question is: where did the dimension 9 come from? We could have used 6 or 16, say. A basis with dimension 6 will have less flexibility than one with 9,

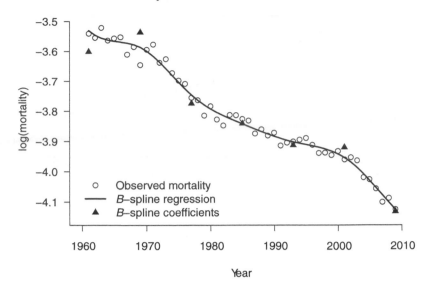

Figure 11.5 *B*-spline regression with fitted mortality and fitted regression coefficients; basis dimension 9. Data: USA females age 70 from 1961 to 2009 from Human Mortality Database.

while one of dimension 16 will have more. What is the preferred dimension? One possibility is simply to search through the possible bases and choose either by eye or, more satisfactorily, by some goodness-of-fit criterion for an optimal size for the basis.

Eilers and Marx (1996) came up with another idea. First, they chose a basis with a large dimension; with 49 data points they might use a basis of dimension 20. Presumably, the resulting *B*-spline regression is too flexible and the fitted curve is inclined to follow the randomness in the data rather than its underlying trend. Figure 11.6 bears out this remark: we see that the fitted curve follows the data rather too slavishly. Eilers and Marx realised that the erratic nature of the curve follows from the erratic nature of the estimated coefficients. Their second idea was to modify the log-likelihood in such a way that the estimated coefficients in Figure 11.6 were less erratic. This is the idea of *penalisation*.

Suppose we denote the regression coefficients $\boldsymbol{\theta} = (\theta_1, \ldots, \theta_c)'$. We define the *roughness penalty*:

$$P(\boldsymbol{\theta}) = (\theta_1 - 2\theta_2 + \theta_3)^2 + \ldots + (\theta_{c-2} - 2\theta_{c-1} + \theta_c)^2. \qquad (11.18)$$

Thus, $P(\boldsymbol{\theta})$ measures the variability of *adjacent* coefficients and the more variable the θ_i are, the larger the value of $P(\boldsymbol{\theta})$; thus $P(\boldsymbol{\theta})$ measures "roughness".

Figure 11.6 *B*-spline regression with fitted mortality and fitted regression coefficients; basis dimension 20. Data: USA females age 70 from 1961 to 2009 from Human Mortality Database.

There are close similarities with Whittaker's definition of smoothness in (11.2). However, we note the following: (a) in (11.2) the fitted values are smoothed directly, whereas (b) in (11.18) the fitted values are smoothed indirectly by smoothing the regression coefficients; the smooth regression coefficients then give the smooth fitted values through the regression equation (11.16).

Expression (11.18) is a second-order penalty, and while other orders are possible (terms like $(\theta_j - \theta_{j+1})^2$ give rise to a first-order penalty, for example) we will deal exclusively with second-order penalties for reasons that will be discussed in Section 13.3, and in particular Figure 13.8.

We can write (11.18) in matrix/vector form in the same way that we wrote the roughness penalty in our discussion of Whittaker smoothing in Section 11.2. Let:

$$\boldsymbol{D} = \begin{bmatrix} 1 & -2 & 1 & 0 & 0 & 0 & \cdots \\ 0 & 1 & -2 & 1 & 0 & 0 & \cdots \\ 0 & 0 & 1 & -2 & 1 & 0 & \cdots \\ \vdots & \vdots & \vdots & \vdots & \vdots & \vdots & \ddots \end{bmatrix}, \; (c-2) \times c \qquad (11.19)$$

be the *second-order difference matrix*. Then $P(\boldsymbol{\theta})$ can be written compactly:

$$P(\boldsymbol{\theta}) = \boldsymbol{\theta}' \boldsymbol{D}' \boldsymbol{D} \boldsymbol{\theta}. \qquad (11.20)$$

The general log-likelihood for Poisson regression at age x is:

$$\ell(\mu_{x+1/2}) = d_x \log(\mu_{x+1/2}) - E^c_x \mu_{x+1/2} \tag{11.21}$$

from (10.37), and, summing over x, we find the full log-likelihood in θ as:

$$\ell(\theta) = \sum_x [d_x \log(\mu_{x+1/2}) - E^c_x \mu_{x+1/2}]; \tag{11.22}$$

here we replace $\mu_{x+1/2}$ with:

$$\log(\mu_{x+1/2}) = \log E^c_x + \sum_{j=1}^{c} \theta_j B_j(x) \tag{11.23}$$

from (11.15). This is the log-likelihood that gives the fits in Figures 11.5 and 11.6. Eilers and Marx (1996) proposed the modified log-likelihood:

$$\ell_p(\theta) = \ell(\theta) - \tfrac{1}{2}\lambda P(\theta). \tag{11.24}$$

The $^1/_2$ is just a convenient scaling factor; λ is known as the *smoothing parameter*. The function $\ell_p(\theta)$ is a *penalised log-likelihood* and, for given λ, we choose θ by maximising $\ell_p(\theta)$ rather than the original likelihood $\ell(\theta)$. The penalised log-likelihood consists of two parts: the log-likelihood, $\ell(\theta)$, and the roughness penalty, $P(\theta)$, so in maximising $\ell_p(\theta)$ we try to maximise the log-likelihood and minimise the roughness penalty; these two aims are contradictory and the role of λ is to balance these conflicting aims. This is precisely Whittaker's idea, which we discussed in Section 11.2 and expression (11.3) in particular. We can get a feel for penalised likelihood if we consider the two extreme cases:

1. If the smoothing parameter $\lambda = 0$, then we are back with ordinary maximum likelihood.
2. If $\lambda \to \infty$, then we maximise $\ell_p(\theta)$ by setting $P(\theta) = 0$, which is achieved with a second-order penalty when the θ_i lie on a straight line. Furthermore, if the θ_i lie on a straight line it can be shown that the resulting fitted regression is also a straight line; i.e. a Gompertz-type law of mortality.

In conclusion, if $\lambda = 0$ we have ordinary maximum likelihood, while if $\lambda = \infty$ we fit a Gompertz-type model. Intermediate values of λ give rise to curves that are smoother (large λ) or rougher (small λ).

We can fit penalised Poisson regression models easily in R with the software package *MortalitySmooth* (Camarda, 2012). Details of how to install a package on your own computer are given in Appendix A. In one dimension we use the function `Mort1Dsmooth()`. A call of `Mort1Dsmooth()` looks like:

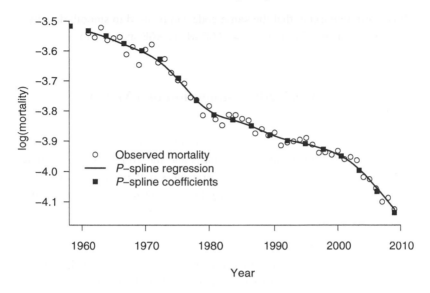

Figure 11.7 *P*-spline fit with fitted mortality and fitted regression coefficients; basis dimension 20. Data: USA females age 70 from 1961 to 2009 from Human Mortality Database.

```
# Using Mort1Dsmooth
Fit = Mort1Dsmooth(x = YEAR, y = Dth, offset = log(Exp),
                    ndx = 17, lambda = 10000, method = 3)
```

The first three arguments specify the year range, the deaths and the offset; the argument ndx specifies 16 internal knots and hence a basis of dimension 20, the argument lambda specifies the value of λ and setting method = 3 forces the fit with this value. The R object Fit contains a wealth of information. The names() function gives the list of objects which includes Fit$logmortality, the fitted log mortality, and Fit$coefficients, the fitted coefficients. R allows unambiguous abbreviations so we can write Fit$logm and Fit$coef. This leads to Figure 11.7; the fitted log mortality tracks the underlying trend well, and the corresponding fitted coefficients have been constrained to lie along the smooth curve. The program Pspline_Regression.r described in Appendix H.4 uses the Mort1Dsmooth() function to fit the *P*-spline model, which gives Figure 11.7. Notice that Figure 11.7 also supports the idea that a regression coefficient has an interpretation as the approximate mean of the observations in the neighbourhood of its corresponding *B*-spline.

It is worth remarking that the same code can be used to smooth data by age for a single year; we simply set x = AGE where AGE specifies the ages.

11.6 Effective Dimension of a Model

We discussed polynomial models in Section 10.10. There the quadratic model has three free parameters and the cubic model has four. In Figure 11.6 we fitted a regression with 20 *B*-splines in the basis and the 20 regression coefficients are chosen freely. The number of free parameters in the model is known as the *dimension of the model* or equivalently the *degrees of freedom of the model*. Thus the quadratic model has dimension 3 while the model in Figure 11.6 has dimension 20. In this section we consider the following question: can we attach a meaning to the dimension of a smooth model?

In our discussion of penalised likelihood on page 196 we saw that when the smoothing parameter $\lambda \to \infty$ we were in effect fitting the Gompertz model, i.e. a model with dimension 2; at the other extreme, when $\lambda = 0$ we have no penalty and ordinary maximum likelihood is being used, i.e. a model with dimension 20 (since there are 20 *B*-splines in the basis and hence 20 free regression coefficients). Evidently, as λ moves from 0 to ∞ the dimension of the model moves from 20 to 2. The intermediate value of the dimension in a smooth model is known as the *effective dimension* or *effective degrees of freedom*; we will often abbreviate effective dimension to ED. We can think of the effective dimension as a measure of the *complexity* of a model.

The technical definition of dimension in a smooth model is beyond the scope of this book; a full discussion can be found in Hastie and Tibshirani (1990) and in Eilers and Marx (1996). However, a call of Mort1Dsmooth() returns the value of the effective dimension, so we can use it to develop a feel for ED. We will also calculate the deviance, Akaike's Information Criterion or AIC, and the Bayesian Information Criterion or BIC; we will refer to these quantities in the next two sections.

```
# Effective dimension, deviance, aic & bic with Mort1Dsmooth
LAMBDA = c(0, 10^2, 10^4, 10^6, 10^8)
ED = NULL; DEV = NULL; AIC = NULL; BIC = NULL
Fit = function(Lambda) Mort1Dsmooth(x = YEAR, y = Dth,
    offset = log(Exp), ndx = 17, lambda = Lambda, method = 3)
for(i in 1:length(LAMBDA)) {
  FIT = Fit(LAMBDA[i])
  ED = c(ED, FIT$df)
```

Table 11.1 *Effective dimension, deviance, AIC and BIC for various smooth models; basis dimension 20. Data: USA females age 70 from 1961 to 2009 from Human Mortality Database.*

λ	ED	Dev	AIC	BIC
0	20	191	231	269
10^2	16.9	195	229	261
10^4	9.0	281	300	317
10^6	3.9	658	666	673
10^8	2.1	1000	1004	1008

```
  DEV = c(DEV, FIT$dev)
  AIC = c(AIC, FIT$aic)
  BIC = c(BIC, FIT$bic)}
cbind(LAMBDA, ED, DEV, AIC, BIC)
```

We define the R function `Fit()` which uses `Mort1Dsmooth()` to fit a smooth model with 20 *B*-splines in the basis; setting `method = 3` forces the fit with given λ. We call `Fit()` in a loop with values of λ supplied by the elements of LAMBDA; the effective dimension, `FIT$df`, the deviance, `FIT$dev`, AIC, `FIT$aic` and BIC, `FIT$bic`, are saved. The output is shown in Table 11.1; in particular, we see that the model in Figure 11.7 has an effective dimension of 9.0. We can think of this model as being very roughly equivalent to fitting a polynomial of degree 8 (with exact dimension 9) while making better use of the available dimension.

The full program `Pspline.r`, of which the above code is a snippet, produces Table 11.1; see Appendix H.4 for details of access to the code. The AIC plot, Figure 11.8, discussed in Section 11.8, and the fitted smooth in Figure 11.10 in Section 11.10 on overdispersion are also produced.

11.7 Deviance of a Model

We saw in Section 11.2 that Whittaker measured the fidelity of a fitted model to the data by the sum of squares of deviations (11.1); this measure is appropriate when the distribution of the random part of the model is normal. However, we are interested in the case when the random part has a Poisson distribution. The appropriate measure is the *deviance*, which we now discuss. In fact, we have already met the deviance in Chapter 6, where it was used for model comparison and tests of goodness-of-fit. Here our focus is on choosing an appropriate level of smoothing, that is, an appropriate value of the smoothing parameter λ.

We follow McCullagh and Nelder (1989, Chapter 2). In general, fitting a model with mean $E[Y] = \mu$ and data y is done by maximising the log-likelihood $\ell(\mu; y)$. Suppose we have a second model with fitted values, $\hat{\mu}$, equal to the data values, y; this would be achieved with a model with the number of parameters equal to the number of data points. The maximum value of the log-likelihood in this case is $\ell(y; y)$, which does not depend on the parameters in the original model. Thus maximising $\ell(\mu; y)$ in our original model is equivalent to minimising $-2[\ell(\mu; y) - \ell(y; y)]$. Let $\hat{\mu}$ be the maximum likelihood estimate of μ; then the minimum value of $-2[\ell(\mu; y) - \ell(y; y)]$ is:

$$\text{Dev} = -2[\ell(\hat{\mu}; y) - \ell(y; y)]. \tag{11.25}$$

This function is known as the *deviance*, which we often denote by Dev.

In the case of normal regression with a single observation y, mean μ and known variance σ^2, we have,

$$f(y; \mu) = \frac{1}{\sqrt{2\pi\sigma^2}} \exp\left(-\frac{(y-\mu)^2}{2\sigma^2}\right) \tag{11.26}$$

$$\Rightarrow \ell(\mu; y) = -\tfrac{1}{2} \log(2\pi\sigma^2) - (y-\mu)^2/(2\sigma^2) \tag{11.27}$$

$$\Rightarrow -2[\ell(\mu; y) - \ell(y; y)] = (y-\mu)^2/\sigma^2, \tag{11.28}$$

since $\ell(y; y) = -\tfrac{1}{2} \log(2\pi\sigma^2)$. Hence, on summing over the sample we find that:

$$\text{Dev} = \sum (y_i - \hat{\mu}_i)^2/\sigma^2. \tag{11.29}$$

Thus, fitting a normal regression model by minimising least squares is equivalent to minimising the deviance.

We turn now to the Poisson case. Let $Y \sim \text{Poisson}(\mu)$. The argument follows the normal case:

$$f(y; \mu) = \exp(-\mu)\mu^y/y! \tag{11.30}$$

$$\Rightarrow \ell(\mu; y) = -\mu + y \log \mu - \log(y!) \tag{11.31}$$

$$\Rightarrow -2[\ell(\mu; y) - \ell(y; y)] = -2[-\mu + y \log \mu + y - y \log y] \tag{11.32}$$

$$= 2[y \log(y/\mu) - (y - \mu)]. \tag{11.33}$$

Thus the deviance in the case of the Poisson distribution is:

$$\text{Dev} = 2 \sum [y_i \log(y_i/\hat{\mu}_i) - (y_i - \hat{\mu}_i)]. \tag{11.34}$$

This formula usually simplifies. All our models include an intercept and in this case $\sum(y_i - \hat{\mu}_i) = 0$. Expression (11.34) becomes:

$$\text{Dev} = 2 \sum y_i \log(y_i/\hat{\mu}_i). \tag{11.35}$$

Table 11.1 shows the deviances computed from (11.35) for the USA female data; the code on page 198 shows how this can be done. We see that as we move from a model with no smoothing ($\lambda = 0$) to one with close to maximum smoothing ($\lambda = 10^8$) the deviance increases from 191 to 1000. We are now ready to discuss smoothing parameter selection.

11.8 Choosing the Smoothing Parameter

Whittaker saw model selection as a trade-off between model fit as measured by the sum of squares of deviations and smoothness as measured by third differences. For given λ the minimum value of his relative balance function (11.3) gave a smooth curve. Trial and error with a small number of test λ gave a satisfactory graduation; the choice of λ is clearly subjective.

In this section we suggest two methods of choosing λ. The modern approach sees model selection as a balance between model fit and model complexity. As model complexity increases the model fit will improve but we run the risks of (a) modelling noise and (b) using a model which is so complex that it fails to explain underlying patterns. Akaike (1973) used information theory to derive a general formula for model selection which now carries his name: Akaike's Information Criterion or AIC. Eilers and Marx (1996), using Akaike's approach, wrote AIC in the following form:

$$\text{AIC} = \text{Dev} + 2\text{ED} \tag{11.36}$$

and chose λ by minimising AIC; here, Dev measures model fit and ED measures model complexity. Again, see Chapter 6 for an earlier discussion of information criteria.

Figure 11.8 shows an AIC plot for the USA female data used in our discussion of *P*-splines. The minimum value of AIC is 228.34 and occurs when $\lambda = 43.7$. The selected smooth curve is shown in Figure 11.9.

Some care is required when the function Mort1Dsmooth() is used to choose the optimal value of λ. We have found that the convergence criterion often needs to be tightened in order to achieve a good optimal value. We can control the convergence criterion with the control option:

```
# Optimizing AIC with Mort1Dsmooth
Opt.aic = Mort1Dsmooth(x = YEAR, y = Dth, offset = log(Exp),
    ndx = 17, method = 2, control = list(TOL2 = 0.01))
summary(Opt.aic)
```

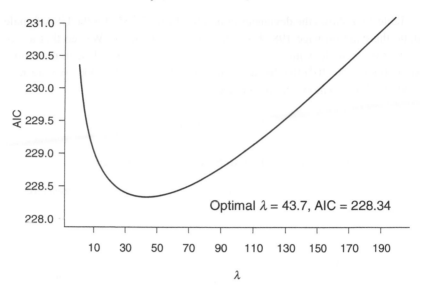

Figure 11.8 AIC against λ. Data: USA females age 70 from 1961 to 2009 from Human Mortality Database.

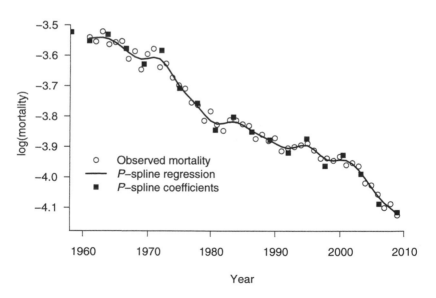

Figure 11.9 *P*-spline fit with fitted mortality and fitted regression coefficients: optimising AIC gives $\lambda = 43.7$; basis dimension 20. Data: USA females age 70 from 1961 to 2009 from Human Mortality Database.

It is instructive to compare Figures 11.7 and 11.9. In Figure 11.7 we forced $\lambda = 10^4$ while in Figure 11.9 we chose $\lambda = 43.7$ by minimising AIC. Many actuaries would be uncomfortable with Figure 11.9, which results when AIC is optimised, and we would agree with this; we would prefer a smoother curve. One approach which deals with this problem, at least in part, is to follow Schwarz (1978) who proposed the Bayesian Information Criterion or BIC:

$$BIC = Dev + \log(n)\,ED, \tag{11.37}$$

where n is the number of observations. This criterion increases the penalty on complexity as the size of the data set increases; in the present case with $n = 49$ we have $\log(49) = 3.89$. We optimise BIC by setting method = 1 in the above code. We find λ is increased from 43.7 with AIC to 120.2 with BIC. However, this apparently large change in λ has very little effect on the effective dimension of the selected model: it reduces from 17.8 with AIC to 16.7 with BIC. The effect on the fitted curve in Figure 11.9 is scarcely visible to the naked eye; our chosen curve is still more flexible than most actuaries would be comfortable with.

The problem is the assumption that our data follow a Poisson distribution. In fact, we have already met this problem in another context: in Section 10.8 we noted the disparity between the standard errors of fitted values under the normal model for the log of observed mortality, $\log(d_x/E_x^c)$, and the Poisson model for observed deaths, d_x. In the next section we will describe an approach which resolves both problems: the choice of smoothing parameter and disparity of the standard errors. The Poisson assumption has been popular in many publications, for example Brouhns et al. (2002), Cairns et al. (2009) and Currie et al. (2004).

11.9 Overdispersion

We suppose that X_i has the Poisson distribution Poisson(λ_i), $i = 1, \ldots, n$. Then a basic property of X_i is that $E[X_i] = Var[X_i] = \lambda_i$, i.e. the mean and variance of X_i are equal. The following calculation is salutary:

$$E\left[\sum (X_i - \lambda_i)^2 / \lambda_i\right] \tag{11.38}$$

$$= \sum E\left[(X_i - \lambda_i)^2\right]/\lambda_i \tag{11.39}$$

$$= \sum Var[X_i]/\lambda_i \tag{11.40}$$

$$= \sum \lambda_i/\lambda_i = n. \tag{11.41}$$

Hence:

$$E\left[\sum (X_i - \lambda_i)^2 / \lambda_i\right]/n = 1. \tag{11.42}$$

We now apply this formula with X_i replaced by our observed deaths, d_i, and λ_i replaced with our estimate of the mean of X_i, i.e. the fitted deaths, $\hat{d}_i$. We will also replace the divisor of n with $n - \text{ED}$ since the estimated λ_i are not independent; a good analogy is the use of a divisor of $n - p$ in the estimate of the variance in a regression model, where p is the dimension of the model. Equation (11.42) tells us that we should find:

$$\left[\sum (d_i - \hat{d}_i)^2 / \hat{d}_i\right]/(n - \text{ED}) \approx 1. \tag{11.43}$$

However, if we use the fitted values from the BIC fit in the previous section we find that the left-hand side of (11.43) evaluates to 6.06. This is a measure of the average of the ratio of the variance to the mean for the distribution of deaths in our example. The Poisson distribution seems a poor assumption, and it is the assumption of the equality of the mean and variance that is letting us down. Can we fit a model which has the variance proportional to the mean? We want:

$$\text{Var}[X_i] = \psi^2 E[X_i]. \tag{11.44}$$

The ratio of the variance to the mean is known as the *dispersion* parameter, which we denote by ψ^2. If the Poisson assumption holds we have $\psi^2 = 1$; in the present case we find:

$$\hat{\psi}^2 = \left[\sum (d_i - \hat{d}_i)^2 / \hat{d}_i\right]/(n - \text{ED}) = 6.06, \tag{11.45}$$

which corresponds to overdispersion; underdispersion, i.e. $\psi^2 < 1$, is also possible, although this is rarely (if ever) encountered with mortality data. An alternative estimate of the dispersion parameter is:

$$\hat{\psi}^2 = \frac{\text{Dev}}{n - \text{ED}} = 6.07. \tag{11.46}$$

A call of `Mort1Dsmooth()` returns an estimate of ψ^2 based on (11.46); for example, the object `Opt.bic` in the previous section gives `Opt.bic$psi2 = 6.07`.

Overdispersion has a profound impact on smoothing parameter selection. An informal discussion is instructive. Under the Poisson assumption the variability about the mean is relatively small, so the chosen curve "believes" the data; the selected smooth curve will favour tracking the data closely. However, if the data are overdispersed, i.e. have a large variance, the selected curve is less inclined to track the data because local patterns are more readily interpreted as noise; the selected curve is smoother.

The main effect of overdispersion is to modify the definitions of AIC and BIC. We find:

$$\text{AIC} = \frac{\text{Dev}}{\psi^2} + 2\text{ED} \qquad (11.47)$$

and

$$\text{BIC} = \frac{\text{Dev}}{\psi^2} + \log(n)\,\text{ED}. \qquad (11.48)$$

Thus if overdispersion is present a smoother curve will be chosen. A careful discussion of smoothing and overdispersion from an actuarial perspective is given in Djeundje and Currie (2011); see also Camarda (2012) for a concise summary. The function Mort1Dsmooth() sets $\psi^2 = 1$ by default, that is, the Poisson assumption with no overdispersion. We modify the call by setting the option overdispersion = TRUE; the code on page 201 becomes:

```
# Optimizing AIC allowing for overdispersion
Opt.aic.over = Mort1Dsmooth(x = YEAR, y = Dth,
    offset = log(Exp), over = TRUE, ndx = 17, method = 2,
    control = list(TOL2 = 0.01))
summary(Opt.aic.over)
```

We find that the smoothing parameter increases from 43.7 to 309, and the ED of the selected model decreases from 17.8 to 11.9. By setting method = 1 we optimise BIC; in this case, the change is even more dramatic: the smoothing parameter increases from 120.2 to 3715, and the ED of the selected model decreases from 16.7 to 7.5. Figure 11.10 shows the fitted curves. Both curves seem satisfactory, although we prefer the stiffer curve given by BIC. As an aside we remark that the trial-and-error method which led to Figure 11.7 gave a curve with ED = 9.0, more or less midway between the fits with AIC and BIC. We recommend setting over = TRUE when the *MortalitySmooth* package is used.

11.10 Dealing with Overdispersion

Overdispersion is extremely common in mortality data. We have seen two serious consequences of ignoring it: underestimating standard errors in Poisson models (Section 10.8) and, in the previous section, undersmoothing in *P*-spline models. We outline an approach which deals with both these problems. We

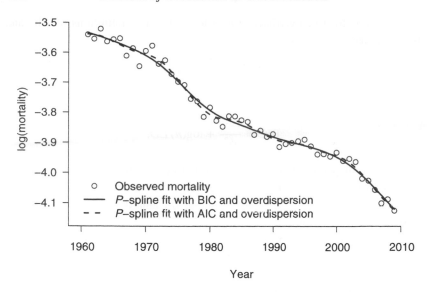

Figure 11.10 *P*-spline fit with fitted mortality: optimising BIC and AIC with overdispersion gives ED = 7.5 and 11.9, respectively; basis dimension 20. Data: USA females age 70 from 1961 to 2009 from Human Mortality Database.

consider a GLM with log-likelihood:

$$\ell(\theta, \phi; y) = [y\theta - b(\theta)]/a(\phi) + c(y, \phi). \qquad (11.49)$$

Then we know from (10.64) that $E[Y] = b'(\theta)$ and from (10.66) that $\text{Var}[Y] = a(\phi)b''(\theta)$. In the case of the Poisson distribution $D \sim \text{Poisson}(E^c\mu)$, expression (11.49) becomes:

$$\ell(\mu, \phi; d) = d \log(\mu) - E^c\mu \qquad (11.50)$$

(see (10.37) for example) and $E[D] = E^c\mu$ and $\text{Var}[D] = E^c\mu = E[D]$. We now suppose that we can modify the log-likelihood (11.50) and use instead:

$$\ell(\mu, \phi; d) = \frac{d \log(\mu) - E^c\mu}{\phi}; \qquad (11.51)$$

this expression still satisfies the definition of a member of the exponential family and we may use the results (10.64) and (10.66) on the mean and variance of a family member again. We find that the mean of D is unaltered, $E[D] = E^c\mu$, but the variance is $\text{Var}[D] = \phi E^c\mu$. The key point is that (11.51) is a log-likelihood which allows inference within the GLM framework and which satisfies the mean/variance relationship (11.44). We refer to this model as a *quasi-Poisson model.*

It is important to note that the log-likelihood (11.51) does not correspond to an actual random variable, i.e. there is no random variable underlying the log-likelihood (11.51). For this reason (11.51) is known a *quasi-log-likelihood*.

It is straightforward to fit the quasi-Poisson model. We modify the code on page 180 as follows:

```
#   Fit the Gompertz model with a quasi-Poisson model
Gompertz.Q = glm(Dth ~ AGE + offset(log(Exp)),
                 family = quasipoisson)
summary(Gompertz.Q)
QuasiP.GLM = predict(Gompertz.Q, se.fit = TRUE)
```

The solutions by weighted least squares in Section 10.3.3 and here by quasi-likelihood are essentially identical: first, fitted log mortalities for weighted least squares, Poisson and quasi-Poisson models are equal; second, the standard errors for fitted log mortalities for weighted least squares and quasi-Poisson models are equal while the standard errors for the Poisson model are much smaller; third, the estimate of the residual variance σ^2 in the weighted least squares model and the dispersion parameter ψ^2 in the quasi-Poisson model are equal.

Overdispersion has received considerable attention from statisticians. Quasi-likelihood was introduced by Wedderburn (1974) and discussed in McCullagh and Nelder (1989, Chapter 8) . Quasi-likelihood and smoothing with P-splines is addressed in Djeundje and Currie (2011). Another approach is to assume that the deaths have a negative binomial distribution, a distribution with variance greater than the mean; see Lawless (1987) for a general introduction, Thurston et al. (2000) for a discussion of smoothing with a negative binomial distribution and Li et al. (2009) for an actuarial application. Computation is usually more challenging since the negative binomial is not in the exponential family. Perperoglou and Eilers (2010) modelled the dispersion directly at the individual cell level; this novel approach is particularly well suited to smoothing with P-splines.

12

Methods of Graduation III: Two-Dimensional Models

12.1 Introduction

In Chapter 10 we described how Gompertz modelled the change in mortality with age; this famous law began the study of mathematical modelling of mortality. Later in the same chapter we used US data to illustrate how mortality has improved over the last 50 years. The US is not alone in this, and Table 12.1 shows how life expectancy at age 65 has increased over the last few decades in both the US and the UK. These are dramatic figures indeed and the implications for society for the provision of pensions and care for the elderly in general are profound.

Table 12.1 *Period life expectancy at age 65 in UK and US. Source: Human Mortality Database.*

	UK		US	
Year	Male	Female	Male	Female
1970	12.1	16.0	13.0	16.9
1980	12.8	16.8	14.1	18.3
1990	14.0	17.9	15.1	19.0
2000	15.7	18.9	16.1	19.1
2010	18.0	20.6	17.9	20.5

Governments have responded by raising, or planning to raise, traditional state pension ages, 65 for males and 60 for females in the UK, by first equalising the pensionable ages for males and females, and then increasing this common age. For example, in the UK under the Pensions Act of 2011, the male and female State Pension ages will equalise at 65 in November 2018 and thereafter increase steadily to 66 by October 2020. The increase to age 67 was initially planned to take place by 2035 but the Pensions Act of 2014 brought this forward to 2028. The further increase to age 68 in 2046 was set by the Pensions Act of 2007; however, in 2017 the UK government proposed that this be brought forward to 2038, although the necessary legislation has been delayed to 2023.

The principal evidence for these changes is twofold. First, Table 12.1 shows life expectancy over the past 40 years has increased sharply. Second, we expect life expectancy to continue to increase in the future; graphs like Figure 10.3 provide compelling evidence that human mortality is improving in time and seems likely to continue to do so.

The purpose of the present chapter is to bring age and time, the two major factors driving these changes in mortality, together in a single model. This is a huge area of current research and we will make no attempt at a comprehensive coverage; rather we will give a flavour of the kinds of models that are studied. All these models have one thing in common: they are designed with forecasting in mind; this will be the subject of our next chapter.

We will consider three distinct models of mortality in two dimensions. The Lee–Carter model (Lee and Carter, 1992) was the first serious attempt to model and then forecast mortality in two dimensions. Cairns et al. (2009) was an influential paper for actuaries that gave a thorough investigation of nine mortality models and their forecasting properties; we will examine one of these models, the Cairns–Blake–Dowd model, in detail. Both the Lee–Carter and the Cairns–Blake–Dowd models are structural models, by which we mean models in which parameters represent age and time effects; forecasting mortality is achieved by forecasting the time effects. Our third and final model was introduced by Currie et al. (2004); here a general smooth surface is fitted and then penalties are used to forecast the surface.

In this chapter we use data on UK males from the Human Mortality Database. We have deaths $d_{x,y}$ at age x in calendar year y for ages 50 to 90 and years 1961 to 2013, and corresponding central exposures $E^c_{x,y}$. In Section 10.2 we gave code to obtain data by age for a single year. Here we want data by both age and year, so we extend the code in Section 10.2 as follows:

```
#   Select male data for ages 50-90 and years 1961 to 2013
Dth = Death.Data$Male.Matrix[(49 < Age) & (Age < 91),
          (1960 < Year) & (Year < 2014)]
Exp = Exposure.Data$Male.Matrix[(49 < Age) & (Age < 91),
          (1960 < Year) & (Year < 2014)]
```

The deaths and exposures are arranged in two matrices, Dth and Exp. We refer to their elements Dth $= (d_{x,y})$ and Exp $= (E^c_{x,y})$, each $n_x \times n_y$, where

$x = x_1, \ldots, x_{n_x}, y = y_1, \ldots, y_{n_y}$. In our example, $x_1 = 50$, $x_{n_x} = 90$, $y_1 = 1961$, $y_{n_y} = 2013$ and so each data matrix is 41×53. We can inspect a portion of these matrices with

```
> Dth[1:2,1:2]
      1961     1962
50 2643.29 2632.89
51 2980.60 3018.17
```

and we have displayed the upper-left 2×2 submatrix of Dth. Conveniently, the rows and columns are automatically correctly labelled by age and year. We can convert a matrix to a vector in R with

```
> c(Dth[1:2,1:2])
[1] 2643.29 2980.60 2632.89 3018.17
```

and it is important to note that the conversion from matrix to vector stacks the *columns* of the matrix on top of each other in column order.

With these data, $\log(d_{x,y}/E^c_{x,y})$ is an estimate of $\log(\mu_{x+1/2,y+1/2})$, where $\mu_{x,y}$ is the force of mortality at exact age x and exact year y. Figure 12.1 shows the raw mortality surface which sits above the age–year plane. It is this surface that we wish both to model and eventually to forecast.

12.2 The Lee–Carter Model

The Lee–Carter model (Lee and Carter, 1992) for mortality in age and time is:

$$\log(\mu_{x+1/2,y+1/2}) = \alpha_x + \beta_x \kappa_y, \quad x = x_1, \ldots, x_{n_x}, y = y_1, \ldots, y_{n_y}. \qquad (12.1)$$

Let $\alpha = (\alpha_{x_1}, \ldots, \alpha_{x_{n_x}})'$, $\beta = (\beta_{x_1}, \ldots, \beta_{x_{n_x}})'$ and $\kappa = (\kappa_{y_1}, \ldots, \kappa_{y_{n_y}})'$ be the vectors of parameters. This model is not *identifiable*, by which is meant that the parameters are not uniquely estimable. For example, if $\hat{\alpha}, \hat{\beta}$ and $\hat{\kappa}$ are any three estimates of α, β and κ then for any constant c we have that (a) $\hat{\alpha} - c\hat{\beta}, \hat{\beta}$ and $\hat{\kappa} + c$ is also a solution and (b) so too is $\hat{\alpha}, c\hat{\beta}$ and $\hat{\kappa}/c$. Following Lee and Carter (1992) we impose location and scale constraints on β and κ, respectively, as follows:

$$\sum \beta_x = 1; \quad \sum \kappa_y = 0. \qquad (12.2)$$

The estimates are then uniquely determined and have the appealing property for each age x that:

$$\sum_y \log(\hat{\mu}_{x+1/2,y+1/2}) = \sum_y (\hat{\alpha}_x + \hat{\beta}_x \hat{\kappa}_y) = n_y \hat{\alpha}_x; \qquad (12.3)$$

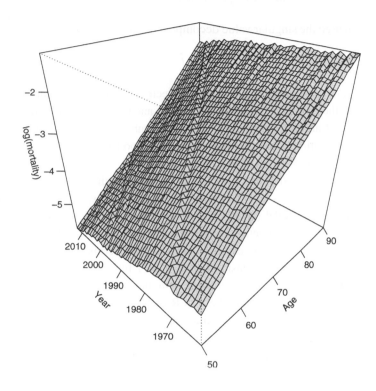

Figure 12.1 Crude mortality rates for UK males. Data: UK males age 50 to 90, years 1961 to 2013 from Human Mortality Database.

thus each $\hat{\alpha}_x$ is simply the average over time of $\log(\hat{\mu}_{x+1/2,y+1/2})$. Other constraint systems are possible: see for example, Richards and Currie (2009) and Girosi and King (2008).

We turn now to estimation in the Lee–Carter model. Lee and Carter (1992) considered the fitting of their model as a matrix approximation problem rather than one of statistical estimation. First, they used (12.3) to estimate α_x as $\sum_y \log(d_{x,y}/E^c_{x,y})/n_y$, that is, as the mean of observed log mortalities over time.

Now define M, $n_x \times n_y$, the matrix of adjusted observed log mortalities:

$$M = (M_{x,y}) = \left(\log(d_{x,y}/E^c_{x,y}) - \hat{\alpha}_x\right). \tag{12.4}$$

They then used the singular value decomposition to approximate M as follows:

$$M \approx \hat{\beta}\hat{\kappa}', \tag{12.5}$$

where $\hat{\beta}$ and $\hat{\kappa}$ are the first left and right vectors appropriately scaled. We do not pursue their method further here; see Lee and Carter (1992) for details.

Brouhns et al. (2002) assumed that the number of deaths $d_{x,y}$ followed a Poisson distribution. They used maximum likelihood and solved the likelihood equations with the Newton–Raphson method.

We will fit the model directly with R. We observe that the difficulty in estimating the parameters in (12.1) is the product term $\beta_x\kappa_y$; the model is not a generalised linear model (GLM) and so the glm() function cannot be used. However, the Mult() function in the gnm() function in the R package *gnm* (Turner and Firth, 2012) does allow such product terms in a model specification. This very useful extension to the glm() function allows us to fit the model with ease, but first we need some data management.

The deaths and exposures are held in R in matrices Dth and Exp, as in the previous section. Like the glm() function, the gnm() function expects its dependent variable to be a vector, so we must convert both Dth and Exp to vectors. On page 210 we used the c() function for this. Thus

```
Dth.V = c(Dth); Exp.V = c(Exp)      # Make Dth & Exp vectors
```

stacks the columns of Dth and Exp on top of each other in column order, and Dth.V is the vector with the deaths in 1961 in ascending age first, followed by the deaths in 1962, etc.

The Lee–Carter model treats age and year as *factors* or categories; that is, each age x and year y has its own parameters, α_x and β_x for age, and κ_y for year. In our example the category or factor *levels* of age are $50, \dots, 90$. Remembering how the vectors Dth.V and Exp.V are stored, we create the required factor variables with the rep() function as follows:

```
AGE = 50:90; YEAR = 1961:2013
Age.F = factor(rep(AGE, n.y))           # Make age a factor
Year.F = factor(rep(YEAR, each = n.x)) # Make year a factor
```

The syntax of the gnm() function is very close to that of the glm() function:

```
LC.Male = gnm(Dth.V ~ -1 + Age.F + Mult(Age.F, Year.F),
              offset = log(Exp.V), family = poisson)
```

We have fitted the Lee–Carter model by maximum likelihood with the gnm() function but we do not have estimates that satisfy our constraints (12.2). The gnm() function returns parameter estimates with a random parameterisation (Turner and Firth, 2012). Hence, two calls of gnm() result in different parameter estimates, although model invariants such as the deviance and the fitted mortality values are the same. We recover the estimates satisfying (12.2) as follows.

Let $\hat{\alpha}_R$, $\hat{\beta}_R$ and $\hat{\kappa}_R$ be any estimates returned by the gnm() function and let $\bar{\kappa}_R = \sum_y \hat{\kappa}_{y,R}/n_y$ and $\bar{\beta}_R = \sum_x \hat{\beta}_{x,R}/n_x$. Then:

$$\hat{\alpha} = \hat{\alpha}_R + \bar{\kappa}_R\hat{\beta}_R \tag{12.6}$$

$$\hat{\kappa} = n_x\bar{\beta}_R(\hat{\kappa}_R - \bar{\kappa}_R\mathbf{1}_{n_y}) \tag{12.7}$$

$$\hat{\beta} = \hat{\beta}_R/(n_x\bar{\beta}_R) \tag{12.8}$$

are the estimates of α, β and κ subject to the constraints (12.2). It is a simple matter to check first that:

$$\hat{\alpha}_x + \hat{\beta}_x\hat{\kappa}_y = \hat{\alpha}_{x,R} + \hat{\beta}_{x,R}\hat{\kappa}_{y,R} \tag{12.9}$$

for all x and y, that is, the fitted log hazard rates are equal; and second that the constraints (12.2) are satisfied by $\hat{\alpha}$, $\hat{\beta}$ and $\hat{\kappa}$.

Figure 12.2 shows the fitted parameter values and the fitted log hazard rates for age 65. The constraints (12.2) allow us to interpret the parameters as follows: the fitted $\hat{\alpha}$ is a description of mortality by age averaged over time; the fitted $\hat{\kappa}$ summarises the change in mortality over time; the fitted $\hat{\beta}$ modulates the change over time by age. The lower-right panel shows the fitted log hazard rates for age 65. An important point is that the trajectory of the fitted log hazard rate curve for any age is exactly the same as that of $\hat{\kappa}$ subject only to a change of location and scale; in other words, if we removed the scale from the left-hand side of the graph all ages would look identical. For this reason the Lee–Carter model is perhaps not an ideal model of mortality, but the model comes into its own when forecasting is required; forecasting is discussed in the next chapter. The program Lee_Carter.r referenced in Appendix H.5 fits the Lee–Carter model and produces Figures 12.1 and 12.2.

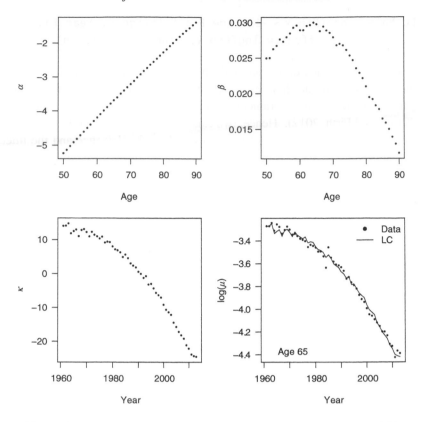

Figure 12.2 Data and parameter estimates in the Lee–Carter model. Data: UK males age 50 to 90, years 1961 to 2013 from Human Mortality Database.

12.3 The Cairns–Blake–Dowd Model

The Cairns–Blake–Dowd or CBD model (Cairns et al., 2006) for mortality in age and time is:

$$\log(\mu_{x+1/2, y+1/2}) = \kappa_y^{(1)} + \kappa_y^{(2)}(x - \bar{x}), \quad x = x_1, \ldots, x_{n_x}, \; y = y_1, \ldots, y_{n_y}, \quad (12.10)$$

where $\bar{x} = \sum x_i / n_x$ is the mean age. Thus the log of the force of mortality is a linear function of age and so if we assume that the number of deaths has the Poisson distribution we will have a GLM. We can think of the CBD model as follows: for each year y, (12.10) defines a Gompertz law of mortality with $\kappa_y^{(1)}$ as the intercept and $\kappa_y^{(2)}$ as the slope. The sets of coefficients $\{\kappa_y^{(1)}, y = 1, \ldots, n_y\}$ and $\{\kappa_y^{(2)}, y = 1, \ldots, n_y\}$ then describe how the Gompertz line changes in time. The idea is to forecast the mortality table by treating these two sets as correlated time series. We discuss this in the next chapter. First, we must fit the model.

The regression coefficients fall into two sets: $\kappa^{(1)} = (\kappa_{y_1}^{(1)}, \ldots, \kappa_{y_{n_y}}^{(1)})'$ and $\kappa^{(2)} = (\kappa_{y_1}^{(2)}, \ldots, \kappa_{y_{n_y}}^{(2)})'$. Let $\kappa = (\kappa^{(1)'}, \kappa^{(2)'})'$ be the full vector of $2n_y$ regression coefficients. Let X be the $n \times 2n_y$ model matrix, where $n = n_x n_y$ is the number of observations. Then we have:

$$X\kappa = \begin{bmatrix} X_1 & X_2 \end{bmatrix} \begin{bmatrix} \kappa^{(1)} \\ \kappa^{(2)} \end{bmatrix} = X_1\kappa^{(1)} + X_2\kappa^{(2)}, \qquad (12.11)$$

where X_1 and X_2 are those parts of the model matrix corresponding to $\kappa^{(1)}$ and $\kappa^{(2)}$, respectively. Let 0 and 1 be the vectors of 0s and 1s, respectively, each of length n_x. Then we can see that X_1 is the block-diagonal matrix:

$$X_1 = \begin{bmatrix} 1 & 0 & \cdots & 0 \\ 0 & 1 & \cdots & 0 \\ \vdots & \vdots & \ddots & \vdots \\ 0 & 0 & \cdots & 1 \end{bmatrix}, \; n \times n_y. \qquad (12.12)$$

The Kronecker product is a convenient notation for many patterned matrices and we can write X_1 compactly as:

$$X_1 = I_{n_y} \otimes 1_{n_x}, \qquad (12.13)$$

where I_{n_y} is the identity matrix of size n_y and we have written $1 = 1_{n_x}$ for emphasis. Appendix G provides a short introduction to the Kronecker product with a simple example. Kronecker products are easily computed in R with the kronecker() function. For example, the Kronecker product of I_2, the identity matrix of size 2, with the column vector $(2, 3)'$ is:

```
kronecker(diag(2), c(2, 3))
     [,1] [,2]
[1,]    2    0
[2,]    3    0
[3,]    0    2
[4,]    0    3
```

whereas the Kronecker product of I_2 with the row vector $(2, 3)$ is:

```
kronecker(diag(2), t(c(2, 3)))
     [,1] [,2] [,3] [,4]
[1,]    2    3    0    0
[2,]    0    0    2    3
```

where $\mathtt{t}()$ is the R function which transposes a vector or matrix.

We now turn to X_2. We define $x_m = (x_1 - \bar{x}, \ldots, x_{n_x} - \bar{x})'$. Then, in the same way that $X_1 = I_{n_y} \otimes 1_{n_x}$, we have:

$$X_2 = I_{n_y} \otimes x_m. \tag{12.14}$$

The CBD model is fitted as follows:

```
X1 = kronecker(diag(n.y), rep(1, n.x))
Age.mean = AGE - mean(AGE)
X2 = kronecker(diag(n.y), Age.mean)
X = cbind(X1, X2)
CBD.Male = glm(Dth.V ~ -1 + X, offset = log(Exp.V),
                family = poisson)
```

and Figure 12.3 summarises the results. We note in particular that the fitted log mortality by age follows the Gompertz law and is linear for each year (lower right); the intercept, κ_1, decreases with year while the slope, κ_2, increases with year. The program CBD.r referenced in Appendix H.5 fits the CBD model and produces Figure 12.3.

There are various extensions to the CBD model. Cairns et al. (2009) added a cohort term to (12.10):

$$\log(\mu_{x+1/2,y+1/2}) = \kappa_y^{(1)} + \kappa_y^{(2)}(x - \bar{x}) + \gamma_{y-x}, \ x = x_1, \ldots, x_{n_x}, \ y = y_1, \ldots, y_{n_y},$$
$$\tag{12.15}$$

and further extended the model by adding a quadratic term in age:

$$\log(\mu_{x+1/2,y+1/2}) = \kappa_y^{(1)} + \kappa_y^{(2)}(x - \bar{x}) + \kappa_y^{(3)}[(x - \bar{x})^2 - \hat{\sigma}_x^2] + \gamma_{y-x},$$
$$x = x_1, \ldots, x_{n_x}, \ y = y_1, \ldots, y_{n_y}, \tag{12.16}$$

where $\hat{\sigma}_x^2$ is an estimate of the variance of age. Both these models are GLMs and are easily fitted by extending the code for CBD. Both models give much improved fits but the forecasting of such a large number of correlated parameters is less straightforward. See Currie (2016) for details on fitting these models and Cairns et al. (2009) for details on forecasting.

12.4 A Smooth Two-Dimensional Model

The previous two sections described the Lee–Carter and CBD models. These models and their extensions describe the mortality surface in terms of interpretable sets of parameters. In this section we take a completely different

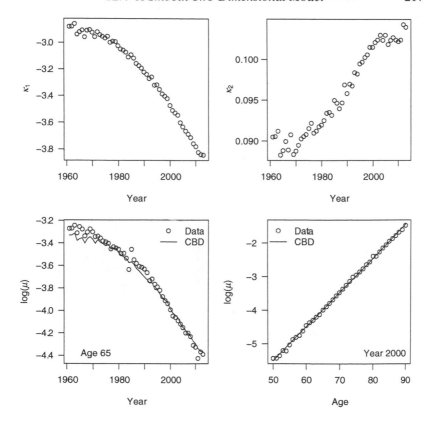

Figure 12.3 Data, parameter estimates and fitted values in the Cairns–Blake–Dowd model. Data: UK males age 50 to 90, years 1961 to 2013 from Human Mortality Database.

approach and describe the mortality surface with a general smooth surface. In Chapter 11 we described the method of *P*-splines and used it to fit general smooth curves to mortality by age (see Figure 11.7) and to mortality by year (see Figure 11.10). The purpose of this section is to extend these models to two dimensions.

Figure 11.3 showed a cubic *B*-spline basis for year with nine *B*-splines in the basis. Suppose we also have a cubic *B*-spline basis for age with eight *B*-splines in the basis. We want to combine these bases to make a two-dimensional basis for the age–year plane. Figure 12.4 shows a reduced two-dimensional basis for ages 50 to 90 and years 1961 to 2013. A useful analogy is to think of the one-dimensional basis in Figure 11.3 as the profile of a mountain range whereas Figure 12.4 shows the full range.

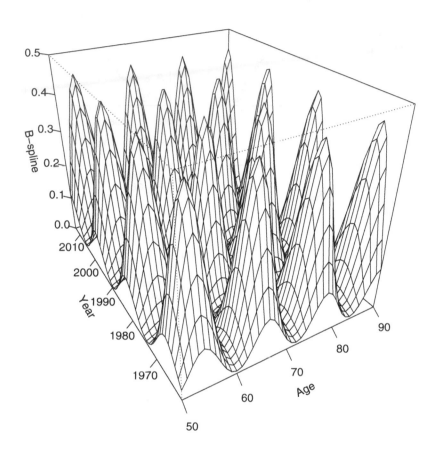

Figure 12.4 Two-dimensional *B*-spline basis.

In Section 11.4 we showed how to construct a regression matrix for a set of ages or years from a given basis. Let B_x, $n_x \times c_x$, and B_y, $n_y \times c_y$, be the regression matrices for age and year, respectively. Then the regression matrix for a two-dimensional *B*-spline model is given by the Kronecker product of B_y and B_x:

$$B = B_y \otimes B_x, \qquad (12.17)$$

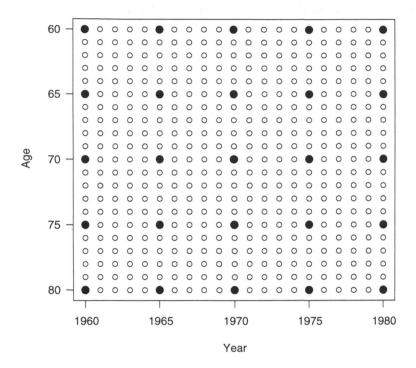

Figure 12.5 Grid of data points, ○, location of regression coefficients, ●.

since the Kronecker product constructs all products of the elements of B_y and B_x, and does so in the correct order. These bases and resulting regression matrices can be quite large. For example, if B_x is $n_x \times c_x = 41 \times 9$ and B_y is $n_y \times c_y = 53 \times 10$, then there are $n_x n_y = 41 \times 53 = 2{,}173$ observations and $c_x c_y = 9 \times 10 = 90$ regression coefficients; B will be $2{,}173 \times 90$.

We are now in a position to fit a surface with a regression model with model matrix B, deaths Dth.V, exposures Exp.V and Poisson errors. The code on page 192 fits the model. If we use a large number of B-splines in the bases for age and year then, just as in the one-dimensional case, the resulting surface will be not be sufficiently smooth. We need a two-dimensional analogue of the penalty (11.18) we used in one dimension.

Figure 12.5 shows the age–year plane for ages 60 to 80 and years 1960 to 1980. We recall that our data matrices, Dth and Exp, are arranged with years in columns and ages in ascending order down a column. Figure 12.5 mirrors this arrangement. In this illustrative example the data are shown as ○ and the summits of the two-dimensional *B*-splines (see Figure 12.4) are shown as •. Each regression coefficient sits on the summit of its associated *B*-spline so we can think of the coefficients as arranged in a matrix, say Θ. In one dimension we used a second-order penalty to penalise the roughness of adjacent coefficients. We bring about smoothness in two dimensions by penalising roughness of adjacent coefficients in the rows of Θ (ensuring smoothness across years for each age) and in the columns of Θ (ensuring smoothness down ages for each year). Let θ be the vector of coefficients obtained by stacking the columns of Θ on top of each other. The roughness penalty that ensures smoothness down ages for each year is:

$$P_x(\theta) = \theta'(I_{n_y} \otimes D'_x D_x)\theta, \tag{12.18}$$

whereas the roughness penalty that ensures smoothness across years for each age is:

$$P_y(\theta) = \theta'(D'_y D_y \otimes I_{n_x})\theta. \tag{12.19}$$

Here D_x and D_y are second-order difference matrices of appropriate size; see (11.19) and (11.6) for examples of second and third-order difference matrices, respectively. The complete penalty matrix is:

$$P(\theta) = \lambda_x P_x(\theta) + \lambda_y P_y(\theta). \tag{12.20}$$

We note that there are separate smoothing parameters for age and year, since there is no reason to suppose that similar amounts of smoothing would be appropriate for age and year. See Currie et al. (2004) for further discussion of penalisation in two dimensions; Richards et al. (2006) also discuss penalisation in two dimensions but from an actuarial perspective.

The function Mort2Dsmooth() in the package *MortalitySmooth* (Camarda, 2012) fits the two-dimensional *P*-spline model. We suppose that the ages are in AGE, the years in YEAR, the deaths in the matrix Dth and the exposures in the matrix Exp. The basic call of Mort2Dsmooth() looks like:

```
# Using Mort2Dsmooth
Fit.2d = Mort2Dsmooth(AGE, YEAR, Dth, offset=log(Exp))
names(Fit.2d)
```

Note in particular that Mort2Dsmooth() expects the deaths and exposures to be supplied as matrices. Figure 12.6 is a graphical summary of the improvement in mortality over time; the figure is produced as follows:

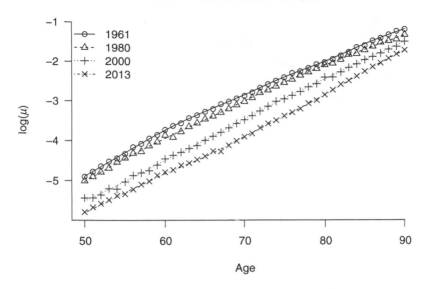

Figure 12.6 Observed and fitted log(μ) for years 1961, 1980, 2000 and 2013 for the two-dimensional *P*-spline model. Data: UK males age 50 to 90, years 1961 to 2013 from Human Mortality Database.

```
logm.hat = Fit.2d$logm[ , c(1, 20, 40, 53)]
matplot(AGE, logm.hat, type = "l", col = 1, xlab = "Age",
        ylab = paste("log(",expression(mu),")"))
matpoints(AGE, Obs[ , c(1, 20, 40, 53)], pch = 1:4,
        col = 1)
legend("topleft", legend = YEAR[c(1, 20, 40, 53)],
        lty = 1:4, pch = 1:4, bty = "n")
```

Figure 12.7, which compares mortality at the start of the period with that at the end, shows how mortality has improved at all ages over time. It is striking that the greatest improvement has taken place around age 65, the pension age in the UK; the smallest improvement is at high ages, a phenomenon often referred to as *mortality convergence*.

There are many options to the `Mort2Dsmooth()` function; for example, the numbers of *B*-splines in the bases can be controlled, overdispersion can be allowed for and weights can be set to remove certain observations from the data. There is also a comprehensive output list; use the `names()` function to access this. Finally, there are examples in the online help for the function and many more in Camarda (2012).

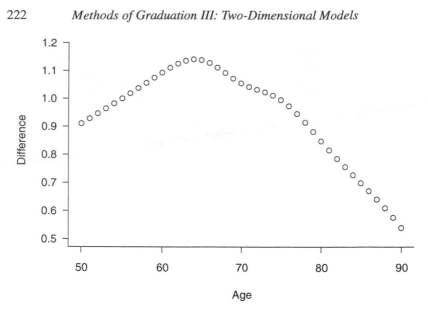

Figure 12.7 Differences in mortality: $\log(\mu_{1961}) - \log(\mu_{2013})$.

Table 12.2 *Comparison of Lee–Carter, Cairns–Blake–Dowd and two-dimensional P-spline models.*

Model	Deviance	Effective dimension	BIC
Lee–Carter	16,032	133	17,054
Cairns–Blake–Dowd	38,748	106	39,563
2-d *P*-spline	9,581	96	10,835

The program `2d_Pspline.r`, referenced in Appendix H.5, fits the two-dimensional *P*-spline model and produces Figures 12.6 and 12.7.

12.5 Comparing Models

We have fitted three models in this chapter: the Lee–Carter model, the Cairns–Blake–Dowd model and finally the two-dimensional *P*-spline model. Our ultimate aim is to forecast mortality but it is of some interest to compare how effective the models are at modelling these data.

Table 12.2 gives deviances, effective dimensions and BIC values for all three models. It is evident that for these data the two-dimensional *P*-spline model is the best fitting model with by far the lowest deviance; furthermore the fit is obtained with the lowest effective dimension. The Cairns–Blake–Dowd model fits a family of time-dependent Gompertz lines; this approach gives a disappointing result for these data. The Lee–Carter model fits a family of curves which are location and scale transforms of an estimate of the time effect; this approach also performs disappointingly when set against the general smooth surface fitted by the two-dimensional *P*-spline model.

There are two cautions we would add to the remarks in the previous paragraph. First, these remarks apply only to the single data set we used here to describe the three models and their fitting. For a more comprehensive study involving six countries (USA, UK, Japan, Australia, Sweden and France), see Currie (2016). Second, our main purpose is to forecast future mortality. The structure of a model may be more important for effective forecasting than the fit of that model to the data. We will discuss forecasting in the next chapter.

13

Methods of Graduation IV: Forecasting

13.1 Introduction

We have already seen compelling evidence that human mortality is improving rapidly. Figure 10.3 shows falling US mortality at age 70 for both males and females while Figure 12.6 shows how mortality has improved for all ages across time for UK male data. Table 12.1 gives values of e_{65} for males and females for selected years for the UK and the USA. Figure 13.1 extends this table and provides further striking evidence; the figure shows life expectancy at age 65, e_{65}, for six developed countries. (To improve clarity we have applied light smoothing to the raw data behind Figure 13.1.) The impact of such improvements on the resources required for the provision of pensions, health services and care of the elderly is obvious and is one of the major challenges facing today's society. One statistic summarises the challenge: the average value of e_{65} in 1961 across the six countries was 12.55 years; this had risen to 18.55 years in 2011, a rise of 6 years in 51 years, or 1 year every 8.5 years. Moreover, Figure 13.1 strongly suggests that increases are likely to continue into the future. Prudent planning should not assume a levelling off any time soon.

It would be a simple matter to forecast the life expectancies in Figure 13.1 but the actuary requires a forecast of the whole mortality table, at least for ages above 40 or 50, say. It is the purpose of this chapter to present methods for forecasting mortality tables. These methods will rely on the two-dimensional models of mortality discussed in the previous chapter. Of course, there are many other approaches to this important problem. Targeting and cause-of-death methods are two such methods; see Booth and Tickle (2008) and Richards (2009) for discussions of these methods. We see two problems with targeting methods: first, the target is itself a forecast and so the method is circular, and second, setting a target in itself reduces the standard error of the forecast.

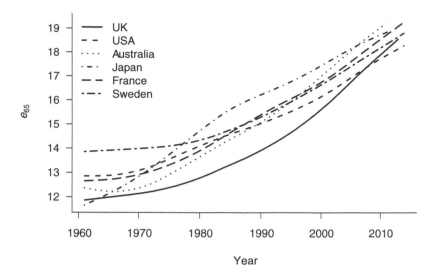

Figure 13.1 Period life expectancy at age 65, e_{65}, for six countries. Data: UK, USA, Australia, Sweden, Japan and France male e_{65} from 1961 to available date from Human Mortality Database.

Forecasting in cause-of-death models is difficult because the various components are highly correlated.

First, we discuss forecasting mortality in one dimension, that is, forecasting in time for a given age. We will divide this discussion into two sections: time series methods used for forecasting for the Lee–Carter and Cairns–Blake–Dowd models, and penalty methods used for the two-dimensional P-spline models.

13.2 Time Series

In Section 12.2 we fitted the Lee–Carter model to UK male data, and Figure 13.2 shows a plot of $\hat{\kappa}$, the estimated value of the time index κ. The idea behind the Lee–Carter model, $\log(\mu_{x+1/2,y+1/2}) = \alpha_x + \beta_x\kappa_y$, is to forecast the mortality table by forecasting $\hat{\kappa}$, while holding the age parameters fixed at their estimated values, $\hat{\alpha}$ and $\hat{\beta}$.

In their original paper, Lee and Carter (1992) modelled κ by a *random walk with drift*. Thus they supposed that successive values of κ_y are generated by:

$$\kappa_y = \mu + \kappa_{y-1} + \epsilon_y, \ y = 2, \ldots, n_y, \tag{13.1}$$

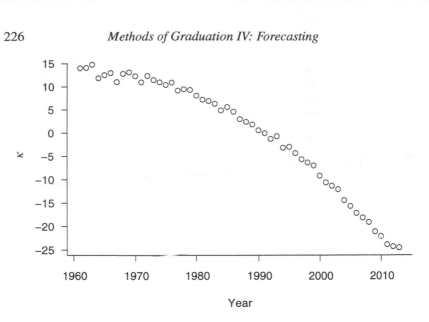

Figure 13.2 Estimated κ in Lee–Carter model. Data: UK males age 50 to 90, years 1961 to 2013 from Human Mortality Database.

where μ is the *drift parameter* and the error terms ϵ_y are independent $N(0, \sigma^2)$ variables; the error terms are often known as *innovations*. We will refer to (13.1) as a *drift model*. Estimation for the drift model is straightforward. Let:

$$\Delta\kappa_y = \kappa_y - \kappa_{y-1}, \ y = 2, \ldots, n_y, \tag{13.2}$$

$$= \mu + \epsilon_y. \tag{13.3}$$

Then $\{\Delta\kappa_y, \ y = 2, \ldots, n_y\}$ is a set of independent $N(\mu, \sigma^2)$ variables. Thus, we estimate μ as the mean of the $\Delta\kappa_y$:

$$\hat{\mu} = \frac{\sum_2^{n_y} \Delta\kappa_y}{n_y - 1} = \frac{\kappa_{n_y} - \kappa_1}{n_y - 1} \tag{13.4}$$

and estimate σ^2 as their sample variance:

$$\hat{\sigma}^2 = \frac{\sum_2^{n_y} (\Delta\kappa_y - \hat{\mu})^2}{n_y - 2}. \tag{13.5}$$

For computation we use standard results for a sample from the normal distribution and we estimate μ and σ^2 with:

```
Mu = mean(diff(Kappa))
Sigma2 = var(diff(Kappa))
```

where the `diff()` function computes first differences of a vector.

Finally, the standard error of $\hat{\mu}$ is:

$$\mathrm{SE}(\hat{\mu}) = \frac{\sigma}{\sqrt{n_y - 1}}, \tag{13.6}$$

since the estimate of μ is based on $n_y - 1$ independent $N(\mu, \sigma^2)$ variables. The estimation error of μ is often known as *parameter error*.

We now turn to forecasting with the drift model. We condition on κ_{n_y}, the final observed value of κ, and set all future innovations ϵ to zero. Then from (13.1) we have:

$$\hat{\kappa}_{n_y+1} = \hat{\mu} + \kappa_{n_y} \tag{13.7}$$

$$\hat{\kappa}_{n_y+2} = \hat{\mu} + \hat{\kappa}_{n_y+1} \tag{13.8}$$

$$= 2\hat{\mu} + \kappa_{n_y}, \tag{13.9}$$

and, in general, the m-step ahead forecast is:

$$\hat{\kappa}_{n_y+m} = m\hat{\mu} + \kappa_{n_y} \tag{13.10}$$

with standard error:

$$\mathrm{SE}(\hat{\kappa}_{n_y+m}) = m\frac{\sigma}{\sqrt{n_y - 1}} \tag{13.11}$$

from (13.6).

Figure 13.3 shows the plausible trajectories for the mean forecast 25 years ahead. We note that this confidence interval allows for a possible error in the estimation of μ; it does not allow for random variation in the actual observations, which are subject to the innovation error variance, σ^2. We discuss the effect of this error in the forecast later in this section.

This is not a book on time series so we will say only enough about this topic to enable us to use R's time series software. The interested reader should consult one of the many specialised books on time series for further information; Shumway and Stoffer (2010) is a comprehensive example and is supported by the R package *astsa* (for Applied Statistical Time Series Analysis); we will make use of this package throughout this chapter.

The drift model can be thought of as a random walk on the differenced series. We now discuss three generalisations of the drift model.

The first generalisation is to allow differencing of a higher order. For example, suppose the series is generated by:

$$\kappa_y = \mu + 2\kappa_{y-1} - \kappa_{y-2} + \epsilon_y, \quad y = 3, \ldots, n_y. \tag{13.12}$$

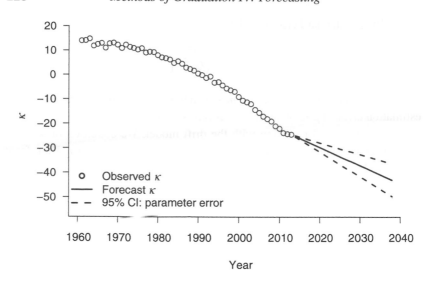

Figure 13.3 Parameter error: estimated and forecast κ in Lee–Carter model together with 95% confidence interval. Data: UK males age 50 to 90, years 1961 to 2013 from Human Mortality Database.

Then:

$$\Delta^2 \kappa_y = \Delta(\kappa_y - \kappa_{y-1}) \tag{13.13}$$

$$= \kappa_y - 2\kappa_{y-1} + \kappa_{y-2} \tag{13.14}$$

$$= \mu + \epsilon_y, \ y = 3, \dots, n_y. \tag{13.15}$$

Again we have a random walk but this time after second-order differencing.

The second generalisation of the random walk with drift is:

$$\kappa_y = \mu + ar_1 \kappa_{y-1} + \epsilon_y, \ y = 2, \dots, n_y. \tag{13.16}$$

This looks like a regression model but with the previous observation as the independent variable; such a model is known as an *autoregressive model* with autoregressive coefficient ar_1. Equation (13.16) defines an autoregressive model of order one or AR(1) model; in general we have an AR(p) model with p autoregressive terms, and p is known as the order of the model.

The third and final generalisation is:

$$\kappa_y = \mu + ma_1 \epsilon_{y-1} + \epsilon_y, \ y = 2, \dots, n_y, \tag{13.17}$$

where we "regress" on the previous innovation. This is known as a *moving average model* of order one or MA(1) model; again, with q moving average terms we have the moving-average model of order q or MA(q) model. Of course, we can have both autoregressive and moving average terms in the same model. For example:

$$\kappa_y = \mu + \text{ar}_1\kappa_{y-1} + \text{ar}_2\kappa_{y-2} + \text{ma}_1\epsilon_{y-1} + \epsilon_y, \; y = 3,\ldots,n_y \qquad (13.18)$$

is a model with two autoregressive terms and one moving average term. The model is known as an *autoregressive moving average model* or ARMA(2,1) model.

Finally, we can combine all three elements to give the *autoregressive integrated moving average* model or ARIMA(p,d,q); here the series of d order differences is an ARMA(p,q) model.

The R function `arima()` can be used both to fit a time series model to data and to forecast future values. This function is not always straightforward to use and we will use the R package *astsa* referred to on the previous page. The package provides the function `sarima()`, a user-friendly front end to R's `arima()` function.

The following code shows how to fit the drift model with the *astsa* package and checks that the answers agree with (13.4) for the drift parameter μ and with (13.5) for the innovation variance σ^2; in the latter case, we must convert the unbiased estimate of σ^2 to the maximum likelihood estimate used by the `sarima()` function. Let `Kappa` contain the estimated values $\hat{\kappa}$.

```
#  Estimate drift parameters with sarima() function
n.y = length(Kappa)
Sarima.out = sarima(Kappa, p=0, d=1, q=0, details = FALSE)
Mu = mean(diff(Kappa))
c(Mu, Sarima.out$fit$coef)
Sigma2 = var(diff(Kappa))
c((n.y-2)/(n.y-1)*Sigma2, Sarima.out$fit$sigma2)
```

We find the drift parameter $\hat{\mu} = -0.742$ and the innovation variance $\hat{\sigma}^2 = 0.946$.

Forecasting is also straightforward with the `sarima.for()` function in the *astsa* package.

```
Forecast = sarima.for(Kappa, n.ahead = 25, p=0, d=1, q=0)
c(Forecast$pred, Kappa[n.y] + Mu*(1:25))
```

in agreement with (13.10). The R object `Forecast` contains two objects: `Forecast$pred` is the forecast given by (13.10) and `Forecast$se` contains the *prediction* or *stochastic error*. We think of the prediction error as follows: the forecast is the best estimate or mean of the future values; the future

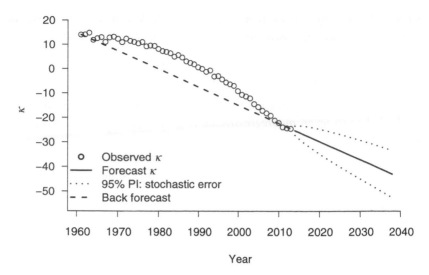

Figure 13.4 Stochastic error: estimated and forecast κ in Lee–Carter model together with 95% prediction interval, PI. Data: UK males age 50 to 90, years 1961
to 2013 from Human Mortality Database.

observed values will also contain innovation error so the m-step ahead forecast
has a predicted value given by (13.10) but the actual observed value will be:

$$\hat{\kappa}_{n_y+m} = m\hat{\mu} + \kappa_{n_y} + \sum_1^m \epsilon_{n_y+j}. \tag{13.19}$$

Thus the predicted value $\hat{\kappa}_{n_y+m}$ has an error variance or prediction error of $m\sigma^2$;
the prediction standard errors, $\sqrt{m}\sigma, m = 1,\ldots,n_f$, where n_f is the length of
the forecast, are contained in `Forecast$se`:

```
c(Forecast$se, sqrt(Sarima.out$fit$sigma2*(1:25)))
```

The 95% prediction interval for the forecast allowing for the prediction error
is $\hat{\kappa}_{n_y+m} \pm 1.96\sqrt{m}\,\hat{\sigma}$ and is computed by:

```
Upper = Forecast$pred + 1.96*Forecast$se
Lower = Forecast$pred - 1.96*Forecast$se
```

from which Figure 13.4 is readily obtained. We emphasise the difference between Figures 13.3 and 13.4: the confidence interval in Figure 13.3 reflects the
possible error in the estimation of μ, i.e. parameter error; the prediction interval
in Figure 13.4 reflects the prediction or stochastic error in the forecast.

From equation (13.4) the forecast for the drift model is the linear extrapolation of the first and last observed values of κ, and Figure 13.4 also shows what we might term the "back" forecast. Evidently, the drift model is a poor model for these data since it fails to pick up the clear curvature in the observations. This is an important observation for actuaries since it means that the drift model will, in all likelihood, underestimate future improvements to mortality.

There are two possible courses of action. We can search through values of p, d and q and select a more promising forecast. The sarima() function provides AIC and BIC values to help in this task. Alternatively, we could forecast κ using the 25 most recent values of κ, say; this is somewhat arbitrary but will produce a more plausible forecast.

This is an appropriate time to sound a note of caution in the forecasting of mortality. It seems self-evident that to forecast mortality 50 or even 100 years ahead is a most challenging task with many possible pitfalls, both known and unknown. Indeed, Richards et al. (2014) list no fewer than five further sources of error, in addition to the parameter and prediction errors noted above. A fuller list of possible errors follows:

- *Model risk.* The risk that we have selected the wrong forecast model. In our example, the drift model clearly performs poorly.
- *Parameter or trend risk.* The risk that our parameter estimates are wrong. Equation (13.6) gives the standard error of $\hat{\mu}$ in the drift model, a measure of the extent of possible error, and Figure 13.3 illustrates this.
- *Stochastic or prediction risk.* The inherent randomness in the process generating mortality as measured by the innovation variance, σ^2, and illustrated by Figure 13.4.
- *Basis risk.* The risk that our data set does not reflect the mortality of our portfolio. Our example uses UK mortality data but national mortality will often be heavier than the mortality of private annuitants or pensioners.
- *Volatility.* Often, the actuary requires forecasts over a small number of years; for example, Solvency II requires short-term forecasts. Local conditions, such as mild or severe winters, or hot summers, can produce short-term shocks to mortality.
- *Idiosyncratic risk.* Random variation in the mortality of the portfolio. Small portfolios are particularly vulnerable to this risk since the mortality of a few high-worth policyholders can have a large impact on the solvency of the whole portfolio. In this case the risk is known as *concentration risk*.
- *Mis-estimation risk.* The richness of the portfolio's underlying data may not support the estimation of the mortality with any degree of confidence.

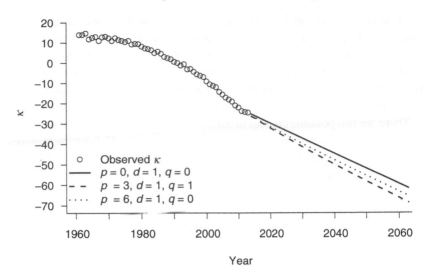

Figure 13.5 Model risk in forecasts of κ in Lee–Carter model for various ARIMA(p, d, q) models. Data: UK males age 50 to 90, years 1961 to 2013 from Human Mortality Database.

Richards and Currie (2009) has a thorough discussion of many of the above risks. As a final illustration of just one of the above possible errors, model risk, we consider forecasting κ 50 years ahead with various ARIMA(p, d, q) models. Calls of the form:

```
sarima.for(kappa, p=p, d=1, q=q, n.head = 50)
```

lead to Figure 13.5.

This discussion of model risk is within the framework of the Lee–Carter model. Of course, model risk is present at this higher level too; we compare forecasts with the two-dimensional P-spline, the Lee–Carter and the Cairns-Blake Dowd models in Section 13.8. First, we discuss forecasting with P-spline models.

13.3 Penalty Forecasting

Forecasting with penalties is very different from forecasting with time series methods. We illustrated the time-series approach in the previous section with a discussion of the Lee–Carter model: $\log \mu_{x+1/2, y+1/2} = \alpha_x + \beta_x \kappa_y$, where $\mu_{x+1/2, y+1/2}$ is the force of mortality at age x in year y. The time series method usually proceeds in two stages:

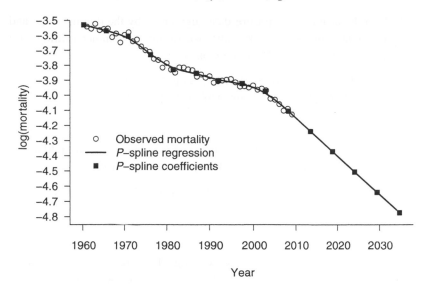

Figure 13.6 *P*-spline fit and forecast with fitted mortality and fitted regression coefficients. Data: USA females age 70 from 1961 to 2009 from Human Mortality Database.

1. estimate the model parameters α, β and κ by fitting to data
2. forecast the estimated κ parameters.

The key point here is that these are two quite separate computations: the method is *fit then forecast*, where the critical word is *then*.

Penalty forecasting is different: the method is *fit and forecast simultaneously*. Thus forecasting is built into the model (unlike Lee–Carter where it is added on to the model). Figure 13.6 shows log hazard rates for USA females age 70 from 1961 to 2009, as discussed in Chapter 11. Forecasting with penalties was introduced by Currie et al. (2004); Richards et al. (2006) describes the method in an actuarial context. We can use the *MortalitySmooth* package (Camarda, 2012) to illustrate forecasting with penalties.

Let us suppose we wish to forecast for 25 years, that is, to 2034. The function MortlDsmooth() allows both interpolation (which can be useful if we have missing or unreliable data) and forecasting. The method depends on a neat trick. We define dummy deaths and exposures for the forecast years; these can be any positive values, but we set future deaths to have the value 1 and future exposures to have the value 2. We also define a weight vector Weight which takes the value 1 for the observed data and the value 0 for the future dummy data. We now apply the MortlDsmooth() function to the years 1961 to 2034, to the death and exposure data augmented by the dummy data, and with the weight option set w = Weight; we strongly recommend fitting with overdispersion switched on. The code looks like:

```
Weight = c(rep(1, length(YEAR)), rep(0,25))
YEAR.NEW = YEAR[1]:(YEAR[length(YEAR)] + 25)
Dth.NEW = c(Dth, rep(1, 25))
Exp.NEW = c(Exp, rep(2, 25))
Smooth.Forecast = Mort1Dsmooth(x = YEAR.NEW, y = Dth.NEW,
    offset = log(Exp.NEW), w = Weight, over = TRUE)
```

and Figure 13.6 shows both the fitted and the forecast log hazard rates, and the estimated regression coefficients. The forecast looks very like a linear extrapolation of the fitted log mortality. How does this come about?

The key idea is that the 0/1 weights ensure that the dummy data have no effect on the fit but do allow the forecasting of the regression coefficients through the penalty. We can see how this works by considering the penalised log-likelihood:

$$\ell_p(\boldsymbol{\theta}) = \ell(\boldsymbol{\theta}) - \tfrac{1}{2}\lambda P(\boldsymbol{\theta}) \qquad (13.20)$$

first discussed in equation (11.24). Now consider the effect of fitting with and without the dummy data. The weight function ensures that the log-likelihood $\ell(\boldsymbol{\theta})$ is unchanged by the introduction of the dummy data. If we can choose the future coefficients such that the penalty $P(\boldsymbol{\theta})$ is also unaltered, then the fit with the original data and the fit with the augmented data will have the same value of $\ell_p(\boldsymbol{\theta})$; this is the optimal fit and the forecast follows from the forecast coefficients. Suppose that the regression coefficients in the model for the data only are $\boldsymbol{\theta} = (\theta_1, \ldots, \theta_c)'$ and that $\hat{\theta}_{c-1}$ and $\hat{\theta}_c$ are the estimates of the final two coefficients. We must choose the first coefficient θ_{c+1} in the forecast region so that the contribution to the penalty is zero; thus, we wish $(\hat{\theta}_{c-1} - 2\hat{\theta}_c + \theta_{c+1})^2 = 0$. Now we see why we have used a second-order penalty: we set $\hat{\theta}_{c+1} = 2\hat{\theta}_c - \hat{\theta}_{c-1}$. This is the condition that $\hat{\theta}_{c-1}$, $\hat{\theta}_c$ and $\hat{\theta}_{c+1}$ lie on a straight line. The second-order penalty implies a linear extrapolation of the coefficients and hence of the log mortality. Figure 13.6 shows that the fitted regression coefficients in the forecast region do lie on a straight line. A more precise check is given by:

```
diff(Smooth.Forecast$coef)
```

which confirms that the forecast coefficients are indeed a linear extrapolation of the data coefficients. We can check this further with:

```
Smooth.Fit = Mort1Dsmooth(x = YEAR, y = Dth,
    offset = log(Exp), df = Smooth.Forecast$df, method = 4)
```

Here we fit to the original data only but force the fitted degrees of freedom to equal the degrees of freedom of the forecast. Now we can compare the two sets of coefficients Smooth.Fit$coef and Smooth.Forecast$coef. We find that in the data region the coefficients agree to within 1%. (The agreement would be exact if we made the *B*-spline basis for the data a subset of the basis for the data augmented by the dummy data. Unfortunately, the *MortalitySmooth* package does not allow the fine tuning of these two bases; see Currie et al. (2004) for a detailed discussion of this point.)

The process we have described for forecasting is a little elaborate. However, if we are only interested in the forecast values of log mortality and their standard errors then the *MortalitySmooth* package provides a painless way of obtaining these. We have data for years 1961 to 2009 stored in the vector YEAR. To forecast for 25 years we define a new vector YEAR.NEW which contains the data and forecast years. The code:

```
YEAR.NEW = 1961:2034
Smooth.Fit = Mort1Dsmooth(x = YEAR, y = Dth,
  offset = log(Exp), over = TRUE)
Smooth.Forecast = predict(Smooth.Fit, newdata = YEAR.NEW,
  se.fit = TRUE)
```

fits the model to the data and then uses the predict() function to obtain both the central forecast in Smooth.Forecast$fit and their standard errors in Smooth.Forecast$se.fit; Figure 13.7 shows the result.

Finally, we demonstrate the effect of changing the order of the penalty on the forecast. The pord option of the Mort1Dsmooth() function has a default value of 2 but we may select other values. For example:

```
Smooth.Fit.1 = Mort1Dsmooth(x = YEAR, y = Dth,
  offset = log(Exp), over = TRUE, pord = 1)
Smooth.Forecast.1 = predict(Smooth.Fit.1,
  newdata = YEAR.NEW)
```

fits with a first-order penalty: $(\theta_1 - \theta_2)^2 + (\theta_2 - \theta_3)^2 + \ldots$; setting pord = 3 gives a third-order penalty: $(\theta_1 - 3\theta_2 + 3\theta_3 - \theta_4)^2 + (\theta_2 - 3\theta_3 + 3\theta_4 - \theta_5)^2 + \cdots$.

Figure 13.8 shows the three forecasts. We see that the first-order penalty gives a constant forecast, the second-order penalty gives a linear forecast and the third-order penalty gives a quadratic forecast. The linear forecast looks by far the most plausible, and we recommend the use of the second-order penalty when forecasting.

The program Pspline_Forecast.r referenced in Appendix H.6 illustrates forecasting with *P*-splines in one dimension, and Figures 13.6 and 13.7 are produced.

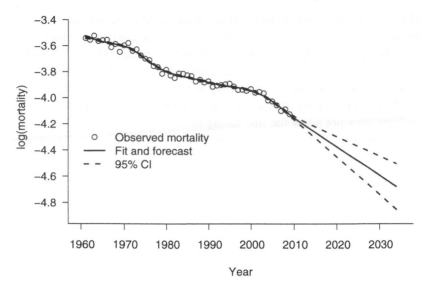

Figure 13.7 *P*-spline fit and forecast with 95% confidence interval. Data: USA females age 70 from 1961 to 2009 from Human Mortality Database.

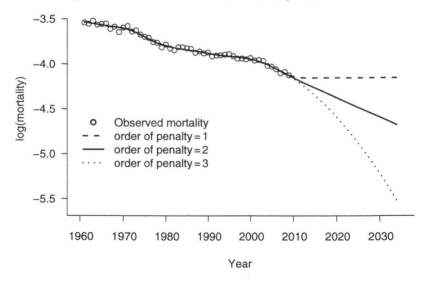

Figure 13.8 *P*-spline fit and forecast with order of penalty 1, 2 and 3. Data: USA females age 70 from 1961 to 2009 from Human Mortality Database.

13.4 Forecasting with the Lee–Carter Model

In Section 12.2 we described the Lee–Carter model:

$$\log(\mu_{x+1/2,y+1/2}) = \alpha_x + \beta_x \kappa_y, x = x_1, \ldots, x_{n_x}, \ y = y_1, \ldots, y_{n_y} \qquad (13.21)$$

and obtained the estimates of the age parameters α and β and the time parameter κ; let us denote these estimates by $\hat{\alpha}$, $\hat{\beta}$ and $\hat{\kappa}$. In Section 13.2 we described the time series approach to forecasting and used the *astsa* package to forecast $\hat{\kappa}$; the forecast also comes with its standard error. For clarity we repeat the code snippet on page 229:

```
Forecast = sarima.for(Kappa, n.ahead = 25, p=0, d=1, q=0)
Central = Forecast$pred; SE = Forecast$se
Z = qnorm(0.975)
Upper = Central + Z*SE; Lower = Central - Z*SE
```

which gives the central forecast and the 95% prediction interval for the drift model; here we have used the qnorm() function to allow a general prediction level. Figure 13.4 summarised these results.

Forecasting with the Lee–Carter model is straightforward. We fix α and β at their estimated values, and set κ equal to its forecast value, κ_f, say. The model gives the forecast of log mortality at age x as:

$$\hat{\alpha}_x 1_f + \hat{\beta}_x \kappa_f, \tag{13.22}$$

where 1_f is the vector of 1s with the same length as κ_f. We can forecast the whole table by using the matrix form:

$$\hat{\alpha} 1'_f + \hat{\beta} \kappa'_f. \tag{13.23}$$

In R this can be written compactly as:

```
Forecast = Alpha.hat + Beta.hat %*% t(Central)
```

where %*% indicates matrix multiplication. Here we have used a neat trick in R. The code Beta.hat %*% t(Central) produces a matrix, M say, of size $n_x \times n_f$, where n_f is the length of the forecast. The vector Alpha.hat has length n_x, *the same as the number of rows* of M. R now repeats Alpha.hat for each column of M, which gives the desired result. In a similar way:

```
Forecast.Upper = Alpha.hat + Beta.hat %*% t(Upper)
```

gives the upper 95% prediction interval for the forecast. Figure 13.9 summarises these results.

An important point is that the intervals in Figure 13.9 only allow for prediction error, not for parameter error. We account for parameter error in Figure 13.10. We compute the appropriate standard error with (13.11) as follows:

```
Sarima.out = sarima(Kappa.hat, p=0, d=1, q=0)
SE = (1:N.Ahead) * sqrt(Sarima.out$fit$sigma2/(n.y-1))
```

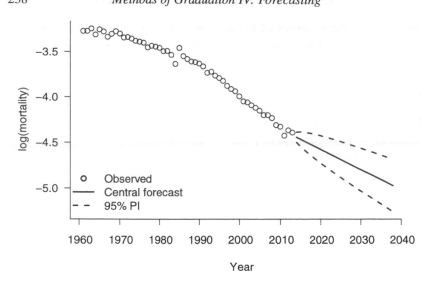

Figure 13.9 Lee–Carter forecast with 95% prediction intervals, PI, for age 65.
Data: UK males ages 50 to 90 from 1961 to 2013 from Human Mortality Database.

With this change to the standard error of the forecast value of κ the above code holds; Figure 13.10 shows the results.

The program `Forecast_LC.r` referenced in Appendix H.6 illustrates forecasting with the Lee–Carter model, and Figures 13.9 and 13.10 are produced.

Although the Lee–Carter model is one of the most popular methods of forecasting mortality it does suffer from one quite serious drawback. The original purpose of the Lee–Carter model (Lee and Carter, 1992) was the forecasting of life expectancy, a more limited objective than forecasting the whole mortality table. If the estimates of the parameter β are at all volatile, the forecasts of mortality at adjacent ages can readily cross over; this would not be acceptable to the practising actuary. Delwarde et al. (2007) addressed this problem by smoothing β; Currie (2013) extended this work and smoothed both α and β. We recommend the use of these later methods in place of Lee and Carter's original proposal.

13.5 Simulating the Future

The requirements of Solvency II demand that the actuary is able to generate plausible future paths of mortality. While Figures 13.9 and 13.10 give a guide

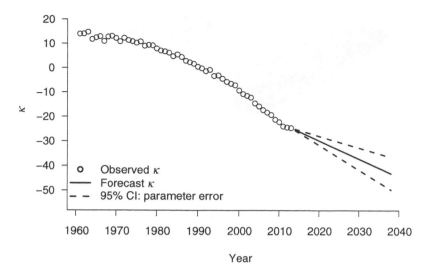

Figure 13.10 Lee–Carter forecast with 95% confidence intervals for age 65. Data: UK males ages 50 to 90 from 1961 to 2013 from Human Mortality Database.

to the possible deviations from the central forecast, they measure average departures. In this section we look at the individual paths that future mortality might take. Our discussion will focus on the Lee–Carter model and in particular the forecasting of κ with the drift model. We will discuss the effects of (a) stochastic or prediction error and (b) estimation or parameter error.

We use equation (13.19) to simulate allowing for stochastic error:

$$\hat{\kappa}_{n_y+m} = m\hat{\mu} + \kappa_{n_y} + \sum_1^m \epsilon_{n_y+j}. \tag{13.24}$$

From page 229, we set `Mu = -0.742` and `Sigma = 0.982`. We have:

```
N.Ahead = 25; Range = 2014:2038; N.Sim = 100
Mat.Sim = matrix(rnorm(N.Sim * N.Ahead, Mu, Sigma),
                 nrow = N.Sim, ncol = N.Ahead)
Stoch = Kappa[n.y] + apply(Mat.Sim, 1, cumsum)
matplot(Range, Stoch, type = "l", col = 1)
```

Simulation is often associated with loops in computing and R does indeed have the control structure `for`, for this purpose. However, in the interests of efficiency, the user is strongly advised to avoid the use of `for` if at all possible. Here, the `apply()` function is helpful and can often be used to avoid explicit

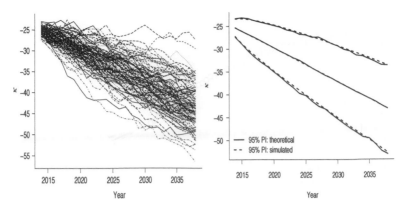

Figure 13.11 Lee–Carter model, stochastic error only. Left: 100 simulations of κ. Right: theoretical and simulated 95% prediction intervals for κ based on 1,000 simulations. Data: UK males ages 50 to 90 from 1961 to 2013 from Human Mortality Database.

loops with `for`. In the above code we apply the `cumsum()` function to the first dimension of the matrix `Mat.Sim`. The object `Stoch` is a matrix whose `N.Sim` columns contain the cumulative sums as required by equation (13.24). Figure 13.11 (left) shows 100 such simulations.

We can also compute quantiles across our simulations. We define a function `Quantile()` to compute the 2.5%, 50% and 97.5% quantiles and once more use the `apply()` function:

```
Quantile = function(x) quantile(x, c(0.025, 0.5, 0.975))
Sim.Q = apply(Stoch, 1, Quantile)
```

The columns of `Sim.Q` now contain the 2.5%, 50% and 97.5% quantiles across our simulated values. Figure 13.11 (right) shows both the simulated and theoretical prediction intervals; the theoretical values are computed as on page 230.

We can repeat the above exercise but this time allowing for parameter error only. For the simulation we need only change the line to compute `Mat.Sim` to:

```
Mat.Sim = matrix(rep(rnorm(N.Sim, Mu, Sigma/sqrt(n.y-1)),
                  N.Ahead), nrow = N.Sim, ncol = N.Ahead)
```

where we have used equation (13.6) for the standard error of μ. Figure 13.12 (left), based on 50 simulations, and Figure 13.12 (right), based on 5,000 simulations, illustrate the results.

Finally, we can allow for both stochastic and parameter error in our simulations. The variance of the forecast allowing for both stochastic and parameter

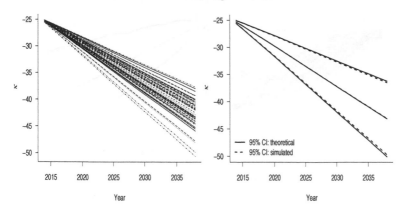

Figure 13.12 Lee–Carter model, parameter error only. Left: 50 simulations of κ. Right: theoretical and simulated 95% confidence intervals for κ based on 5,000 simulations. Data: UK males ages 50 to 90 from 1961 to 2013 from Human Mortality Database.

error follows from equation (13.24) as:

$$\text{Var}(\hat{\kappa}_{n_y+m}) = \text{Var}\left(m\hat{\mu} + \sum_{j=1}^{m} \epsilon_{n_y+j}\right) \tag{13.25}$$

$$= m^2 \text{Var}(\hat{\mu}) + \text{Var}\left(\sum_{j=1}^{m} \epsilon_{n_y+j}\right) \tag{13.26}$$

$$= m^2 \frac{\sigma^2}{n_y - 1} + m\sigma^2 \tag{13.27}$$

by (13.11).

We combine the above code for the stochastic and parameter error with:

```
Mat.Sim.1 = matrix(rnorm(N.Sim * N.Ahead, 0, Sigma),
                   nrow = N.Sim, ncol = N.Ahead)
Mat.Sim.2 = matrix(rep(rnorm(N.Sim, Mu, Sigma/sqrt(n.y-1)),
                   N.Ahead), nrow = N.Sim, ncol = N.Ahead)
Mat.Sim = Mat.Sim.1 + Mat.Sim.2
```

and Figures 13.13 (left) and 13.13 (right) follow as in the earlier examples; the theoretical quantiles are computed from equation (13.27).

In this section we have explained the importance of (a) stochastic or prediction error, and (b) parameter or estimation error. Our discussion has focused on the forecasting of κ in the Lee–Carter model. What are the consequences of these errors for the forecasting of mortality? There are three sets of parameters

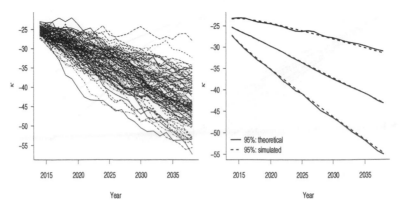

Figure 13.13 Lee–Carter model, stochastic and parameter error. Left: 100 simulations of κ. Right: theoretical and simulated 95% intervals for κ based on 1,000 simulations. Data: UK males ages 50 to 90 from 1961 to 2013 from Human Mortality Database.

in the Lee–Carter model: α, β and κ. We have shown how to measure the prediction and parameter errors for κ. The assumption in the Lee–Carter model is that the age parameters, α and β, are fixed in time and so by assumption are not subject to prediction error; they are however subject to estimation error. In their original paper, Lee and Carter (1992) argued strongly that the effect of the estimation error in α and β was small compared to the errors in estimating and forecasting κ. With this assumption we can use equation (13.23):

$$\hat{\alpha} 1'_f + \hat{\beta} \kappa'_f \tag{13.28}$$

to produce forecasts with both prediction and confidence intervals. Figure 13.14 shows how the uncertainty in forecasting κ transfers to uncertainty in forecasting mortality. Twenty-five years ahead is not so far into the future for the actuary, yet even with this relatively near horizon, the width of these intervals is salutary.

The ideas of stochastic and parameter error apply equally to ARIMA models, and the principles which we have laid down for the drift model can be extended to ARIMA models. The R function `arima.sim()` is helpful for simulations which allow for stochastic uncertainty. For parameter uncertainty, one approach is to use bootstrapping. Richards (2008) is a brief introduction to bootstrapping in an actuarial setting while Kleinow and Richards (2016) is a thorough discussion of bootstrapping in our present application. The book by Efron and Tibshirani (1993) is a good general introduction.

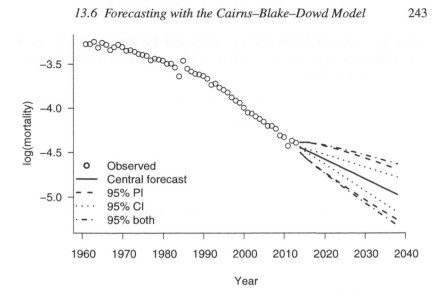

Figure 13.14 Interval forecasts for age 65 mortality with Lee–Carter model; intervals allow for stochastic error, PI, parameter error, CI, and both. Data: UK males ages 50 to 90 from 1961 to 2013 from Human Mortality Database.

The program `Simulation.r` referenced in Appendix H.6 illustrates simulation of the Lee–Carter model with the drift model; Figures 13.11 to 13.13 are produced.

13.6 Forecasting with the Cairns–Blake–Dowd Model

In Section 12.3 we discussed the Cairns–Blake–Dowd or CBD model:

$$\log(\mu_{x+1/2,y+1/2}) = \kappa_y^{(1)} + \kappa_y^{(2)}(x - \bar{x}), \quad x = x_1, \ldots, x_{n_x}, \ y = y_1, \ldots, y_{n_y}, \quad (13.29)$$

and described the fitting of the model as a generalised linear model. Figure 12.3 showed the fitted parameters (and the fits at age 65 and in year 2000). There are two sets of parameters, $\kappa^{(1)}$ and $\kappa^{(2)}$, and Cairns et al. (2006) suggested that these should be forecast as a bivariate random walk with drift. We have already described estimating and forecasting the simple random walk with drift in connection with the Lee–Carter model in Section 13.4, and our development here follows this closely.

Let $\kappa_y = (\kappa_y^{(1)}, \kappa_y^{(2)})'$ be the observed values at time y. The bivariate random walk with drift says that successive values are generated by:

$$\kappa_y = \mu + \kappa_{y-1} + \epsilon_y, \ y = 2, \ldots, n_y, \quad (13.30)$$

where $\mu = (\mu_1, \mu_2)'$ is the drift parameter, and the innovation terms ϵ_y are independent bivariate normal $N(0, \Sigma)$ variables. Estimation follows the simple random walk with drift. Let:

$$\Delta\kappa_y = \kappa_y - \kappa_{y-1}, \quad y = 2, \ldots, n_y \tag{13.31}$$

$$= \mu + \epsilon_y \tag{13.32}$$

and $\Delta\kappa_y, y = 2, \ldots, n_y$ be a set of independent $N(\mu, \Sigma)$ variables, where:

$$\Sigma = \begin{bmatrix} \sigma_1^2 & \sigma_{1,2} \\ \sigma_{1,2} & \sigma_2^2 \end{bmatrix} \tag{13.33}$$

is the variance-covariance matrix. Estimation of μ and Σ corresponds respectively to (13.4) and (13.5) in the one-dimensional case. The estimate of μ is:

$$\hat{\mu} = \frac{\sum_2^{n_y} \Delta\kappa_y}{n_y - 1} = \frac{\kappa_{n_y} - \kappa_1}{n_y - 1} \tag{13.34}$$

and its distribution is:

$$\hat{\mu} \sim N\left(\mu, \frac{1}{n_y - 1}\Sigma\right); \tag{13.35}$$

the estimate of Σ is

$$\hat{\Sigma} = \frac{1}{n_y - 2} \sum_2^{n_y} (\Delta\kappa_y - \hat{\mu})(\Delta\kappa_y - \hat{\mu})'. \tag{13.36}$$

With our running example on UK male data we find:

$$\hat{\mu} = \begin{bmatrix} -0.01857 \\ 0.00026 \end{bmatrix}, \quad \hat{\Sigma} = \begin{bmatrix} 0.000625 & 0.000013 \\ 0.000013 & 0.000009 \end{bmatrix}, \quad \hat{\rho} = 0.553. \tag{13.37}$$

There is substantial correlation between the components of ϵ_y.

We now turn to forecasting. The m-step ahead forecast for κ corresponds to equation (13.10) and is:

$$\hat{\kappa}_{n_y+m} = \kappa_{n_y} + m\hat{\mu}, \tag{13.38}$$

with variance matrix:

$$\mathrm{Var}\left(\hat{\kappa}_{n_y+m}\right) = \frac{m^2}{n_y - 1}\Sigma. \tag{13.39}$$

Let us suppose that we wish to forecast $n_f = 25$ years ahead. Let $\hat{\kappa}_f^{(1)}$ and $\hat{\kappa}_f^{(2)}$ be the forecast values of $\kappa^{(1)}$ and $\kappa^{(2)}$, respectively, and $c_f = (1, 2, \ldots, n_f)'$. Then by (13.38) we have:

$$\hat{\kappa}_f^{(1)} = \kappa_{n_y}^{(1)} 1_f + \hat{\mu}_1 c_f \tag{13.40}$$

$$\hat{\kappa}_f^{(2)} = \kappa_{n_y}^{(2)} 1_f + \hat{\mu}_2 c_f, \tag{13.41}$$

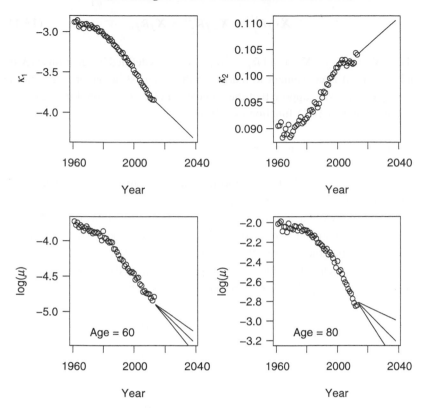

Figure 13.15 Forecasting mortality with the CBD model. Forecasts of $\kappa^{(1)}$ and $\kappa^{(2)}$, and $\log(\mu)$ for ages 60 and 80 with 95% confidence intervals. Data: UK males ages 50 to 90 from 1961 to 2013 from Human Mortality Database.

where $\mathbf{1}_f$ is a vector of 1s of length n_f. The upper panels of Figure 13.15 show the forecast values of $\kappa^{(1)}$ and $\kappa^{(2)}$.

We can now use the CBD model to forecast the log mortality. Let $\boldsymbol{X}_{1,f}$ and $\boldsymbol{X}_{2,f}$ be the parts of the model matrix for the CBD model corresponding to $\kappa^{(1)}$ and $\kappa^{(2)}$ as in Section 12.11. Then:

$$\boldsymbol{X}_{1,f} = \boldsymbol{I}_{n_f} \otimes \mathbf{1}_{n_x} \tag{13.42}$$

and

$$\boldsymbol{X}_{2,f} = \boldsymbol{I}_{n_f} \otimes \boldsymbol{x}_m, \tag{13.43}$$

as in equations (12.13) and (12.14). The central forecast for the log mortality now follows from the model assumption as:

$$X_{1,f}\hat{\kappa}_f^{(1)} + X_{2,f}\hat{\kappa}_f^{(2)} = X_f\hat{\kappa}_f, \qquad (13.44)$$

where $X_f = [X_{1,f} : X_{2,f}]$ and $\hat{\kappa}_f = \left(\hat{\kappa}_f^{(1)\prime}, \hat{\kappa}_f^{(2)\prime}\right)'$ is defined in equations (13.40) and (13.41). It is convenient to rearrange this forecast in vector form into an $n_x \times n_f$ matrix. We suppose that the forecast in (13.44) is stored in `Forecast` in R and use the `dim()` function as follows:

```
dim(Forecast) = c(n.x, n.f)
```

The forecast for age x is the appropriate row of this matrix. The lower panels of Figure 13.15 show the central forecasts for ages 60 and 80.

We deal next with measuring (a) the stochastic or prediction error and (b) the estimation or parameter error associated with the CBD model. We start with the estimation or parameter error. We compute the variance-covariance matrix of the forecast, which will enable us to compute the required confidence intervals. We condition on κ_{n_y}, the final observed value of κ. Then:

$$
\begin{aligned}
\mathrm{Var}\left(X_f\hat{\kappa}_f\right) &= X_f\mathrm{Var}\left(\hat{\kappa}_f\right)X_f' \quad \text{by Appendix E} \\[4pt]
&= X_f\mathrm{Var}\begin{bmatrix} \hat{\mu}_1 c_f \\ \hat{\mu}_2 c_f \end{bmatrix}X_f' \quad \text{by (13.40) and (13.41)} \\[4pt]
&= X_f\begin{bmatrix} c_f\mathrm{Var}\left(\hat{\mu}_1\right)c_f' & c_f\mathrm{Cov}\left(\hat{\mu}_1,\hat{\mu}_2\right)c_f' \\ c_f\mathrm{Cov}\left(\hat{\mu}_2,\hat{\mu}_1\right)c_f' & c_f\mathrm{Var}\left(\hat{\mu}_2\right)c_f' \end{bmatrix}X_f' \\[4pt]
&= X_f\left[\mathrm{Var}\left(\hat{\mu}\right)\otimes c_f c_f'\right]X_f' \quad \text{by Appendix G} \\[4pt]
&= \frac{1}{n_y - 1}X_f\left[\Sigma \otimes c_f c_f'\right]X_f' \qquad (13.45)
\end{aligned}
$$

by (13.35); here we set $\Sigma = \hat{\Sigma}$ by (13.36). Expression (13.45) gives the full variance-covariance matrix of the forecasts for all ages. We require the variances only, that is, the diagonal elements of $\mathrm{Var}\left(X_f\hat{\kappa}_f\right)$. We can compute these neatly in R: we suppose the full variance-covariance matrix has been computed and stored in `Var.Mat.Forecast`. We extract the diagonal elements with the `diag()` function, re-dimension the vector `Var.Forecast` into the $n_x \times n_f$ matrix whose elements match the matrix of forecast mortalities, and finally obtain the standard errors required for the confidence intervals:

```
Var.Forecast = diag(Var.Mat.Forecast)
dim(Var.Forecast) = c(n.x, n.f)
SD.Forecast = sqrt(Var.Forecast)
```

The 95% confidence intervals for the log mortality for ages 60 and 80 have been added to Figure 13.15.

The prediction or stochastic error is computed from the bivariate equivalent of (13.19):

$$\hat{\kappa}_{n_y+m} = m\hat{\mu} + \kappa_{n_y} + \sum_1^m \epsilon_{n_y+j}, \qquad (13.46)$$

with prediction variance $m\Sigma$. The prediction variance of the m-step ahead forecast mortality at age x follows as:

$$
\begin{aligned}
\mathrm{Var}&\left(\hat{\kappa}^{(1)}_{n_y+m} + \hat{\kappa}^{(2)}_{n_y+m}(x - \bar{x})\right)\\
&= \mathrm{Var}\left([1 : x - \bar{x}]\begin{bmatrix} \hat{\kappa}^{(1)}_{n_y+m} \\ \hat{\kappa}^{(2)}_{n_y+m} \end{bmatrix}\right)\\
&= \mathrm{Var}(\tilde{x}'\hat{\kappa}_{n_y+m}) \quad \text{where } \tilde{x} = (1, x - \bar{x})'\\
&= \tilde{x}'\mathrm{Var}(\hat{\kappa}_{n_y+m})\tilde{x}\\
&= \tilde{x}'m\Sigma\tilde{x}, \quad m = 1,\ldots,n_f. \qquad (13.47)
\end{aligned}
$$

Hence, the variances of the forecast n_f years ahead at age x are:

$$(\tilde{x}'\Sigma\tilde{x})\,c_f. \qquad (13.48)$$

Figure 13.16 shows the central forecasts together with the 95% prediction intervals for ages 50, 65, 80 and 90. It is striking that the interval at age 50 is much narrower than that at age 90. The width of the prediction interval depends on the behaviour of $\tilde{x}'\Sigma\tilde{x}$ as a function of x. The substantial positive correlation (0.55) between $\Delta\kappa^{(1)}$ and $\Delta\kappa^{(2)}$ means that $\tilde{x}'\Sigma\tilde{x}$ is monotonic increasing in age above about age 55. Conversely, if the correlation were -0.55 it would be the prediction intervals at lower ages that would be larger.

We do not discuss simulating from the CBD model here. The methods described in Section 13.5 on simulating from the drift model can be extended to the bivariate drift model used in the CBD model. Simulating from a bivariate normal distribution can be achieved with the mvrnorm() function available in the MASS library.

The program Forecast_CBD.r referenced in Appendix H.6 illustrates forecasting with the CBD model; Figure 13.15 is produced.

13.7 Forecasting with the Two-Dimensional P-Spline Model

In Section 13.3 we saw how to use the *MortalitySmooth* package to forecast with P-splines in one dimension. In two dimensions the method is very similar,

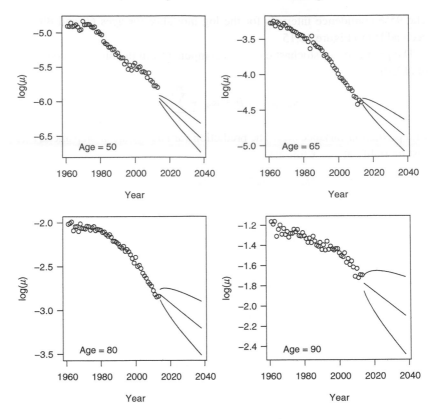

Figure 13.16 Forecasting mortality with the CBD model. Forecasts of $\log(\mu)$ for ages 50, 65, 80 and 90 with 95% prediction intervals. Data: UK males ages 50 to 90 from 1961 to 2013 from Human Mortality Database.

although there are some important pitfalls to be aware of. We continue with our example of forecasting 25 years ahead for UK male data for ages 50 to 90 and years 1961 to 2013. We use the Mort2Dsmooth() function together with the predict() function just as on page 235:

```
Fit.2d = Mort2Dsmooth(AGE, YEAR, Dth, offset=log(Exp),
                     over = TRUE)
c(Fit.2d$lambda, Fit.2d$df, Fit.2d$dev, Fit.2d$psi2)
YEAR.NEW = 1961:2038; NEW.DATA = list(x = AGE, y = YEAR.NEW)
Forecast.2d = predict(Fit.2d, newdata = NEW.DATA)
```

The above code fits the model with overdispersion switched on and then forecasts the table to 2038. Figure 13.17 shows the forecasts for ages 60, 75

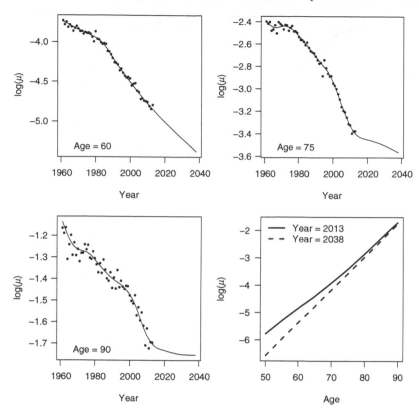

Figure 13.17 Forecasting mortality with two-dimensional *P*-spline model: $\lambda_x = 1000$, $\lambda_y = 316$, ED = 44.8, $\psi^2 = 4.9$. Data: UK males ages 50 to 90 from 1961 to 2013 from Human Mortality Database.

and 90: while the forecast at age 60 looks reasonable, the forecasts at ages 75 and 90 both indicate a very rapid deceleration in the improvements to mortality with time. The approximately linear forecasts we saw in one dimension hold here only for the younger ages. The problem is that the relatively strong penalty across age ($\lambda_x = 1,000$ while $\lambda_y = 316$) undermines this linearity in time and replaces it with a linear function (exact Gompertz) across age at the forecast horizon, as the lower-right panel in Figure 13.17 shows.

One solution is to fit a model with approximately the same degrees of freedom as the optimal model but with $\lambda_x < \lambda_y$. For example, if we take $\lambda_y = 2\lambda_x$, then a search over the λ values quickly arrives at:

```
Fit.2d = Mort2Dsmooth(AGE, YEAR, Dth, offset=log(Exp),
  method = 3, lambdas = c(2000, 4000))
Forecast.2d = predict(Fit.2d, newdata=NEW.DATA, se.fit=TRUE)
```

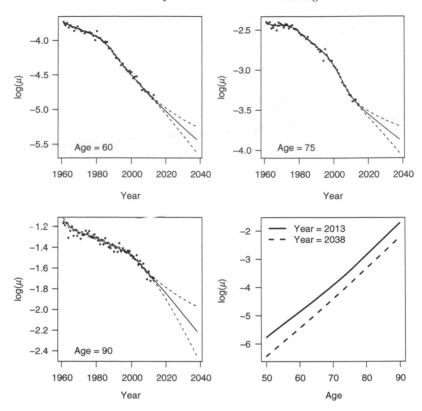

Figure 13.18 Forecasting mortality with two-dimensional *P*-spline model: $\lambda_x =$ 2000, $\lambda_y = 4000$, ED $= 44.7$, $\psi^2 = 5.0$. Data: UK males ages 50 to 90 from 1961 to 2013 from Human Mortality Database.

which has degrees of freedom 44.71 compared to 47.75 for the original model, and overdispersion parameter 4.97 compared to 4.93. Of course, we have sacrificed fit in this exercise and the deviance has gone up from 10,491 for the original model to 10,577 for the new model.

Figure 13.18 summarises the results. The forecasts now look reasonable and the hazard rate at the forecast horizon is nearly parallel to that for 2013, the last data year. One point of detail in Figure 13.18 is that the standard errors are computed from:

```
SE = sqrt(Fit.2d$psi2) * Forecast.2d$se.fit
```

since the `Mort2Dsmooth()` function under `method = 3` has overdispersion switched off; the resulting standard errors must be inflated by ψ as above.

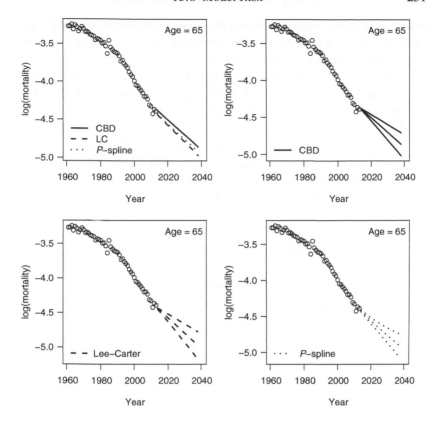

Figure 13.19 Model risk: forecasting with the CBD, Lee–Carter and *P*-spline models at age 65 with 95% confidence intervals. Data: UK males ages 50 to 90 from 1961 to 2013 from Human Mortality Database.

We do not describe simulation of sample paths for the two-dimensional *P*-spline model here; see Richards et al. (2014) for a full discussion. The program `2d_Pspline_Forecast.r` referenced in Appendix H.6 illustrates forecasting with two-dimensional *P*-splines; Figures 13.17 and 13.18 are produced.

13.8 Model Risk

In this chapter we have concentrated on two sources of error in forecasting mortality: prediction or stochastic error, and estimation or parameter error. Prediction error is measured by the width of the prediction intervals while parameter error is measured by the width of the confidence intervals. Both sources of error can, and one supposes usually do, operate simultaneously, as discussed in Section 13.5 and Figure 13.14 in particular. We close this chapter with a discussion of a further source of error, *model risk*.

We have already discussed model risk within the framework of a single model, the Lee–Carter model. Here model risk depends on which ARIMA model is chosen to forecast the κ term in the model; see Section 13.2 and Figure 13.5. Model risk arises more generally with the choice of the basic model of mortality. Figure 13.19 shows the forecasts at age 65 for the CBD, Lee–Carter and P-spline models. The upper-left panel shows the impact that model choice has on the forecasts; the forecasts of the log hazard rate at age 65 are -4.86, -4.97 and -4.92 at the forecast horizon, 2038, for the CBD, Lee–Carter and P-spline models, respectively. The remaining panels demonstrate that the width of the confidence intervals also varies among these models. Again at the forecast horizon, the corresponding widths of the 95% confidence intervals are 0.31, 0.39 and 0.34. Thus the precision of the forecast is also model-dependent.

Defined-benefit pension schemes and annuity portfolios are exposed to longevity risk, so the forecasting of mortality is important to actuaries. Richards and Currie (2009) gives a thorough discussion of the financial implications of model risk within a family of Lee–Carter models. Short-term mortality forecasts are also important for actuaries in, for example, a one-year, value-at-risk view of reserves for Solvency II calculations; Richards et al. (2014) investigate the financial implications of model choice for these kinds of calculations.

PART THREE

MULTIPLE-STATE MODELS

14

Markov Multiple-State Models

14.1 Insurance Contracts beyond "Alive" and "Dead"

Many forms of insurance based on a person's life history involve events more complicated than moving from being "alive" to being "dead". Some examples are the following:

- An active member of a pension scheme may cease to be an active member in more than one way. Typically, they might:
  - retire at a normal retirement age in good health (note that "normal" retirement may fall into some range of ages, for example 55 to 75)
  - retire early because of poor health
  - die before reaching retirement
  - leave the scheme because they change employer.
- A critical illness insurance policy (also known as dread disease insurance) will pay a sum assured upon diagnosis or occurrence of a defined list of diseases. The list generally includes heart attacks, stroke, malignant tumours and other less common but serious conditions.
- A life insurance policy may pay the death benefit before death actually occurs, if the policyholder is suffering from a terminal condition. This is called *benefit acceleration*.
- Disability insurance (also known as income protection insurance) pays an income to the policyholder if he or she is unable to work because of illness or injury. In particular it is sometimes necessary to model a reversible event, such as falling ill and then recovering to good health.
- Long-term care (LTC) insurance pays for the cost of care, either in the policyholder's own home or in a residential home, if the policyholder becomes unable to look after themselves. Usually this occurs in old age. The criteria for claiming may be based on physical infirmity (often expressed as the loss

255

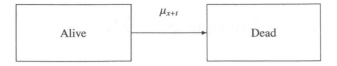

Figure 14.1 A two-state model of mortality.

of ability to carry out a specified number of *activities of daily living* (ADLs) such as washing, dressing, feeding or using the toilet) or it may be based on loss of cognitive capacity. A person may pass through several stages of infirmity, each more severe than the last (and more costly), so a model suitable for use with LTC insurance must allow for transfer between successively worse states of disability, as well as death.

14.2 Multiple-State Models for Life Histories

These and other forms of insurance mean that we must look for models of *life histories* including many possible events besides death. A natural starting point is the alternative formulation of a model for an individual's future lifetime that we introduced in Section 3.5. For convenience, we reproduce the figure from that section here, as Figure 14.1. The idea is that, at any particular time, an individual occupies one of a set of *states* that describe their current status, for example: healthy, ill, retired normally, retired because of ill health, requiring care at home, requiring care in a nursing home, or dead. The collection of states is called the *state space*. The individual's life history will be represented by time spent in states and movements between states. Any such model is called a *multiple-state model*, and Figure 14.1 shows the simplest non-trivial example. Figure 14.2 shows the only other two-state model. Figure 14.3 shows a three-state example, with applications to life insurance, while Figure 14.4 shows a three-state example with applications to disability insurance. The new feature introduced by Figures 14.2 and 14.4 is that of life histories that can both enter and leave the same state, possibly repeatedly.

The life history represented by a multiple-state model can be described in several ways. We suppose that there are $M + 1$ states labelled $0, 1, \ldots, M$, and we consider the ith of n individuals.

- The simplest, but least useful, formulation is to define a process $J_i(x + t)$ to be the label of the state occupied at age $x + t$.

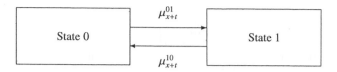

Figure 14.2 A two-state model with reversible transition.

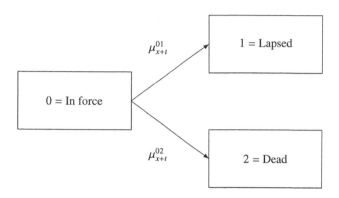

Figure 14.3 A three-state model of mortality and lapse of life insurance policies.

- A variation of the state occupied is to define indicator processes of state occupancy. For $j = 0, 1, \ldots, M$ define $Y_i^j(t) = 1$ if the state occupied immediately before age $x + t$ by the ith individual is state j, and $Y_i^j(t) = 0$ otherwise. The time "immediately before" age $x + t$ is a technicality which we met in Chapter 8 in the denominators of non-parametric estimators, and to which we will return in Chapter 17. It is denoted by $x+t^-$. These indicator processes will be very useful indeed in later chapters.

- Alternatively, instead of tracking state occupancy, we could count transitions. For each pair $j \neq k$ of state labels, define $N_i^{jk}(t)$ to be the number of transitions from state j to state k made by the ith individual up to and including age $x + t$.

In Chapter 17, the last two of these will be central to the theory of counting processes

Our first task, along now-familiar lines, is to specify a mathematical model of these life histories. To do so, we extend the ideas behind Figure 14.1 to allow for more than two states, and hazard rates governing transitions between

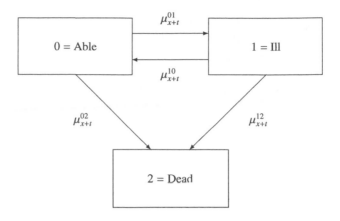

Figure 14.4 A model of illness and death.

states. This we will do in this chapter. The resulting model is probabilistic
but describes complete life histories without reference to observed data, much
as Chapter 3 introduced a mathematical model of human lifetimes. Then, in
Chapter 15 we will consider what data we may observe, allowing for left-
truncation and right-censoring, and formulate a probabilistic model capable of
generating the observed data (compare with Chapter 4).

14.3 Definitions

In Chapter 3, we specified a model in terms of the random lifetime T_x and its
distribution $F_x(t)$, and defined the hazard rate μ_{x+t}. Given $F_x(t)$ we can find
μ_{x+t} (equation (3.18)) and given μ_{x+t} we can find $F_x(t)$ (equation (3.21)), so
either serves to fully specify the model. In Section 3.5 we saw how the hazard
rate could specify a model of transition from the state of being alive to the state
of being dead. The model in Chapter 3 was so simple that it hardly mattered
whether we took $F_x(t)$ or μ_{x+t} as our starting point. When we turn to more
general models, in particular models with reversible transitions, it matters a
great deal.

We will motivate the development of this chapter by using as an example the
illness–death model shown in Figure 14.4. which might be used to model disability insurance. The basic idea is that the policyholder would pay premiums

while able, would receive a regular income while ill (the benefit) and might also receive a death benefit.

In general, suppose there are $M + 1$ states ($M \geq 1$) labelled $0, 1, \ldots, M$. (We could also label states in other ways, such as "a" for alive, "i" for ill and "d" for dead. Labelling states using numbers makes it easier to write down sums of quantities over states, which we will need often.) Figure 14.4 shows a model with three states labelled 0, 1 and 2. Transitions between pairs of states are, where possible, governed by a suitable hazard rate, in this context usually called a *transition intensity*. For example:

- It is possible for someone in state 0 to move to state 1, so the figure depicts this by an arrow, and shows the relevant transition intensity which at age $x + t$ is denoted by μ^{01}_{x+t}.
- It is not possible for someone in state 2 to move to state 1, so the figure omits the corresponding arrow and transition intensity. An alternative way to think about this is that the arrow is there but the transition intensity μ^{21}_{x+t} is zero for all $x + t$.

We begin by generalising the traditional $_tp_x$ and μ_{x+t} notations to any multiple-state model. Consider an individual at age x.

Definition 14.1 Define $_tp^{ij}_x$ to be the probability that an individual in state i at age x will be in state j at age $x + t$.

The probabilities $_tp^{ij}_x$ are called the *occupancy probabilities* of the model. If $i = j$, then $_tp^{ii}_x$ is *not* in general the probability that the individual never leaves state i, only that they were in state i at the given times.

Definition 14.2 Define $_tp^{\overline{ii}}_x$ to be the probability that an individual in state i at age x will remain in state i until at least age $x + t$.

If it is impossible to return to state i after leaving it, then $_tp^{ii}_x = {}_tp^{\overline{ii}}_x$. Otherwise, as long as $_tp^{\overline{ii}}_x < 1$ we must have $_tp^{\overline{ii}}_x < {}_tp^{ii}_x$, since remaining in state i is then just one of several ways to be in state i at ages x and $x + t$. The probabilities $_tp^{ii}_x$ and $_tp^{\overline{ii}}_x$ are analogues of the traditional $_tp_x$, while the probabilities $_tp^{ij}_x$ with $i \neq j$ are analogues of the traditional $_tq_x$.

Definition 14.3 For $i \neq j$, define μ^{ij}_{x+t} to be the transition intensity from state i to state j at age $x + t$, as:

$$\mu^{ij}_{x+t} = \lim_{dt \to 0} \frac{_{dt}p^{ij}_{x+t}}{dt}. \tag{14.1}$$

It is interpreted exactly as μ_{x+t} was: for small dt, $_{dt}p^{ij}_{x+t} = \mu^{ij}_{x+t}dt + o(dt)$.

We will always assume that the occupancy probabilities are well behaved so that this limit exists. This is not a serious restriction.

In Chapter 3, the multiplicative property of the survival function, equation (3.6), was important. Among other things, it meant that knowledge of $S_x(t)$ meant that we also knew $S_y(t)$ for all $y > x$. The corresponding property for multiple-state models belongs not to individual occupancy probabilities but to the matrix of occupancy probabilities. Define the $(M+1) \times (M+1)$ matrix $\boldsymbol{P}_x(t)$ as follows:

$$
\begin{bmatrix}
{}_tp_x^{00} & {}_tp_x^{01} & \cdots & {}_tp_x^{0M} \\
{}_tp_x^{10} & {}_tp_x^{11} & \cdots & {}_tp_x^{1M} \\
\cdots & \cdots & \cdots & \cdots \\
{}_tp_x^{M0} & {}_tp_x^{M1} & \cdots & {}_tp_x^{MM}
\end{bmatrix}.
\tag{14.2}
$$

Then the analogue of equation (3.6) is the following, known as the Chapman–Kolmogorov equations:

$$
\boldsymbol{P}_x(t+s) = \boldsymbol{P}_x(t)\,\boldsymbol{P}_{x+t}(s),
\tag{14.3}
$$

in which the order of multiplication matters because matrix multiplication is non-commutative: in general $\boldsymbol{P}_{x+t}(s)\,\boldsymbol{P}_x(t) \neq \boldsymbol{P}_x(t)\,\boldsymbol{P}_{x+t}(s)$. Thus, knowledge of $\boldsymbol{P}_x(t)$ means that we also know $\boldsymbol{P}_y(t)$ for all $y > x$.

14.4 Examples

The single-decrement model illustrated in Figure 14.1 has two states, so:

$$
\boldsymbol{P}_x(t) = \begin{bmatrix} {}_tp_x^{00} & {}_tp_x^{01} \\ {}_tp_x^{10} & {}_tp_x^{11} \end{bmatrix}.
\tag{14.4}
$$

Consideration of impossible transitions shows that in fact:

$$
\boldsymbol{P}_x(t) = \begin{bmatrix} {}_tp_x^{00} & {}_tp_x^{01} \\ 0 & 1 \end{bmatrix}.
\tag{14.5}
$$

Applying equation (14.3), we get:

$$
\boldsymbol{P}_x(t+s) = \boldsymbol{P}_x(t)\,\boldsymbol{P}_{x+t}(s) = \begin{bmatrix} {}_tp_x^{00} & {}_tp_x^{01} \\ 0 & 1 \end{bmatrix} \begin{bmatrix} {}_sp_{x+t}^{00} & {}_sp_{x+t}^{01} \\ 0 & 1 \end{bmatrix}
\tag{14.6}
$$

$$
= \begin{bmatrix} {}_tp_x^{00}\,{}_sp_{x+t}^{00} & ({}_tp_x^{00}\,{}_sp_{x+t}^{01} + {}_tp_x^{01}) \\ 0 & 1 \end{bmatrix},
\tag{14.7}
$$

which just expresses the elementary relationships of the life table.

Next, consider the two-decrement model shown in Figure 14.3. The insurance context is that life insurance policies are said to be "in force" while the policyholder continues to pay premiums and the insurer remains at risk in the event of death. Policies may cease to be in force before the policy term has expired because the policyholder either dies, or ceases to pay premiums and the policy *lapses*. Lapse rates are usually very much higher than mortality rates. This model has three states, so:

$$
\boldsymbol{P}_x(t) = \begin{bmatrix} {}_tp_x^{00} & {}_tp_x^{01} & {}_tp_x^{02} \\ {}_tp_x^{10} & {}_tp_x^{11} & {}_tp_x^{12} \\ {}_tp_x^{20} & {}_tp_x^{21} & {}_tp_x^{22} \end{bmatrix} \tag{14.8}
$$

and consideration of impossible transitions shows that in fact:

$$
\boldsymbol{P}_x(t) = \begin{bmatrix} {}_tp_x^{00} & {}_tp_x^{01} & {}_tp_x^{02} \\ 0 & 1 & 0 \\ 0 & 0 & 1 \end{bmatrix}. \tag{14.9}
$$

Finally, consider the two-state model shown in Figure 14.2. We might suppose, for example, that this represents periods spent in good health (state 0) and in poor helath (state 1), ignoring the possibility of death. The transition probability matrix for this model is the same as in equation (14.4), but this time the simplification to equation (14.5) does not happen, unless $\mu_{x+t}^{10} = 0$ for all t.

Now, if we apply equation (14.3), we get:

$$
\boldsymbol{P}_x(t+s) = \begin{bmatrix} ({}_tp_x^{00}\,{}_sp_{x+t}^{00} + {}_tp_x^{01}\,{}_sp_{x+t}^{10}) & ({}_tp_x^{00}\,{}_sp_{x+t}^{01} + {}_tp_x^{01}\,{}_sp_{x+t}^{11}) \\ ({}_tp_x^{10}\,{}_sp_{x+t}^{00} + {}_tp_x^{11}\,{}_sp_{x+t}^{10}) & ({}_tp_x^{10}\,{}_sp_{x+t}^{01} + {}_tp_x^{11}\,{}_sp_{x+t}^{11}) \end{bmatrix}. \tag{14.10}
$$

It is easy to see from this that in the general multiple-state model with $M + 1$ states, the Chapman–Kolmogorov equations imply the following identity for the ijth element of the transition probability matrix $\boldsymbol{P}_x(t + s)$:

$$
{}_{t+s}p_x^{ij} = \sum_{k=0}^{M} {}_tp_x^{ik}\,{}_sp_{x+t}^{kj}, \tag{14.11}
$$

which amounts to conditioning on the state occupied at time t.

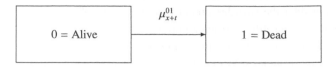

Figure 14.5 A two-state model of mortality, specified in terms of its intensity.

14.5 Markov Multiple-State Models

We will now properly define our multiple-state model, and state something we have glossed over until now, namely that we are restricting our attention to Markov multiple-state models. Suppose there are $M + 1$ states, labelled $0, 1, \ldots, M$, and that transitions between states represent particular events. The occupancy probabilities $_t p_x^{ij}$ and transition intensities are defined as in Section 14.3. The properties of the model then follow from these three assumptions:

Assumption 14.4 The model has the *Markov* property. This means that the probabilities of events after time t depend only on the state occupied at time t and not on any other information known at time t. In other words, we may know the entire past history of transitions between states, and possibly other information too, but none of it matters except the state occupied at time t.

Assumption 14.5 For every pair of distinct states $i \neq j$, occupancy probabilities and transition intensities are related as follows: $_{dt} p_{x+t}^{ij} = \mu_{x+t}^{ij} dt + o(dt)$. Alternatively, assume the limits $\lim_{dt \to 0^+}\ _{dt} p_{x+t}^{ij} / dt$, $(i \neq j)$ exist and are finite; they then define the intensities μ_{x+t}^{ij}.

Assumption 14.6 The probability that two or more transitions take place in time dt is $o(dt)$.

Figure 14.5 shows the single-decrement model, specified in terms of its intensity. Clearly Assumption 14.6 is not needed here. Note that in Chapter 3 we must at some point have introduced assumptions equivalent to Assumption 14.4 above, since we did end up with a Markov model. In fact it was the consistency condition in equation (3.3), which led to equation (3.6), which in turn generalises to the Chapman–Kolmogorov equations (14.3). The latter are a defining characteristic of Markov models.

The Markov property is not trivial even in this simple model; it means that no information except the age and continuing survival of the subject is relevant to the probability that they might die. This is reasonable if no other information is available, but perhaps not if we know about the person's health, habits or occupation. In general, there are two approaches to incorporating other information. One is to sub-divide the population being studied into more homogeneous subgroups, within each of which the Markov assumption is more reasonable. This is a form of stratification. The other is to make the transition intensities depend on the additional information in ways that preserve the computational simplicity of the Markov model. For example, if the hazard of death is a specified function of smoking status, then, conditional on knowing the smoking status, all calculations go through as easily as in the Markov model. This leads to exactly the same choices as were discussed in Chapter 7.

We like the Markov property because it leads to much simpler mathematics, not because it is difficult to specify intensities that depend on factors other than age. A common example is when the intensities depend on age and on the time spent in the current visit to the currently occupied state. Such a model is called *semi-Markov*, and and we will discuss these briefly in Section 14.10. Luckily, the Markov assumption is often found to be consistent with the data, but this can be decided *only* by examining the data. It may be tempting to go straight for a Markov model because it is easiest, but if the data suggest strong dependence on information other than the currently occupied state, the results could be quite misleading. Actuaries have long been accustomed to dealing with this: the select life table is a discrete time semi-Markov model.

It is worth looking at this point in slightly more detail. The Markov model as specified here is a probabilistic model capable of generating the observed data. This means that:

- any life history that we could conceivably observe can happen in the model; and

- all life histories that are impossible (for example, return from the dead) cannot happen in the model.

However, this does not say that a Markov model can be parameterised in a way consistent with the observations. That is, it may be that *no* set of transition intensities μ_{x+t}^{ij} depending only on age and the state currently occupied can be found, under which the probability assigned to the observed data is acceptably high. If so, this will be discovered at the inference stage, by attempting to fit a Markov model and testing its goodness-of-fit. This highlights again that the probabilistic and the inferential (or statistical) aspects of the modelling process are distinct.

14.6 The Kolmogorov Forward Equations

Associated with the simple alive–dead model is the differential equation (3.19). Given that $_tp_x + {}_tq_x = 1$, all non-trivial aspects of the model are represented by the following pair of differential equations (in the notation of multiple-state models):

$$\frac{\partial}{\partial t}{}_tp_x^{00} = -{}_tp_x^{00}\mu_{x+t}^{01} \tag{14.12}$$

$$\frac{\partial}{\partial t}{}_tp_x^{01} = {}_tp_x^{00}\mu_{x+t}^{01}. \tag{14.13}$$

There are also two trivial differential equations, making four in all:

$$\frac{\partial}{\partial t}{}_tp_x^{10} = \frac{\partial}{\partial t}{}_tp_x^{11} = 0. \tag{14.14}$$

The last two are self-evident, because the "dead" state is *absorbing* (also called a *coffin state*) and in future we will omit differential equations for absorbing states.

This system of differential equations generalises to any Markov multiple-state model. A model with $M + 1$ states defines $(M + 1)^2$ possible occupancy probabilities, and therefore a system of $(M + 1)^2$ differential equations, one for each occupancy probability. In most examples of actuarial interest, many of these will be trivial because of impossible transitions, as in the alive–dead model above. The system of equations is called the *Kolmogorov forward equations*.

The Kolmogorov equations are useful because they form a system of linear ordinary differential equations (ODEs) specifying occupancy probabilities in terms of transition intensities. So if a model has been defined in terms of transition intensities, but occupancy probabilities are needed in applied work, they are the necessary link. It may not be possible to solve them explicitly, but linear ODEs are usually straightforward to solve numerically.

In this section, we will derive the Kolmogorov forward equations, first for the alive–dead model to illustrate the technique, and then in the general case. Before considering methods of solving the Kolmogorov equations, in Section 14.7 we compare them with the alternative approach we might consider for obtaining occupancy probabilities. By analogy with equations (3.18) and (3.21) the choices we have are:

- integrate over the joint density of random times, representing the times of movements between states;

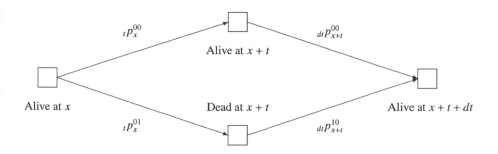

Figure 14.6 Possible paths from "alive" at age x to "alive" at age $x + t + dt$.

- formulate and solve differential equations for the occupancy probabilities.

We shall see that the first of these can be much harder than the second; that is why we focus on the Kolmogorov equations.

14.6.1 The Single-Decrement Model

We now have to derive equations (14.12) and (14.13) using the assumptions of the multiple-state model (we ignore their two trivial companions). It will be enough to derive equation (14.12), because $_tp_x^{00} + {_tp_x^{01}} = 1$.

Proposition 14.7 *In the single-decrement Markov model, equation (14.12) holds.*

Proof: Notice first that $_tp_x^{00}$ is well defined because of the Markov property. That is, how the life got to be in state 0 at age x is irrelevant. We consider the interval of age $x + t$ to $x + t + dt$, and ask: how can a life alive at age x become a life alive at age $x + t + dt$? Figure 14.6 shows all possible routes through the states that could be occupied at age $x + t$.

Figure 14.6 appears to show a possible transition from dead to alive. We could have omitted this. It is better not to at this stage, for two reasons:

- Direct transitions from state i to state j that are impossible will automatically be taken care of when the transition intensity $\mu_{x+t}^{ij} = 0$ appears in the derivation.
- In more complicated models, indirect transition between two states may be possible via other states, even if direct transition is not possible.

Conditioning on the state occupied at age $x + t$ (the Chapman–Kolmogorov equations):

$$
\begin{aligned}
_{t+dt}p_x^{00} &= (_tp_x^{00} \times {_{dt}}p_{x+t}^{00}) + (_tp_x^{01} \times {_{dt}}p_{x+t}^{10}) \\
&= {_t}p_x^{00} \times (1 - {_{dt}}p_{x+t}^{01}) + (_tp_x^{01} \times 0) \\
&= {_t}p_x^{00} \times (1 - \mu_{x+t}^{01}\,dt + o(dt)) \quad \text{by Assumption 14.5.}
\end{aligned}
$$

Thus:

$$
_{t+dt}p_x^{00} - {_t}p_x^{00} = -{_t}p_x^{00}\,\mu_{x+t}^{01}\,dt + o(dt).
$$

Now divide both sides by dt and let $dt \to 0$. We find:

$$
\lim_{dt \to 0^+} \frac{_{t+dt}p_x^{00} - {_t}p_x^{00}}{dt} = -{_t}p_x^{00}\,\mu_{x+t}^{01} + \lim_{dt \to 0^+} \frac{o(dt)}{dt}
$$

$$
\Rightarrow \frac{\partial}{\partial t}{_t}p_x^{00} = -{_t}p_x^{00}\,\mu_{x+t}^{01}. \tag{14.15}
$$

□

Apart from the change in notation, this is the same as equation (3.19), so the same method of solution gives:

$$
_tp_x^{\overline{00}} = {_t}p_x^{00} = \exp\left(-\int_0^t \mu_{x+s}^{01}\,ds\right). \tag{14.16}
$$

The presence of $_tp_x^{\overline{00}}$ on the left of this equation is advance warning of the fact that, in more general models with reversible transitions, it generalises to the probabilities $_tp_x^{\overline{ii}}$ and *not* to the probabilities $_tp_x^{ii}$. In the single-decrement model it makes no difference.

14.6.2 The General Multiple-State Model

Now consider any multiple-state model, with $M + 1$ states and arbitrary allowable transitions. We need not distinguish between possible and impossible transitions in proving any results; an impossible transition is simply one for which the corresponding intensity is zero. First we prove our main result, the Kolmogorov forward equations.

Proposition 14.8 *For any states i and j, not necessarily distinct, in the general Markov multiple-state model:*

$$\frac{\partial}{\partial t}{}_t p_x^{ij} = \sum_{k \neq j} {}_t p_x^{ik} \mu_{x+t}^{kj} - {}_t p_x^{ij} \sum_{k \neq j} \mu_{x+t}^{jk}. \tag{14.17}$$

Proof: We just follow the same steps as in Proposition 14.7. Consider ${}_{t+dt}p_x^{ij}$, and condition on the state occupied at age $x+t$. This could be any of $0, 1, \ldots, M$, so:

$$
\begin{aligned}
{}_{t+dt}p_x^{ij} &= \sum_{k=0}^{M} {}_t p_x^{ik} \, {}_{dt}p_{x+t}^{kj} \\
&= \sum_{k \neq j} {}_t p_x^{ik} \, {}_{dt}p_{x+t}^{kj} + {}_t p_x^{ij} \, {}_{dt}p_{x+t}^{jj} \\
&= \sum_{k \neq j} {}_t p_x^{ik} \, {}_{dt}p_{x+t}^{kj} + {}_t p_x^{ij} \left(1 - \sum_{k \neq j} {}_{dt}p_{x+t}^{jk} \right) \\
&= {}_t p_x^{ij} + \sum_{k \neq j} {}_t p_x^{ik} (\mu_{x+t}^{kj} \, dt + o(dt)) - {}_t p_x^{ij} \sum_{k \neq j} (\mu_{x+t}^{jk} \, dt + o(dt)). \tag{14.18}
\end{aligned}
$$

Therefore:

$$
\begin{aligned}
\frac{\partial}{\partial t}{}_t p_x^{ij} &= \lim_{dt \to 0^+} \frac{{}_{t+dt}p_x^{ij} - {}_t p_x^{ij}}{dt} \tag{14.19} \\
&= \sum_{k \neq j} {}_t p_x^{ik} \mu_{x+t}^{kj} - {}_t p_x^{ij} \sum_{k \neq j} \mu_{x+t}^{jk} + \lim_{dt \to 0^+} \frac{o(dt)}{dt} \\
&= \sum_{k \neq j} {}_t p_x^{ik} \mu_{x+t}^{kj} - {}_t p_x^{ij} \sum_{k \neq j} \mu_{x+t}^{jk},
\end{aligned}
$$

as required. $\square$

Next, we generalise equation (14.16). We need a preliminary result. In general, ${}_t p_x^{ii} \neq {}_t p_x^{\overline{ii}}$, but over short time intervals they are almost the same, thanks to Assumption 14.6.

Proposition 14.9 *For any state i in the general Markov multiple-state model:*

$$_{dt}p_x^{ii} = {}_{dt}p_x^{\overline{ii}} + o(dt). \tag{14.20}$$

Proof: If the subject is in state i at ages x and $x + dt$ then either there were no transitions in between, with probability ${}_{dt}p_x^{\overline{ii}}$, or there were at least two transitions in between, with probability that is $o(dt)$, and these two possibilities are mutually exclusive. $\square$

Proposition 14.10 *For any state i in the general Markov multiple-state model:*

$$_tp_x^{\overline{ii}} = \exp\left(-\int_0^t \sum_{j\ne i}\mu_{x+s}^{ij}ds\right).$$
(14.21)

Proof: We use the same conditioning argument as before: consider $_{t+dt}p_x^{\overline{ii}}$, and condition on the fact that the state occupied at age $x+t$ was state i.

$$
\begin{aligned}
_{t+dt}p_x^{\overline{ii}} &= {_tp_x^{\overline{ii}}} \times {_{dt}p_{x+t}^{\overline{ii}}}\\
&= {_tp_x^{\overline{ii}}} \times ({_{dt}p_{x+t}^{ii}} + o(dt)) \qquad \text{by Proposition 14.9}\\
&= {_tp_x^{\overline{ii}}} \times \left(1 - \sum_{j\ne i}{_{dt}p_{x+t}^{ij}}\right) + o(dt).
\end{aligned}
$$
(14.22)

Therefore:

$$
\begin{aligned}
\frac{\partial}{\partial t}{_tp_x^{\overline{ii}}} &= \lim_{dt\to 0^+}{_tp_x^{\overline{ii}}}\frac{_{t+dt}p_x^{\overline{ii}} - {_tp_x^{\overline{ii}}}}{dt}\\
&= {_tp_x^{\overline{ii}}}\lim_{dt\to 0^+}\frac{1}{dt}\left(-\sum_{j\ne i}{_{dt}p_{x+t}^{ij}} + o(dt)\right)\\
&= {_tp_x^{\overline{ii}}}\lim_{dt\to 0^+}\frac{1}{dt}\left(-\sum_{i\ne i}(\mu_{x+t}^{ij}dt + o(dt)) + o(dt)\right)\\
&= -{_tp_x^{\overline{ii}}}\sum_{j\ne i}\mu_{x+t}^{ij} + \lim_{dt\to 0^+}\frac{o(dt)}{dt}\\
&= -{_tp_x^{\overline{ii}}}\sum_{j\ne i}\mu_{x+t}^{ij}.
\end{aligned}
$$
(14.23)

(14.24)

This is an ODE of the same form as equation (3.19), with boundary condition $_0p_x^{\overline{ii}} = 1$, so the same method of solution gives:

$$_tp_x^{\overline{ii}} = \exp\left(-\int_0^t \sum_{j\ne i}\mu_{x+s}^{ij}ds\right)$$
(14.25)

as required.

14.7 Why Multiple-State Models and Intensities?

We have set out notation which extends the ordinary life table to more general settings, and found the analogues of the basic life table relationships in the form of matrix multiplication. We have yet to justify the claim that the transition intensities (hazards), rather than the occupancy probabilities, should be taken to be the elementary quantities in the model.

Here we give the first justification (the second, ease of statistical inference, is the subject of Chapters 15 and 16). We ask what are the analogues, in the multiple-state model, of calculations such as equations (3.18) and (3.14) based on the ordinary life table? As noted before, it is often necessary to be able to calculate probabilities in terms of hazards and *vice versa*, or, in the terminology of multiple-state models, occupancy probabilities in terms of transition intensities and *vice versa*.

Consider the two-state model shown in Figure 14.2, and let us try to write down directly the analogue of equation (3.18), which for convenience is reproduced below:

$$t q_x = \int_0^t {}_s p_x \, \mu_{x+s} \, ds. \tag{14.26}$$

Consider the possible life histories that the individual may experience in the interval $[0, t]$. We may express them as a series of random times at which transitions occur. The key point is that there is no limit to the possible number of transitions in any finite time, so we have:

- histories with no transitions;
- histories with one transition, and one random time, denoted by T^1;
- histories with two transitions, and two random times, denoted by T^1, T^2;
- in general, for all $N > 0$, histories with N transitions, and N random times, denoted by $T^1, T^2, \ldots, T^N$.

There will also be time spent in the final state between the last transition and the end of observation, which we denote by T^∞. The enormous increase in complexity, for the addition of just one transition to the alive–dead model, is already apparent. To write down the analogue of equation (14.26) for ${}_t p_x^{01}$, we have to allow for all possible life histories in which the individual was in state 0 at time 0 and in state 1 at time t; in other words, all life histories with an odd number of transitions. Each such possibility contributes a term, of which we display just the first two below:

$$_tp_x^{01} = \int_0^t {}_sp_x^{\overline{00}}\, \mu_{x+s}^{01}\, {}_{t-s}p_{x+s}^{\overline{11}}\, ds$$

$$+ \int_0^t {}_sp_x^{\overline{00}}\, \mu_{x+s}^{01} \int_s^t {}_rp_{x+s}^{\overline{11}}\, \mu_{x+s+r}^{10} \int_{s+r}^t {}_xp_{x+s+r}^{\overline{00}}\, \mu_{x+s+r+v}^{01}\, {}_{t-s-r-v}p_{x+s+r+v}^{\overline{11}}\, dv\, dr\, ds$$

+ infinitely many other terms ...   (14.27)

The first term above is obtained by integrating over the distribution of T^1, in the case that there is only one transition. The reasoning is as follows:

- There was only one transition, so it must have been at some time s between 0 and t.
- That event is described by the individual:
  - not leaving state 0 between times 0 and s;
  - moving to state 1 at time s; then
  - not leaving state 1 between times s and t.

  The probability of this event is ${}_sp_x^{\overline{00}}\, \mu_{x+s}^{01}\, {}_{t-s}p_{x+s}^{\overline{11}}\, ds$.
- Now integrate over all possible times s.

The second term above is obtained by integrating over the joint distribution of T^1, T^2 and T^3. The third term, which we have omitted, would be obtained by integrating over the joint distribution of T^1, T^2, T^3, T^4 and T^5, and so on.

Thus, the analogue of equation (14.26) is an infinite series, the nth term of which is a multiple integral of dimension $2n - 1$. The numerical solution of high-dimensional integrals is not trivial, so even evaluating the first few terms of this series quickly becomes impractical. However, corresponding to equation (3.19) are the Kolmogorov forward equations, which as a further example we display in full:

$$\frac{\partial}{\partial t}{}_tp_x^{00} = {}_tp_x^{01}\, \mu_{x+t}^{10} - {}_tp_x^{00}\, \mu_{x+t}^{01}$$

$$\frac{\partial}{\partial t}{}_tp_x^{01} = {}_tp_x^{00}\, \mu_{x+t}^{01} - {}_tp_x^{01}\, \mu_{x+t}^{10}$$

$$\frac{\partial}{\partial t}{}_tp_x^{10} = {}_tp_x^{11}\, \mu_{x+t}^{10} - {}_tp_x^{10}\, \mu_{x+t}^{01}$$

$$\frac{\partial}{\partial t}{}_tp_x^{11} = {}_tp_x^{10}\, \mu_{x+t}^{01} - {}_tp_x^{11}\, \mu_{x+t}^{10}.$$ (14.28)

The appropriate boundary conditions are ${}_0p_x^{00} = {}_0p_x^{11} = 1$ and ${}_0p_x^{01} = {}_0p_x^{10} = 0$.

In fact we could dispense with two of these equations, because $_tp_x^{00} + _tp_x^{01} = _tp_x^{10} + _tp_x^{11} = 1$. The system is linear and first-order, and its numerical solution is straightforward, for any given intensities (see Section 14.8.2).

Clearly the problems of evaluating horrors like equation (14.27) will just get worse in more complicated models. Not so with the Kolmogorov equations: such a system exists for *any* multiple-state model with *any* number of states and transitions, and solving it numerically in complex models is not more difficult than in simple models. This is a strong reason to take the intensities as the fundamental quantities in the model, rather than the analogues of random lifetimes – the times between transitions – and their distributions. It is also the case (see Chapter 15) that intensities are usually simpler to estimate with the kind of data often available in actuarial work. Other, more subtle, complications with random lifetimes will appear in Chapter 16.

14.8 Solving the Kolmogorov Equations

14.8.1 Explicit Solutions with Simple Intensities

The Kolmogorov equations are linear first-order ODEs for the occupancy probabilities, in which the intensities appear as coefficients. The solution of such equations is routine. Sometimes, if the intensities are simple, they can be solved explicitly. Since they can always be solved numerically, however, there is no need always to look for simple intensities. We give an example of explicit solutions here, and some general numerical methods in Section 14.8.2.

Consider the modelling of life insurance policies subject to lapse risk as well as mortality risk. Figure 14.7 shows a three-state model of this process, similar to Figure 14.3 but in which the intensities are constant, not depending on age. The non-trivial Kolmogorov forward equations are found to be:

$$\frac{\partial}{\partial t}{}_tp_x^{00} = -{}_tp_x^{00}(\mu^{01} + \mu^{02}) \tag{14.29}$$

$$\frac{\partial}{\partial t}{}_tp_x^{01} = {}_tp_x^{00}\mu^{01} \tag{14.30}$$

$$\frac{\partial}{\partial t}{}_tp_x^{02} = {}_tp_x^{00}\mu^{02}, \tag{14.31}$$

with boundary conditions $_0p_x^{00} = 1$ and $_0p_x^{01} = {}_0p_x^{02} = 0$. Return to state 0 is impossible, so $_tp_x^{00} = {}_tp_x^{\overline{00}}$ and by Proposition 14.10 $_tp_x^{00} = \exp(-\int_0^t(\mu^{01} + \mu^{02})ds) = \exp(-(\mu^{01} + \mu^{02})t)$. Substitute this into equation (14.30) and:

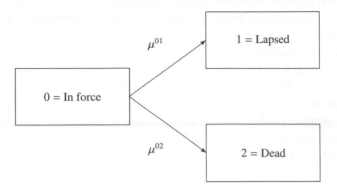

Figure 14.7 A three-state model of mortality and lapse of life insurance policies, with constant intensities.

$$
\begin{aligned}
{}_tp_x^{01} &= \int_0^t \frac{\partial}{\partial s} \, {}_sp_x^{01} \, ds \qquad\qquad (14.32)\\[2mm]
&= \int_0^t {}_tp_x^{00} \mu^{01} \, ds\\[2mm]
&= \int_0^t e^{-(\mu^{01}+\mu^{02})s} \mu^{01} \, ds\\[2mm]
&= \frac{\mu^{01}}{\mu^{01}+\mu^{02}} \left(1 - e^{-(\mu^{01}+\mu^{02})t}\right) + c,
\end{aligned}
$$

and the initial condition shows that $c = 0$. Intuitively this makes sense: $1 - \exp(-(\mu^{01}+\mu^{02})t)$ is the probability of having left state 0, and given that that has happened, the probability of being in state 1 or state 2 is proportional to the magnitude of the two intensities.

However, explicit solutions with simple intensities are the exception rather than the rule, especially if the intensities must be functions of age. Fortunately, numerical solution of the Kolmogorov equations with *any* intensities, as long as they are reasonably well behaved, is straightforward.

14.8.2 Numerical Solutions in the General Case

Solution of ODEs is a standard topic of numerical analysis, and the simplest methods are enough for the Kolmogorov equations. We shall just sketch the ideas and refer to Conte and de Boor (1981) or Press et al. (1986) for details.

The simplest method of all is an *Euler scheme*. Given an ODE of the form:

$$\frac{df(t)}{dt} = g(t, f(t)) \qquad \text{or} \qquad df(t) = g(t, f(t))dt, \qquad (14.33)$$

we discretise it by choosing a small time step Δt and putting:

$$f(t + \Delta t) \approx f(t) + df(t)\Delta t = f(t) + g(t, f(t))\Delta t. \qquad (14.34)$$

For the Kolmogorov equations, we start with an initial condition and $\Delta t > 0$. If we have a system of ODEs, we advance the solution of all of them simultaneously over each time step. As an example, the first two steps in the approximate solution of equations (14.28), with boundary conditions $_0p_x^{00} = _0p_x^{11} = 1$ and $_0p_x^{01} = _0p_x^{10} = 0$, are:

$$\Delta t p_x^{00} \approx _0p_x^{00} + (_0p_x^{01}\mu_x^{10} - _0p_x^{00}\mu_x^{01})\Delta t = 1 - \mu_x^{01}\Delta t$$
$$\Delta t p_x^{01} \approx _0p_x^{01} + (_0p_x^{00}\mu_x^{01} - _0p_x^{01}\mu_x^{10})\Delta t = 0 + \mu_x^{01}\Delta t$$
$$\Delta t p_x^{10} \approx _0p_x^{10} + (_0p_x^{11}\mu_x^{10} - _0p_x^{10}\mu_x^{01})\Delta t = 0 + \mu_x^{10}\Delta t$$
$$\Delta t p_x^{11} \approx _0p_x^{11} + (_0p_x^{10}\mu_x^{01} - _0p_x^{11}\mu_x^{10})\Delta t = 1 - \mu_x^{10}\Delta t \qquad (14.35)$$

and:

$$2\Delta t p_x^{00} \approx \Delta t p_x^{00} + (\Delta t p_x^{01}\mu_{x+\Delta t}^{10} - \Delta t p_x^{00}\mu_{x+\Delta t}^{01})\Delta t$$
$$2\Delta t p_x^{01} \approx \Delta t p_x^{01} + (\Delta t p_x^{00}\mu_{x+\Delta t}^{01} - \Delta t p_x^{01}\mu_{x+\Delta t}^{10})\Delta t$$
$$2\Delta t p_x^{10} \approx \Delta t p_x^{10} + (\Delta t p_x^{11}\mu_{x+\Delta t}^{10} - \Delta t p_x^{10}\mu_{x+\Delta t}^{01})\Delta t$$
$$2\Delta t p_x^{11} \approx \Delta t p_x^{11} + (\Delta t p_x^{10}\mu_{x+\Delta t}^{01} - \Delta t p_x^{11}\mu_{x+\Delta t}^{10})\Delta t. \qquad (14.36)$$

The Euler scheme is not very efficient, however, and there are faster and more accurate methods. One that is still simple is a *fourth-order Runge–Kutta* algorithm. In moving from time t to $t + \Delta t$, Euler's scheme uses only function values at time t. Runge–Kutta methods first estimate function values on the interval $[t, t + \Delta t]$, then use these to obtain an improved estimate of $f(t + \Delta t)$. This particular scheme makes four recursive estimates of the increment in f per time step (note that the superscripts on f in what follows do not denote powers of f):

$$\Delta f^1 = g(t, f(t))\Delta t$$
$$\Delta f^2 = g(t + \Delta t/2, f(t) + \Delta f^1/2)\Delta t$$
$$\Delta f^3 = g(t + \Delta t/2, f(t) + \Delta f^2/2)\Delta t$$
$$\Delta f^4 = g(t + \Delta t, f(t) + \Delta f^3)\Delta t, \qquad (14.37)$$

and then takes as the next function value:

$$f(t + \Delta t) \approx f(t) + \frac{\Delta f^1 + 2\Delta f^2 + 2\Delta f^3 + \Delta f^4}{6}. \qquad (14.38)$$

Taking equation (14.12) as an example, and supposing that we have just reached time t, the next step would be:

$$\Delta p^1 = -{}_t p_x^{00} \mu_{x+t}^{01} \Delta t$$
$$\Delta p^2 = -({}_t p_x^{00} + \Delta p^1/2) \mu_{x+t+\Delta t/2}^{01} \Delta t$$
$$\Delta p^3 = -({}_t p_x^{00} + \Delta p^2/2) \mu_{x+t+\Delta t/2}^{01} \Delta t$$
$$\Delta p^4 = -({}_t p_x^{00} + \Delta p^3) \mu_{x+t+\Delta t} \Delta t \qquad (14.39)$$

and then:

$$_{t+\Delta t} p_x^{00} \approx {}_t p_x^{00} + \frac{\Delta p^1 + 2\Delta p^2 + 2\Delta p^3 + \Delta p^4}{6}. \qquad (14.40)$$

Given a system of ODEs, we advance the solution of all of them simultaneously, not only over each time step as before, but also over each step in the Runge–Kutta algorithm. The Runge–Kutta algorithm ought usually to be sufficient in practice, but many refinements can be made, and the books mentioned above are good starting points.

14.9 Life Contingencies: Thiele's Differential Equations

If life insurance mathematics is first approached via a life table giving values of l_x or ${}_t p_0$ at integer ages, it is natural to develop expected present values (EPVs) of cash-flows arising at these discrete ages. This ignores realities such as deaths taking place at any time, or premiums being paid monthly, that lead to approximate adjustments being made to the EPVs. Almost the whole subject can be summed up by the recursive relation between policy values at integer durations, which is in fact a difference equation. Let $_t V_x$ be the prospective policy value at age $x + t$ under a life insurance contract that commenced at age

x, where t is an integer duration. At the start of the year $[t, t + 1)$, a premium of P_t will be paid if the person is alive, and at the end of the year a sum assured S_t will be paid if they died during the year. Let i be the annual rate of interest. Then:

$$({}_tV_x + P_t)(1 + i) = q_{x+t}\,S_t + p_{x+t}\,{}_{t+1}V_x. \tag{14.41}$$

In words, the policy value plus the premium just received, accumulated to the year-end, is sufficient to pay the expected death benefits and to set up the next year's policy values for the expected survivors.

This can be solved backwards using the fact that the policy value at expiry is equal to any maturity benefit of pure-endowment type then payable. By suitable choice of payments, all EPVs used in practice can be obtained in this way. Since payments need not be constant, and can even depend on the current level of the policy value, this is a much more general approach than using tabulated EPVs of level assurance and annuity payments, and it is trivial to use with a spreadsheet.

Rearranging equation (14.41) makes it clear that it is a difference equation:

$$_{t+1}V_x - {}_tV_x = i\,{}_tV_x + (1 + i)P_t - q_{x+t}(S_t - {}_{t+1}V_x). \tag{14.42}$$

In words, the increment in the policy value per expected survivor is the accumulated premium, plus the interest earned on the policy value, minus the expected loss on deaths, allowing the policy value to offset the payment of the death benefit. The quantity $(S_t - {}_{t+1}V_x)$ is called the *sum-at-risk*.

Equation (14.42) is the link between discrete time and continuous time models, because it generalises to *Thiele's differential equation*. Suppose that premiums are paid continuously at rate P per annum and interest is earned at force of interest δ per annum (in fact both of these can be functions of time). Then:

$$_{t+dt}V_x - {}_tV_x = \delta\,{}_tV_x\,dt + (1 + \delta\,dt)P\,dt - \mu_{x+t}\,dt\,(S_t - {}_{t+dt}V_x). \tag{14.43}$$

Upon dividing both sides by dt and taking the limit, we obtain *Thiele's differential equation*:

$$\frac{d}{dt}{}_tV_x = \delta\,{}_tV_x + P - \mu_{x+t}\,(S_t - {}_tV_x). \tag{14.44}$$

In the general Markov model with $M + 1$ states, suppose that a premium will be paid continuously at rate b_t^i while the life is in state i at time t, and a benefit of b_t^{ij} will be paid immediately upon a transition from state i to state j at time t. Let $_tV_x^i$ be the prospective policy value to be held if the life is in state i at time t, and let the force of interest at time t be δ_t. Then Thiele's differential equations are:

$$\frac{\partial}{\partial t}{_tV_x^i} = \delta_t\,{_tV_x^i} + b_t^i - \sum_{j\neq i}\mu_{x+t}^{ij}(b_t^{ij} + {_tV_x^j} - {_tV_x^i}), \qquad (14.45)$$

one for each state. The sum-at-risk now allows for the possible need to set up a different policy value, not necessarily zero, after a transition to another state.

This is another system of linear first-order ODEs, this time with boundary conditions given by the maturity benefit, if any, in each state. They can be solved backwards using the same numerical methods as we used to solve the Kolmogorov equations, this time with $\Delta t < 0$. In particular, there is no need to restrict models to have simple forms of intensities (constant, piecewise constant, exponential and so on) for use in applications. Periodic payments, rather than continuously paid cash-flows, can easily be handled; the policy value changes by the amount of the payment when it is paid.

Good references for life insurance mathematics are Gerber (1990) and Dickson et al. (2013).

14.10 Semi-Markov Models

The defining feature of a Markov process is that the only part of its history up to time t relevant for the probabilities of events after time t is the state occupied at time t. All other history, such as the nature and times of past transitions, is irrelevant. This is the source of the Markov model's simplicity and tractability. However, it is reasonable to adopt a Markov modelling framework only if the data support this assumption. Sometimes they do, but sometimes they do not.

An example is disability insurance based on life histories of the kind illustrated in Figure 14.4, where recovery from disability is possible. The CMI introduced such a model in the UK for disability insurance sold to individuals (Continuous Mortality Investigation, 1991). It was found reasonable to model the intensities of transitions out of the "able" state as functions of age alone, consistent with a Markov model. However, the intensities of transitions out of the "ill" state showed stronger dependence on duration of illness than on age, although the latter was clear enough that it could not be ignored. Thus, part of the past history – duration since falling ill – and not just occupancy of the "ill" state, had to be included. A model in which at least one transition intensity depends on the state occupied and the duration of that occupation is called a *semi-Markov model*.

Semi-Markov models are less tractable computationally than Markov models, but not so much that they cannot be used. The main complication is that the calculation of certain probabilities, and likewise premium rates and policy values, requires double rather than single integration, or alternatively the solution of integro-differential equations rather than ordinary differential equations. For example, in the Markov model of Figure 14.4 we could write the following expression for $_t p_x^{00}$:

$$_t p_x^{00} = {}_t p_x^{\overline{00}} + \int_0^t {}_s p_x^{01} \, \mu_{x+s}^{10} \, {}_{t-s} p_{x+s}^{\overline{00}} \, ds. \tag{14.46}$$

Intuitively, the first term is the probability of never leaving the "able" state, and the second term is the probability of having left and returned to the "able" state, where the integration is over the time s of the last return from the "ill" state (which must exist).

If we now model the transition intensity from "ill" to "able" as a function of age $x + t$ and duration u since last entering the "ill" state, denoted $\mu_{x+t,u}^{10}$, the corresponding expression is:

$$_t p_x^{00} = {}_t p_x^{\overline{00}} + \int_0^t \left(\int_0^s {}_{s,u} p_x^{01} \, \mu_{x+s,u}^{10} \, du \right) {}_{t-s} p_{x+s}^{\overline{00}} \, ds, \tag{14.47}$$

where $_{s,u} p_x^{01}$ is the probability (strictly, a density in the variable u) that a life in the "able" state at age x should be in the "Ill" state at age $x + s$ with duration of current illness u. A variety of numerical schemes are suggested in Continuous Mortality Investigation (1991) to meet the needs of actuaries using this disability model.

A semi-Markov model does not present any special problems when estimating intensities over single years of age and duration, along the lines of Section 4.6. The number of events in a given range of age and duration, denoted by $D_{x,u}$, say, and the person-years of exposure at those same ranges, denoted by $E_{x,u}^c$, say, furnish the usual occurrence-exposure rate (mortality ratio if the event is death):

$$\hat{r}_{x,u} = \frac{D_{x,u}}{E_{x,u}^c}, \tag{14.48}$$

which has exactly the same statistical properties as its counterparts in the Markov model (see Chapter 15). If it is required to smooth or graduate the resulting intensities, however, the smoothing must be done in two dimensions, which can be difficult. We refer to Continuous Mortality Investigation (1991) for examples.

14.11 Credit Risk Models

In recent years, a new financial application of multiple-state models has arisen, namely the so-called *reduced form credit risk model,* The credit risk of a firm or government is expressed as a *credit rating* issued by a credit rating agency such as Standard & Poor's. Large volumes of financial contracts such as credit derivatives are traded, whose value depends on the *future* credit rating of the firm or government. Thus future changes in credit rating, such as a downgrade from AAA to AA, can have large financial consequences.

The reduced form credit risk model represents an entity's credit rating as presence in a state, and a change in credit rating as a movement between states, governed by a transition intensity. Thus it is identical in form to the models considered in this chapter. The methods used to estimate the intensities have much in common with those to be discussed in Chapter 15, except that in financial economics there is the additional complication of the difference between "real-world" and "risk-neutral" probabilities. We refer to Bielecki and Rutkowski (2002) for details.

15

Inference in the Markov Model

15.1 Introduction

In Chapters 4 and 5 we set out two approaches to modelling survival data. First, we fitted models to complete lifetime data. Then, we estimated the hazard rate for single years of age, and graduated (smoothed) these estimates. In both cases we used parametric models fitted using maximum likelihood. In Section 5.8 we noted a close relationship between the two approaches, deriving from the fact that likelihoods for survival data are functionally similar to likelihoods for Poisson-distributed observations. This is not a surprise because the Poisson distribution is derived as a model of counting events happening as time passes, and in a survival study we are counting events (deaths) as time passes (people get older). That story will be reiterated here in the setting of multiple-state models. Indeed in his classic work Feller (1950) remarked that:

The true nature of the Poisson distribution will become apparent only in connection with the theory of stochastic processes . . .

We will first look at estimating intensities at single years of age, where the assumption that the intensities are constant over single years of age will be needed. Then we will look at fitting parametric models to complete life history data. In both cases the likelihoods will be closely related to the Poisson distribution, for the same reason as stated above: we are counting events during periods of time.

The basic idea is simple. If we consider an "episode" in the life history of the ith individual to be entering state j at age x_i, staying there for some length of time r, then leaving it for state k, we see that each such "episode" contributes a factor of the form:

$$P[\text{``episode''}] = \exp\left(-\int_0^r \sum_{l \neq j} \mu_{x_i+s}^{jl} \, ds\right) \mu_{x_i+r}^{jk} \qquad (15.1)$$

to the likelihood. The exponentiated term is the probability of remaining in state j, and the last term is the intensity responsible for exit from state j. The total likelihood is the product of such terms from all the "episodes" that make up the life history, and, if the life history ends by right-censoring, also a final term of the form:

$$P[\text{final ``episode''}] = \exp\left(-\int_0^r \sum_{l \neq j} \mu_{x_i+s}^{jl} \, ds\right). \qquad (15.2)$$

Simple though this idea is, writing down a probabilistic model capable of describing all possible observed life histories, as we did in Chapters 4 and 5, requires some development of our notation. Part of the reason is that the number of possible transitions may be unbounded. This leads us to specify a probabilistic model using *counting processes*.

15.2 Counting Processes

We have to define quantities that describe the observations we can make on the ith of n individuals. Note that we did not do so in Chapter 14, because that chapter described a purely mathematical model of a single life history. For inference, as before, we need a probabilistic model capable of generating the data we may observe, which consists of individuals making transitions between states at random times, subject to left-truncation and right-censoring.

Suppose we have a multiple-state model with $M + 1$ states, labelled $0, 1, \ldots, M$. Suppose we have n individuals under observation, perhaps at different times. Then for the ith individual define a set of processes $N_i^{jk}(t)$ as follows, for all distinct j and k in $\{0, 1, \ldots, M\}$:

$N_i^{jk}(t) =$ number of *observed* transitions from state j to state k made by

the ith individual up to and including age t. $\qquad (15.3)$

Each $N_i^{jk}(t)$ is a *counting process*. It counts the number of transitions the ith individual makes from state j to state k up to and including age t. By convention, we include transitions made at exact age t. Therefore $N_i^{jk}(t)$ takes non-negative integer values, is non-decreasing, and by our convention is right-continuous.

We will also assume that $N_i^{jk}(t)$ has limits from the left, so its sample paths are *càdlàg* (an acronym of the French *continue à droite, limite à gauche*). It is a stochastic process, because the times of transitions are not known in advance.

We assume that all $N_i^{jk}(0) = 0$. This is equivalent to defining $N_i^{jk}(x + t)$ to be the number of observed transitions in time t since entering observation at age x, if $x = 0$. We will see shortly that entry into observation at non-zero ages (left-truncation) is taken care of in a new way that allows us to drop the age x at left-truncation from the notation.

Because, for given i, the $N_i^{jk}(t)$ all describe events that happen to the same individual, collectively they comprise a *multivariate counting process* that we may denote by $N_i(t)$, with $N_i^{jk}(t)$ being the *jk*th component of $N_i(t)$. For technical reasons, we impose the assumption that no two components of a multivariate counting process can jump simultaneously.

In order for the *i*th individual to be observed to jump from state j to state k at age t, he or she must actually be under observation in state j just before age t, which time we denote by t^-. In Section 5.8 we defined an indicator function $Y_i(x)$, indicating that the *i*th individual was alive at age x. We extend that idea by defining a set of indicator functions as follows, for $j = 0, 1, \ldots, M$:

$$Y_i^j(t) = 1 \text{ if the } i\text{th individual is } observed \text{ in state } j \text{ at age } t^-; \text{ or}$$
$$Y_i^j(t) = 0 \text{ otherwise.} \tag{15.4}$$

At any age t, if the *i*th individual is under observation at age t^-, precisely one of these indicator functions has value 1, and all the others have value 0. If the *i*th individual is not being observed, because of left-truncation or any form of censoring, all $Y_i^j(t) = 0$. This is why we can suppress any age x at left-truncation from the notation, because entry into observation is taken care of by these indicator functions. These indicator functions are also stochastic processes, since their values depend on the times of any transitions, or random censorings, which are not known in advance.

Given all the $N_i^{jk}(t)$, we can reconstruct the $Y_i^j(t)$ and *vice versa*, so either specifics the *i*th individual's observed life history. It is necessary in what follows to have both (see Chapter 17).

The key difference between the processes defined here and those described loosely in Section 14.2 is the presence here of the word *observed*. Chapter 14 set out a probabilistic model of a life history without consideration of what parts of that life history might happen to be observed, much as Chapter 3 set out a probabilistic model of a human lifetime. To proceed to inference, we define what data may be observed, and use the model from Chapter 14 to define a probabilistic model capable of generating those observed data.

In what follows, when we refer to a counting process making a jump, we implicitly assume that it is under observation at the time. Jumps made while not under observation are not counted.

15.3 An Example of a Life History

To make some expressions clearer, we adopt in this chapter a functional notation $\mu^{jk}(t)$ for intensities rather than the subscript notation μ_{x+t}^{jk}.

As an example, we will use the following life history, based on the three-state illness–death model of Figure 14.4:

- An individual enters observation at age $x = 0$.
- She remains healthy until age 40, then falls ill.
- She is ill for two years before recovering at age 42.
- She stays healthy until falling ill again at age 50.
- She remains ill until she dies at age 60.

The processes $N_i^{jk}(t)$ and $Y_i^j(t)$ evolve jointly and dynamically as follows:

- On the age interval $[0, 40]$, the indicator $Y_i^0(t) = 1$ and $Y_i^1(t) = Y_i^2(t) = 0$. Note that the interval is closed on the right because the indicator shows presence in a state (or not) at time t^- just before time t. On the interval $[0, 40)$ all the counting processes are zero: $N_i^{01}(t) = N_i^{10}(t) = N_i^{02}(t) = N_i^{12}(t) = 0$ as no transitions have occurred. Note that the interval is open on the right, because we defined counting processes to have right-continuous paths. The individual is subject to the intensities $\mu^{01}(t)$ and $\mu^{02}(t)$.
- On the age interval $(40, 42]$ the indicator $Y_i^1(t) = 1$ and $Y_i^0(t) = Y_i^2(t) = 0$. On the age interval $[40, 42)$ the counting process $N_i^{01}(t) = 1$ while $N_i^{10}(t) = N_i^{02}(t) = N_i^{12}(t) = 0$. The individual is subject to the intensities $\mu^{10}(t)$ and $\mu^{12}(t)$.
- On the age interval $(42, 50]$ the indicator $Y_i^0(t) = 1$ and $Y_i^1(t) = Y_i^2(t) = 0$. On the age interval $[42, 50)$ the counting processes $N_i^{01}(t) = N_i^{10}(t) = 1$ while $N_i^{02}(t) = N_i^{12}(t) = 0$. The individual is again subject to the intensities $\mu^{01}(t)$ and $\mu^{02}(t)$.
- On the age interval $(50, 60]$ the indicator $Y_i^1(t) = 1$ and $Y_i^0(t) = Y_i^2(t) = 0$. On the age interval $[50, 60)$ the counting process $N_i^{01}(t) = 2$ while $N_i^{10}(t) = 1$ and $N_i^{02}(t) = N_i^{12}(t) = 0$. The individual is again subject to the intensities $\mu^{10}(t)$ and $\mu^{12}(t)$.

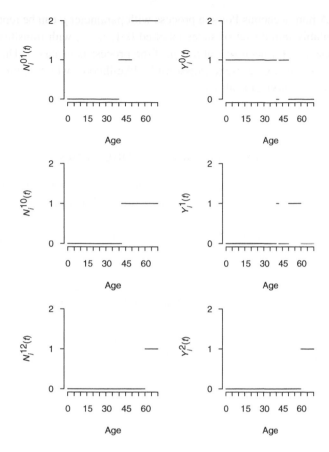

Figure 15.1 Illustration of sample paths of $N_i^{01}(t)$, $N_i^{10}(t)$ and $N_i^{12}(t)$ (left side) and $Y_i^0(t)$, $Y_i^1(t)$ and $Y_i^2(t)$ (right side) for the example in Section 15.3. $N_i^{02}(t)$ is zero for all t and is omitted.

- Finally, on the age interval $(60, \infty)$ the indicator $Y_i^2(t) = 1$ and $Y_i^0(t) = Y_i^1(t) = 0$. On the age interval $[60, \infty)$ the counting process $N_i^{01}(t) = 2$ while $N_i^{10}(t) = N_i^{12}(t) = 1$ and $N_i^{02}(t) = 0$. The individual is subject to no intensities.

Figure 15.1 illustrates these sample paths, omitting $N_i^{02}(t)$ which is zero everywhere. Had these observations been left-truncated, say at age 30, then on the age interval $[0, 30]$ the indicator $Y_i^0(t)$ would have been zero as well as the others. Right-censoring can be accommodated similarly.

The canonical example of a counting process is, not surprisingly, the Poisson process. A homogeneous Poisson process with parameter μ can be represented as a countably infinite set of states labelled $0, 1, 2, \ldots$, with transitions from state j to state $j + 1$ occurring at rate μ, if the process is in state j. This is why counting processes give rise to Poisson-like likelihoods, which in turn feature prominently in survival analysis.

15.4 Jumps and Waiting Times

In Chapter 4, our observations consisted of the random variables D_i, an indicator of observed death, and T_i, the time spent under observation. What are their counterparts in the counting process model just defined?

It is easily seen that the counterpart of T_i is the total time spent by the ith individual in each state, which for state j is:

$$\text{Time spent by } i\text{th individual in state } j = \int_0^\infty Y_i^j(t)\, dt. \qquad (15.5)$$

In stochastic process terminology, time spent in a state is called a *waiting time*. Equally simply, the total time spent by the ith individual in state j between any specified times or ages x_i and x_i' is:

$$\text{Time spent by } i\text{th individual in state } j \text{ between ages } x_i \text{ and } x_i' = \int_{x_i}^{x_i'} Y_i^j(t)\, dt.$$
$$(15.6)$$

The counterpart of D_i, the indicator of observed death of the ith individual, is the collection of jumps or *increments* in the processes N_i^{jk}. Since our convention means that the paths of a counting process are right-continuous, the increment of $N_i^{jk}(t)$ at time t is defined as:

$$dN_i^{jk}(t) = \lim_{dt \to 0} (N_i^{jk}(t) - N_i^{jk}(t - dt)), \qquad (15.7)$$

with the convention that $dN_i^{jk}(0) = 0$. If the ith individual jumps from state j to state k at age t, then $dN_i^{jk}(t) = 1$, otherwise $dN_i^{jk}(t) = 0$. Under some simple conditions on the transition intensity $\mu_i^{jk}(t)$ (Karr, 1991), which we assume are always satisfied, it can be shown that the number of jumps of $N_i^{jk}(t)$ in any finite time is sufficiently well behaved that we can recover the process $N_i^{jk}(t)$ by integrating its increments:

$$N_i^{jk}(t) = \int_0^t dN_i^{jk}(s). \tag{15.8}$$

So the increments of $N_i^{jk}(t)$ give us yet another complete specification of the life history. The integral in equation (15.8) is in fact a sum, because the integrand has value one at the times of any jumps and zero everywhere else:

$$\int_0^t dN_i^{jk}(s) = \sum_{\text{jump times}} 1. \tag{15.9}$$

Using the example in Section 15.3 again, the only non-zero increments of the counting processes are the following:

$$dN_i^{01}(40) = dN_i^{10}(42) = dN_i^{01}(50) = dN_i^{12}(60) = 1 \tag{15.10}$$

and the total waiting times in the "alive" states were:

$$\int_0^\infty Y_i^0(t)\,dt = 48 \text{ years} \tag{15.11}$$

$$\int_0^\infty Y_i^1(t)\,dt = 12 \text{ years}. \tag{15.12}$$

15.5 Aalen's Multiplicative Model

The new idea is as follows. Instead of the transition intensity between states j and k at age t being the deterministic function $\mu^{jk}(t)$ for the ith individual, it is the stochastic process $Y_i^j(t)\mu^{jk}(t)$. The set of counting processes (15.3) and these stochastic intensities give us a probabilistic model "capable of generating any observations" defined in Section 15.4.

This formulation is known as *Aalen's multiplicative model*, because the deterministic hazard rate $\mu^{jk}(t)$ is multiplied by the stochastic indicator $Y_i^j(t)$ and hence is nullified when a transition from state j to state k is impossible. Simple though it is, it is really this idea that accounts for the power and elegance of counting process models in survival analysis.

Equation (3.16) expressed the fundamental link between life-table probabilities and the hazard rate. For counting processes, the equivalent link, and hence interpretation of the intensities, is the following:

$$P\left[\int_t^{t+dt} dN_i^{jk}(s) = 1\right] = P[N_i^{jk}(t+dt) - N_i^{jk}(t) = 1] = Y_i^j(t)\,\mu^{jk}(t)\,dt + o(dt).$$

(15.13)

The left-hand side of equation (15.13) is the probability that the counting process jumps between states j and k in the next dt after age t. It is the direct analogue of the left-hand side of equation (3.16). The right-hand side of equation (15.13) has exactly the same interpretation as in equation (3.16), except that the indicator function $Y_i^j(t)$ is needed to take care of times when the ith individual is not present in state j.

In the simple survival model, equation (3.16) was what allowed us to write down the probabilities of observed events entirely in terms of the hazard function. Similarly, equation (15.13) is the link that allows us to write down the probabilities of observed life histories, hence likelihoods, in the Markov model entirely in terms of the transition intensities.

15.6 The Likelihood for Single Years of Age

We now follow in the path of the Poisson model in Section 4.8 and assume that all intensities are constant over a single year of age. We suppose that the year of age spans x to $x + 1$ and write, for all i, j, k:

$$\mu^{jk}(t) = \mu_x^{jk}, \text{ a constant, for } x < t \leq x + 1.$$

(15.14)

The probabilistic model in Chapter 4 consisted of the observations (D_i, T_i), representing an event and a time. We now have observations represented by the counting process. The events are recorded as the counts $N_i^{jk}(t)$.

The hazard rate of leaving state j, while the ith individual is in state j, is $\sum_{k \neq j} \mu_x^{jk}$. So the total probability of the ith individual spending the observed periods of time in state j between ages x and $x + 1$ is:

$$P[\text{time spent by } i\text{th individual in state } j] = \exp\left(-\int_x^{x+1} Y_i^j(t) \sum_{k \neq j} \mu_x^{jk}\, dt\right)$$

$$= \exp\left(-\sum_{k \neq j} \mu_x^{jk} \int_x^{x+1} Y_i^j(t)\, dt\right).$$

(15.15)

Compare this with $\exp\left(-\int E_x^c \mu\right)$ in the simple survival model. Then, each time a transition from state j to state k is observed, say at time t, that event contributes probability $\mu_x^{jk} \, dt$ to the likelihood. Since $N_i^{jk}(x+1) - N_i^{jk}(x)$ such transitions are observed, the total contribution is $(\mu_x^{jk})^{(N_i^{jk}(x+1)-N_i^{jk}(x))}$ (suppressing the dts which we will drop from the likelihood anyway). For brevity, define:

$$\Delta N_i^{jk}(x) = N_i^{jk}(x+1) - N_i^{jk}(x).\tag{15.16}$$

Thus the likelihood of the whole life history – time spent in states and transitions between states – is:

$$L_i(\mu) = \prod_{j=0}^{M} \exp\left(-\sum_{k \neq j} \mu_x^{jk} \int_x^{x+1} Y_i^j(t) \, dt\right) \prod_{k \neq j} (\mu_x^{jk})^{\Delta N_i^{jk}(x)}\tag{15.17}$$

and the likelihood over all lives observed is:

$$L(\mu) = \prod_{i=1}^{n} \prod_{j=0}^{M} \exp\left(-\sum_{k \neq j} \mu_x^{jk} \int_x^{x+1} Y_i^j(t) \, dt\right) \prod_{k \neq j} (\mu_x^{jk})^{\Delta N_i^{jk}(x)}.\tag{15.18}$$

If we again consider an "episode" in the life history to be entering state j, staying there for some length of time r, then leaving it for state k, we see that each such "episode" contributes a factor:

$$P[\text{"episode"}] = \exp\left(-\int_0^r \sum_{l \neq j} \mu_x^{jl} \, ds\right) \mu_x^{jk}\tag{15.19}$$

to the likelihood, exactly as in equation (5.10). The apparent complexity of the likelihood (15.18) is really nothing more than the book-keeping needed to account for all possible life histories, bearing in mind that the number of transitions possible in any finite period need not be bounded. Even simpler, since:

$$\exp\left(-\sum_{k \neq j} \mu_x^{jk} \int_x^{x+1} Y_i^j(t) \, dt\right) = \prod_{k \neq j} \exp\left(-\mu_x^{jk} \int_x^{x+1} Y_i^j(t) \, dt\right),\tag{15.20}$$

we see that the likelihood factorises into terms contributed by each possible transition:

$$L(\mu) = \prod_{i=1}^{n} \prod_{j=0}^{M} \prod_{k \neq j} \left[\exp\left(-\mu_x^{jk} \int_x^{x+1} Y_i^j(t) \, dt\right) (\mu_x^{jk})^{\Delta N_i^{jk}(x)}\right].\tag{15.21}$$

Each term inside the square brackets is identical to the form of the likelihood obtained in the simple survival model. Inference in the Markov model is thus straightforward. Inference on each transition can proceed separately from all the others, just as in the simple survival model, because the likelihood factorises as above.

15.7 Properties of the MLEs for Single Ages

It follows immediately from the factorisation of the likelihood (15.21) that the maximum likelihood estimate of μ_x^{jk} is the ratio:

$$\hat{r}_x^{jk} = \frac{\sum_{i=1}^{n} \Delta N_i^{jk}(x)}{\sum_{i=1}^{n} \int_x^{x+1} Y_i^j(t)\,dt}, \tag{15.22}$$

which is identical in form to the mortality ratio introduced in Chapter 1. As in Chapter 4 and its sequel, we would usually assume that this is an estimate of the intensity at age $x + 1/2$.

We are interested in the statistical sampling properties of $\hat{r}_x^{jk}$. Since $\hat{r}_x^{jk}$ is an MLE, we can use some standard results about MLEs to deduce its properties. This approach is outlined in Appendix B, including the required likelihood theory. Or, we can just note that, because each estimator $\hat{r}_x^{jk}$ is of the same form as those in the Poisson model in Section 10.4, the same methods will give the asymptotic sampling distribution:

$$\hat{r}_x^{jk} \to \text{normal}\left(\mu_{x+s}^{jk}, \frac{\mu_{x+s}^{jk}}{\mathrm{E}\left[\sum_{i=1}^{n} \int_x^{x+1} Y_i^j(t)\,dt \right]} \right), \tag{15.23}$$

where $0 < s < 1$, and we usually assume $s = 1/2$. With some more work, which requires more theory of counting processes than we have covered so far, we can also show that all these estimates are asymptotically uncorrelated.

We saw in Chapter 14 that probability calculations in any Markov multiple-state model could be carried out by solving systems of ordinary differential equations. We now see also that MLEs for constant intensities over single years of age are simple occurrence-exposure rates with simple asymptotic sampling properties. This makes these models very attractive for many areas of actuarial work.

15.8 Estimation Using Complete Life Histories

The probabilistic model defined in Section 15.2 describes complete observed life histories. In particular, by defining the indicator functions $Y_i^j(t)$ we have implicitly taken care of left-truncation and right-censoring. Note that transition out of a state is a form of right-censoring of observation in that state, even if the individual may return to it.

The main change on fitting a survival model to complete observed life histories is that we assume the intensities to be parametric functions of age, here denoted $\mu^{jk}(t)$, instead of constants over years of age. To transform the likelihood (15.21) into a likelihood based on complete observed life histories, we do two things:

- Instead of integrating between ages x and $x+1$, we simply integrate between 0 and ∞. The indicator functions $Y_i^j(t)$ take care of when each individual is under observation in each state.

- If the ith individual jumps from state j to state k at age t, a factor $\mu^{jk}(t)$ must appear in the likelihood. Noting that $dN_i^{jk}(t) = 1$ precisely at such jump times, and $dN_i^{jk}(t) = 0$ at all other times, a neat way to accommodate the intensities at the jump times is by means of a *product integral*. A product integral, as its name suggests, is the product of a set of function values defined on an interval. If the function takes on the value 1 at all but a finite number of points on the interval, it is simply a product of a finite number of terms, in the same way as the integral in equation (15.9) was in fact a finite sum. Thus if we form the product integral, between 0 and ∞, of the function $(\mu^{jk}(t))^{dN_i^{jk}(t)}$, denoting this by:

$$\text{product integral} = \prod_{t \in [0,\infty]} (\mu^{jk}(t))^{dN_i^{jk}(t)}, \qquad (15.24)$$

we see that it does exactly what we need.

Thus, the likelihood for complete observed life histories can be written:

$$L(\mu) = \prod_{i=1}^{n} \prod_{j=0}^{M} \exp\left(- \int_0^\infty Y_i^j(t)\, \mu^{jk}(t)\, dt\right) \prod_{\substack{t \in [0,\infty] \\ k \neq j}} (\mu^{jk}(t))^{dN_i^{jk}(t)}. \qquad (15.25)$$

Using the example from Section 15.3 again, the likelihood (15.25) in full is:

$$L_i(\mu) = \exp\left(-\int_0^{40} \mu^{01}(t)\,dt\right) \exp\left(-\int_0^{40} \mu^{02}(t)\,dt\right) (\mu^{01}(40))^{dN_i^{01}(40)}$$

$$\times \exp\left(-\int_{40}^{42} \mu^{10}(t)\,dt\right) \exp\left(-\int_{40}^{42} \mu^{12}(t)\,dt\right) (\mu^{10}(42))^{dN_i^{10}(42)}$$

$$\times \exp\left(-\int_{42}^{50} \mu^{01}(t)\,dt\right) \exp\left(-\int_{42}^{50} \mu^{02}(t)\,dt\right) (\mu^{01}(50))^{dN_i^{01}(50)}$$

$$\times \exp\left(-\int_{50}^{60} \mu^{10}(t)\,dt\right) \exp\left(-\int_{50}^{60} \mu^{12}(t)\,dt\right) (\mu^{12}(60))^{dN_i^{12}(60)}.$$

$$(15.26)$$

It is easily seen how this decomposes into components of the form in equation (15.1), one component for each possible transition.

15.9 The Poisson Approximation

We can now explain why the Poisson distribution is so useful in the actuarial analysis of survival data.

- In Section 4.8 we argued heuristically that observing a collection of individuals for a total time E_x^c person-years, each one subject to a constant hazard rate μ, was analogous to observing a Poisson process with parameter μ for E_x^c years. Thus the number of deaths observed, d_x, should be a Poisson($E_x^c\mu$) random variable. We noted that this could only be approximate since the Poisson random variable would have positive probability of an impossible event, namely observing more than n deaths among n individuals.

  We imagined that observation of the ith individual was like observing a Poisson process with parameter μ and immediately ceasing observation at the first jump of the process, if that occurred before right-censoring. In the notation used there, the ith individual contributed a factor $\exp(-\mu(x_i'-x_i))\mu^{d_i}$ to the likelihood, and the total likelihood was proportional to the Poisson likelihood $\exp(-E_x^c\mu)\mu^{d_x}$.

- In Chapter 10 this Poisson model formed the basis of several approaches to graduating mortality ratios, taken to be estimates of $\mu_{x+1/2}$ at single years of age.

- In Chapter 5, in which we fitted parametric models to complete observed lifetimes, we made no assumption about a Poisson distribution (or any other distribution). The random lifetime model in Chapter 3, applied to the observations defined in Section 4.5, led to the following contribution to the likelihood by the ith individual in Section 5.3:

$$P[(d_i, t_i)] \propto \exp\left(-\int_0^{t_i} \mu_{x_i+t} \, dt\right) \mu_{x_i+t_i}^{d_i} \qquad (15.27)$$

and a total likelihood as in equation (5.12):

$$L(\mu) \propto \exp\left(-\sum_{i=1}^n \int_0^{t_i} \mu_{x_i+t} \, dt\right) \prod_{d_i=1} \mu_{x_i+t_i}^{d_i}. \qquad (15.28)$$

This likelihood is exact, not approximate. If the hazard rate is assumed to be a constant μ it takes the familiar form $\exp(-E_x^c \mu_x)\mu^{d_x}$ which, superficially, looks like the likelihood of a Poisson random variable. The right-hand side of equation (15.28) does not, however, define a Poisson random variable because it is random, not deterministic; it depends on the random times t_i. If hazard rates are very small, the great majority of individuals survive to be right-censored, so the variation in the total exposure times attributable to deaths will be very small. This is why the Poisson model gives such good results if we replace the random total exposure time with the observed E_x^c as if the latter was a deterministic constant. Sverdrup (1965) observed that if the hazard rate is very small, most of the information is in the number of deaths, while if the hazard rate is very large, most of the information is in the times of the deaths. This suggests that estimating q_x using binomial-type models is unwise at very high ages, since information about the times of deaths is lost in the process.

In Section 5.7 we took the likelihood (15.28) and replaced the hazard function μ_{x+s} between integer ages x and $x + 1$ with the hazard rate $\mu_{x+1/2}$, and got the likelihood from the Poisson model for single years of age. This showed links between the Poisson model, graduating mortality ratios at single years of age and fitting a survival model to complete observed lifetimes. It should not be overlooked that it is the likelihood (15.28) that is exact.

• In this chapter we have generalised all the above to a Markov model, obtaining Poisson-like likelihoods just like equation (15.28) for each possible transition. The key invention was the definition of the indicator processes $Y_i^j(t)$ as part of the probabilistic model of the observations, which unified observations over single years of age, complete observed lifetimes, left-truncation and right-censoring.

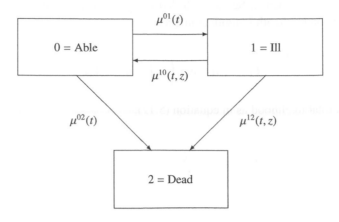

Figure 15.2 A semi-Markov model of illness and death, specified in terms of its intensities, which may depend on the duration of illness z as well as current age $x + t$.

Our exact models always lead to likelihoods that are functionally similar to those of a Poisson random variable, and as long as hazards or intensities are not too large, the fact that exposure times in states are random variables, not constants, does not matter very much.

15.10 Semi-Markov Models

Despite its mathematical convenience, a Markov model will not suffice if the process being modelled generates data that appear not to possess the Markov property. Data generated by disability insurance policies often clearly do not display the Markov property. In particular, transition intensities out of the "ill" state often depend much more strongly on the duration since the inception of sickness than on the current age. This leads to a semi-Markov model; see Section 14.10. It is enough that just one intensity out of one state depends on duration since entering the state; the Markov property does not hold.

Figure 15.2 shows a semi-Markov illness–death model in which the intensities out of the "ill" state depend on duration since entering the "ill" state, denoted by z, as well as current age t (in the notation of this chapter).

Without the Markov property, the Kolmogorov differential equations introduced in Chapter 14 do not hold. In their place, we can derive a system of integro-differential equations which are, except in special circumstances, harder to solve. The CMI in the UK has analysed the model shown in Figure 15.2 extensively (see Waters, 1984 and Continuous Mortality Investigation, 1991) and practical numerical methods for actuarial calculations have been devised. These are beyond the scope of this book. The key point is that using a Markov model because of its simplicity, when the data clearly lack the Markov property, is likely to be a gross model error.

Estimation of transition intensities in a semi-Markov model is more complicated if the data are complete life histories, because at least some of the intensities are functions of two variables instead of one. Thus, we may have to fit two-dimensional surfaces rather than one-dimensional curves. However, if we model single years of age and duration, assuming a constant hazard between ages x and $x + 1$ and durations z and $z + 1$, the semi-Markov model is no more difficult to handle than the Markov model. MLEs are the familiar occurrence-exposure rates and their properties are exactly the same as in Section 15.7.

15.11 Historical Notes

The generality and simplicity of Markov multiple-state models for actuarial work was set out in the landmark paper by Sverdrup (1965), which used as an example the illness–death model shown in Figure 14.4. It is not surprising that this development appeared in Scandinavia, because the education of actuaries in Scandinavian countries emphasised the continuous time model parameterised by the force of mortality μ_{x+t}, and Thiele's differential equation. Erling Sverdrup (1917–1994) was Professor of Actuarial Mathematics and Mathematical Statistics at the University of Oslo, and his work had far-reaching influence on both subjects. One of his doctoral students, Jan Hoem, introduced Markov models for general life and health insurance contracts (Chapter 14). Another, Odd Aalen, introduced the counting process models that now dominate survival analysis (Chapter 17) and applied them to the eponymous Nelson–Aalen estimator (Chapter 8). See Hoem (1969), Hoem (1988), Hoem and Aalen (1978), Aalen (1978) and Andersen et al. (1993).

16

Competing Risks Models

16.1 The Competing Risks Model

One of the most common extensions of the basic single-decrement model is called the *multiple-decrement model* by actuaries, or the *competing risks model* by statisticians (we will use both freely). It is important enough to deserve a short chapter to itself, not least because it is enlightening to compare the modern presentation and the traditional actuarial approach.

Figure 16.1 shows again the double-decrement model of death and lapses among life insurance policyholders. This makes the name "competing risks" clear: in any such model there are several decrements, or reasons to exit the starting state, and at most we observe the first to occur. It is possible, of course, that none of them occur. There is no transition from the "lapsed" state to the "dead" state. Someone who lapses a life insurance policy will certainly die but we will not observe the death. The model in Figure 16.1 is, intentionally, limited to describing what we can observe.

The treatment of this model in the Markov framework with the two transition intensities shown was covered completely in Chapters 14 and 15 and we have nothing to add here. In this chapter we describe alternative approaches, motivated mainly by the following question:

- Suppose the decrements are not independent of each other. For example, people in poor health may be less likely to lapse a life insurance policy. Or, two decrements may have a causal factor in common; for example, death from lung cancer and death from heart disease have smoking habits as a common factor.

- Suppose we could alter one of the decrements, for example eliminate deaths from lung cancer.

- What then would be the effect on the other decrements?

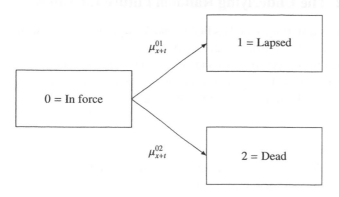

Figure 16.1 A three-state model of mortality and the lapse of a life insurance policy.

Our first observation is that Markov models have nothing to say about this question.

16.1.1 What Multiple-State Models Can't Tell Us

Multiple-state models, so useful elsewhere, cannot answer this question. A transition intensity in a multiple-state model, by definition, represents the rate of movement between two states in the presence of all the other transitions in the model. What happens if any of the intensities changes is not specified; the model assumptions do not lead to any answer at all. In a sense, the model does not have enough structure to let us do so. So, if we set $\mu_{x+t}^{01} = 0$ in Figure 16.1, the best we can do is to make an educated guess about what would happen to μ_{x+t}^{02}.

Clearly, it would be useful to have a model that could tell us what would happen to μ_{x+t}^{02}, and that is the motivation for the classical competing risks model. If we think of rates of decrement or intensities "depending" on each other, we soon realise that we lack a proper mathematical definition for such "dependencies". We do, however, have a proper mathematical definition for dependencies among random variables. Perhaps this offers a way forward. Instead of specifying a multiple-decrement model by its intensities, the "classical" approach turns back to the starting point of Chapter 3, and specifies the model in terms of random future lifetimes. In the following, we introduce this model, ask whether it can answer the question that the Markov model could not, and compare also the probabilistic and traditional actuarial approaches to such a model.

16.2 The Underlying Random Future Lifetimes

Suppose there is a starting state labelled 0, and M decrements represented by states as in Figure 16.1, labelled $1, 2, \ldots, M$.

First consider the observations we may make. In the absence of right-censoring we can define these by a pair of random variables T and J. The random variable T records the time spent under observation in the starting state, and J is a random variable taking one of the values $1, 2, \ldots, M$, indicating which decrement was observed.

As in Chapter 4, we need to propose a probabilistic model capable of generating these data. The "classical" competing risks model does so by associating a random lifetime with each decrement. In the following, we suppose that we are observing a single individual, and subscripts on random variables represent decrements.

Definition 16.1   For $j = 1, 2, \ldots, M$, define T_j to be a random variable representing the time from the start of observation until exit from state 0 because of the jth decrement.

These lifetimes are often called the *latent lifetimes* or *latent failure times*. The idea is that if we could observe the jth decrement alone, we would observe T_j just as in Chapter 3. However, all we can possibly observe is the smallest of all the T_j, and maybe not even that if right-censoring occurs.

To allow for right-censoring, define C to be the time when right-censoring happens, if it happens, and $C = \infty$ otherwise, and allow J also to take the value 0, indicating that observation was ended by right-censoring. Then we observe:

$$T = \min(T_1, T_2, \ldots, T_M, C) \quad \text{and} \quad J. \tag{16.1}$$

Sometimes, the lifetimes $T_1, T_2, \ldots, T_M$ do have a realistic physical interpretation. The time to failure of a machine with several components, or the time to the first death of several lives, might naturally be modelled as the minimum of several lifetimes. But if, as here, the $T_1, T_2, \ldots, T_M$ are "lifetimes" in respect of different events befalling one person, the model may include events that are physically impossible. In Figure 16.1, for example, if $J = 1$ represents lapse and $J = 2$ represents death, the event $T_2 < T_1$ is physically impossible. This is not necessarily a fatal blow, as long as impossible events do not have positive probability, so we would require $P[T_2 < T_1] = 0$ in this example.

16.2.1 Dependent and Independent Hazards

Associated with each decrement $(j = 1, 2, \ldots, M)$ are two hazard rates. Recall the definition of the hazard rate, given the future lifetime T, in the single-decrement case (Definition 3.6, but here to avoid clutter we have suppressed age x and just retained time t):

$$\mu_t = \lim_{dt \to 0^+} \frac{P[T \le t + dt \mid T > t]}{dt}. \tag{16.2}$$

This definition is still perfectly valid here, but it now represents the hazard of exit from the starting state by all causes combined. In the context of a multiple-decrement model, we shall call it the *total* hazard rate.

Definition 16.2 The dependent hazard rate associated with the jth decrement is:

$$(a\mu)_t^j = \lim_{dt \to 0^+} \frac{P[T \le t + dt, J = j \mid T > t]}{dt}. \tag{16.3}$$

The symbol $(a\mu)_t^j$ is the traditional actuarial notation. This hazard is also known as the *crude or gross hazard rate* to statisticians. Comparing this with the multiple-state framework, we recognise the identity:

$$P[T \le t + dt, J = j \mid T > t] = {}_{dt}p_t^{0j}, \tag{16.4}$$

so the dependent hazard rate is exactly the same as the transition intensity in the Markov model; that is, $(a\mu)_t^j = \mu_t^{0j}$. (We will generally avoid the multiple-state notation in this chapter, to avoid confusion; we will introduce it only in order to make a few comparisons.)

What is new, and not available in the multiple-state framework, is the following definition of independent hazard rate:

Definition 16.3 The independent hazard rate associated with the jth cause of exit is:

$$\mu_t^j = \lim_{dt \to 0^+} \frac{P[T_j \le t + dt \mid T_j > t]}{dt}. \tag{16.5}$$

This is known as the *net hazard rate* to statisticians. Its interpretation is that the independent (net) hazards correspond to each decrement acting alone, while the dependent (gross) hazards correspond to each decrement acting in the presence of the others.

We have seen that, when the decrements represent different events befalling a single individual, we may have strong grounds to suppose that they influence each other; some examples were given in Section 16.1.1. In other words, we should *not* expect dependent and independent hazards to be equal. However, unless dependent and independent hazards *are* assumed to be equal, the mathematics quickly becomes intractable. Most of the theory of competing risks (certainly that part which has appeared in actuarial science) starts by assuming dependent and independent hazards to be equal. In one direction, there is an important result that appears to give some support for this:

Proposition 16.4 *Suppose the latent lifetimes $T_1, T_2, \ldots, T_M$ are independent random variables. Then $\mu_t^j = (a\mu)_t^j$ ($j = 1, 2, \ldots, M$).*

We omit the rather long proof. A proof using the same notation as used here can be found in Macdonald (1996). Note the unfortunate clash of terminology. The independence of the lifetimes in Proposition 16.4 means independence of random variables in its usual sense. Calling hazard rates "dependent" and "independent" is just a naming convention and should not be interpreted as any kind of probabilistic statement.

In the other direction, however, the converse of Proposition 16.4 is false: equality of dependent and independent hazard rates does *not* imply that the random variables $T_1, T_2, \ldots, T_M$ are independent.

16.3 The Unidentifiability Problem

The task of inference, given this model, is to estimate the *joint* distribution of $T_1, T_2, \ldots, T_M$. But the fact that we observe only T and J leads to a major problem; the joint distribution of $T_1, T_2, \ldots, T_M$ cannot be identified from these data. This is perhaps not surprising, since we have specified a model in terms of random variables that we cannot observe directly. Only the dependent hazard rates are observable, even in theory; the independent hazards are intrinsically unobservable.

We will outline the reason for the problem. Define the survival function for T to be:

$$S_T(t) = P[T > t] = P[T_1 > t \text{ and } T_2 > t \text{ and } \ldots \text{ and } T_M > t]. \qquad (16.6)$$

Proposition 16.5 *Assuming no two of the latent lifetimes are equal, the overall hazard rate associated with $S_T(t)$ is $(a\mu)_t^1 + \cdots + (a\mu)_t^M$.*

Proof: From equation (16.2), the hazard rate associated with $S_T(t)$ is:

$$\mu_t = \lim_{dt \to 0^+} \frac{P[T \le t + dt \mid T > t]}{dt}. \tag{16.7}$$

Since one and only one of the decrements must have occurred at time T, this is:

$$\mu_t = \lim_{dt \to 0^+} \frac{1}{dt} \sum_{j=1}^{j=M} P[T \le t + dt, J = j \mid T > t]$$

$$= \sum_{j=1}^{j=M} \lim_{dt \to 0^+} \frac{P[T \le t + dt, J = j \mid T > t]}{dt}$$

$$= (a\mu)_t^1 + \cdots + (a\mu)_t^M. \tag{16.8}$$

$\square$

Upon integrating we have:

$$S_T(t) = \prod_{j=1}^{j=M} \exp\left(- \int_0^t (a\mu)_s^j ds \right). \tag{16.9}$$

Note that this is formally equivalent to equation (14.25), only the notation is different.

Now comes the crucial point. We have not assumed that $T_1, T_2, \ldots, T_M$ are independent. They may be dependent, and in general will be. Now *define* a set of latent lifetimes $U_1, U_2, \ldots, U_M$ that are independent by assumption and have the same survivor functions as $T_1, T_2, \ldots, T_M$:

$$S_{U_j}(t) = \exp\left(- \int_0^t (a\mu)_s^j ds \right) \qquad (j = 1, 2, \ldots, M). \tag{16.10}$$

Then these new latent lifetimes define a competing risks model. What we will be able to observe are the minimum lifetime $U = \min(U_1, U_2, \ldots, U_M)$ and the cause of exit $K = (k : U_k = U)$. However, by construction (equations (16.9) and (16.10)) U and K have the same joint distribution as T and J. Therefore, we cannot tell *from any amount of data* whether the process being observed is represented by independent or dependent competing risks.

This is the *unidentifiability problem*, described by Tsiatis (1975). It gets worse; Crowder (1991) showed that, even if it were possible to observe each cause of exit in the absence of the others as well as in their presence, the model would still be unidentifiable (he referred to "the identifiability crisis").

Any question about how the various decrements might interact with each other is a question about the probabilistic model that is supposed to generate the observations. The multiple-state model specified by its intensities did not suggest any such mechanism, so we provided one, namely the latent lifetimes. However, when these lifetimes are all mutually censoring, we can never observe the full probabilistic model in action. With hindsight, it is hardly surprising that real observations should not allow us to identify the model.

16.4 A Traditional Actuarial Approach

In this section, we look briefly at the traditional actuarial treatment of multiple decrements, which has two important features:

- It is not based explicitly on a probabilistic model.
- The key quantities are probabilities corresponding to the single-decrement probabilities q_x. As with the actuarial estimate of Section 10.5, these probabilities are the target of estimation, rather than the hazard rates.

16.4.1 Multiple-Decrement Probabilities

We introduce the traditional actuarial terminology (see Neill, 1986 or Benjamin and Pollard, 1980). First, supposing we start with a person age x, the overall survival function, denoted $S_T(t)$ before, is now denoted $_t(ap)_x$. Then define the following probabilities:

Definition 16.6 In the multiple-decrement model in which the dependent hazards are $(a\mu)_t^1, \ldots, (a\mu)_t^M$:

- The dependent probability that a person alive at age x will exit by decrement j by age $x + t$ is:

$$_t(aq)_x^j = \int_0^t {}_s(ap)_x (a\mu)_{x+s}^j ds = \int_0^t e^{-\int_0^s (a\mu)_{x+r}^1 + \cdots + (a\mu)_{x+r}^M dr} (a\mu)_{x+s}^j ds. \quad (16.11)$$

- The total dependent probability that a person alive at age x will exit by age $x + t$ is:

$$_t(aq)_x = {}_t(aq)_x^1 + {}_t(aq)_x^2 + \cdots + {}_t(aq)_x^M. \quad (16.12)$$

Of course $_t(aq)_x^j$ is the same as the multiple-state occupancy probability $_t p_x^{0j}$. If $t = 1$, we usually just write $(aq)_x^j$ and so on. Note that there are no dependent survival probabilities $_t(ap)_x^j$ associated with each decrement, as the very idea contradicts itself.

Dependent probabilities represent each decrement acting in the presence of every other decrement. We may hypothesise, however, that we could isolate each decrement so that it acted alone. This hypothesis is usually approached through the so-called independent hazards as follows.

Definition 16.7 In the multiple-decrement model in which the independent hazards are $\mu_t^1, \ldots, \mu_t^M$:

- The independent probability that a person alive at age x will survive decrement j until at least age $x + t$ is:

$$_t p_x^j = \exp\left(-\int_0^t \mu_{x+s}^j ds\right). \qquad (16.13)$$

- The independent probability that a person alive at age x will exit by decrement j by age $x + t$ is:

$$_t q_x^j = \int_0^t {_s p_x^j} \mu_{x+s}^j ds. \qquad (16.14)$$

If $t = 1$, we usually just write q_x^j and so on. In this setting, independent survival probabilities $_t p_x^j = 1 - {_t q_x^j}$ make sense, and they define the single-decrement life table associated with the jth decrement. These single-decrement tables are just as meaningful, or meaningless, as the case may be, as the independent hazards from which they were constructed.

16.4.2 Changing Rates of Decrement

We now turn to the question posed at the start of this chapter: if we change one of a set of dependent hazards, what will be the effect on the others? It would be common, for example, to take a multiple-decrement table based on one pension scheme and try to adjust it to suit the circumstances of another pension scheme. For the sake of example, suppose we wish to double the rates of ill-health retirement at all ages.

The starting point in this problem is a table of dependent probabilities $(aq)_x^j$, and the end-point is another such table, say $(aq)_x^{j*}$, where the asterisk denotes the desired probabilities. Let $j = 1$ denote death, $j = 2$ normal retirement and $j = 3$ ill-health retirement. The general idea is this: when we say that rates of ill-health retirement are doubled, we mean that the *independent* probabilities q_x^3 should be doubled, because the meaning of this is not complicated by any dependencies between ill-health retirement and the other decrements. Simply doubling the dependent probabilities $(aq)_x^3$ would ignore such dependencies. This leads to the following three steps:

(i) Deduce relationships between dependent and independent probabilities of decrement, such that we can find q_x^1, q_x^2 and q_x^3 from the given $(aq)_x^1$, $(aq)_x^2$ and $(aq)_x^3$.

(ii) Then put $q_x^{1*} = q_x^1$, $q_x^{2*} = q_x^2$ and $q_x^{3*} = 2q_x^3$.

(iii) Then invert the relationship to get new dependent rates $(aq)_x^{1*}$, $(aq)_x^{2*}$ and $(aq)_x^{3*}$.

We will not illustrate this process in detail, but give an example of the third step above, whose real purpose is to tease out the kind of assumptions that are being made. Suppose we have, after the first two steps, obtained a set of three independent probabilities of decrement, q_x^{1*}, q_x^{2*} and q_x^{3*}. Can we construct a dependent set of probabilities $(aq)_x^{1*}$, $(aq)_x^{2*}$ and $(aq)_x^{3*}$? The answer is yes, it can be done in several ways, depending on the assumptions we make, but the key step is to assume that $(a\mu)_{x+t}^{j*} = \mu_{x+t}^{j*}$. If we do so, then:

$$_tp_x^{1*}\,_tp_x^{2*}\,_tp_x^{3*} = \exp\left(-\int_0^t (a\mu)_{x+s}^{1*}\,ds\right)\exp\left(-\int_0^t (a\mu)_{x+s}^{2*}\,ds\right)\exp\left(-\int_0^t (a\mu)_{x+s}^{3*}\,ds\right)$$

$$= \exp\left(-\int_0^t (a\mu)_{x+s}^{1*} + (a\mu)_{x+s}^{2*} + (a\mu)_{x+s}^{3*}\,ds\right) = {}_t(ap)_x^*. \qquad (16.15)$$

Next, in each of the associated single-decrement tables, assume that the decrement is uniformly distributed between integer years of age. Then ${}_tq_x^{j*} = tq_x^{j*}$, and ${}_tp_x^{j*}\mu_{x+t}^{j*}$ is constant and equal to q_x^{j*}, for $0 \leq t < 1$ (see Dickson et al., 2013, Chapter 3). So, taking $j = 1$ as an example:

$$(aq)_x^{1*} = \int_0^1 {}_t(ap)_x^* (a\mu)_{x+t}^{1*} \, dt = \int_0^1 {}_tp_x^{1*} \, {}_tp_x^{2*} \, {}_tp_x^{3*} \mu_{x+t}^{1*} \, dt$$

$$= q_x^{1*} \int_0^1 (1 - tq_x^{2*})(1 - tq_x^{3*}) \, dt$$

$$= q_x^{1*} \left(1 - \frac{q_x^{2*} + q_x^{3*}}{2} + \frac{q_x^{2*} q_x^{3*}}{3} \right). \quad (16.16)$$

Different assumptions will lead to slightly different approximations; usually the discrepancies are of no practical importance. If we choose assumptions that make the first step easy (find the q_x^j in terms of the $(aq)_x^j$), then usually the third step is hard (find the $(aq)_x^j$ in terms of the q_x^j) and *vice versa*.

The key assumption, as always, was that $(a\mu)_{x+t}^j = \mu_{x+t}^j$. Although we may imagine the existence of the independent hazards μ_{x+t}^j, in practice we cannot make progress without assuming them to be equal to the dependent hazards $(a\mu)_{x+t}^j$, which we recall are just the intensities of a multiple-state model in a different notation. The elaborate and somewhat artificial scaffolding of "dependent" and "independent" probabilities tends to hide this vital point.

In the multiple-state framework, all our assumptions are in the open, we do not have to try to estimate anything that is intrinsically unobservable, and when we enter the realm of the hypothetical ("what would happen to the other decrements if this decrement were changed?") there is no pretence that the model will be more objective than our own educated guess. And of course, multiple-state models are not confined to representing decrements alone.

16.4.3 Estimation in the Binomial Model Revisited

Since the traditional approach, rooted in life tables and the binomial model, concentrates on estimating probabilities, we have to re-examine the methods of Section 10.5 to see how we might estimate q_x^j or $(aq)_x^j$. Our starting point is to consider the interpretation of the actuarial estimate, equation (10.42), when there is more than one decrement.

Suppose we introduce another decrement, such as lapsing a life insurance policy. As well as observing d_x deaths, we observe w_x lapses. For the purpose of labelling probabilities, suppose $j = 1$ represents lapsing and $j = 2$ represents death. The calculation of the waiting time or central exposed-to-risk E_x^c is unchanged; the question is how it should be adjusted in the denominator of the actuarial or binomial-type estimate.

We can always recover a single-decrement model from a multiple-decrement model by combining all the decrements together. It follows that:

$$\frac{w_x + d_x}{E_x^c + (w_x + d_x)/2} \tag{16.17}$$

is the actuarial estimate of $(aq)_x$. From this, it is reasonably obvious that:

$$\frac{w_x}{E_x^c + (w_x + d_x)/2} \quad \text{and} \quad \frac{d_x}{E_x^c + (w_x + d_x)/2} \tag{16.18}$$

are actuarial estimates of $(aq)_x^1$ and $(aq)_x^2$, respectively. Quantities of the same form as equation (10.42), namely:

$$\frac{w_x}{E_x^c + w_x/2} \quad \text{and} \quad \frac{d_x}{E_x^c + d_x/2}, \tag{16.19}$$

can now be interpreted as actuarial estimates of q_x^1 and q_x^2, respectively. This involves selecting one decrement as the target, then following the analysis in Section 10.5, treating all other decrements as censored observations. It can be summed up by these rules:

- Treat all cases of the decrement of interest as exposed until the end of the year, assuming that the decrement occurs at age $x + 1/2$.
- Treat all censored observations, including other decrements, as exposed until the time of censoring.

See Benjamin and Pollard (1980) for a relatively recent treatment along these lines.

16.5 Are Competing Risks Models Useful?

16.5.1 The Statistical Critique

In specific cases, such as a machine whose components we can observe, competing risks models might continue to be useful. When the physical set-up allows us to see the mechanism that generates the data, then a model describing that mechanism makes sense. In the vast majority of actuarial, medical or demographic applications, however, this is not the case; the suggested mechanism (latent lifetimes) is intrinsically unobservable.

This approach has attracted considerable criticism in the statistical literature. Aalen (1987) described the approach as leading to "distortion of the statistical analysis, and to artificial problems, like the question of identifiability". He also pointed out the infeasibility of a competing risks approach to more general transitions. Prentice et al. (1978) said (we have added the words in italics):

It therefore seems important to concentrate on the $\lambda_j(t, \mathbf{z})$ functions [*the dependent hazards*] for statistical modelling as they lead to procedures that have a clear interpretation regardless of the interrelation between causes of failure and yet are identical with the more traditional results, based upon independent latent failure times, in circumstances in which an independence assumption is justifiable. .... It is perhaps surprising that this approach has received so much attention in the literature.

For inference, it is sensible to restrict attention to what is observable, and estimate the dependent forces of decrement $(a\mu)_t^j$, unless there is some pressing reason to do otherwise. This leads us back to multiple-state models, specified by their intensities. If we must compute unobservable quantities, it seems better to treat the resulting calculations as hypothetical.

16.5.2 Actuarial Mortality Analysis

It used to be part of actuarial student folklore, at least in the UK, that the analysis of mortality was a matter especially obscure and difficult, despite the high standards set by some of the other subjects. It is worth considering why this might have been so:

- The available data would be grouped by age. Analysis of data on individual lives would have been infeasible with the computing power available, even as late as when Benjamin and Pollard (1980) was written.
- The aim was to estimate probabilities q_x, and the estimate to be used was most often the actuarial estimate, even if it was not called that by actuaries. The central exposed-to-risk E_x^c could usually be found quite easily, so the main question was what adjustments to make to E_x^c in the denominator of the actuarial estimate, to obtain the appropriate initial exposed-to-risk.
- Ostensibly, single decrements were treated before multiple decrements. The first task, therefore, was to estimate a single-decrement life table. In a practical setting, however, other decrements were usually present, such as lapsing of life insurance policies, *which were allowed for in the estimation*. See Benjamin and Pollard (1980), for example.
- The outcome was that the quantity being estimated right from the outset was not, in fact, the q_x of the single-decrement model, but one of the "independent" q_x^j of the multiple-decrement model, even though that model had not yet been presented.

- Probabilistic models would appear in a minor role only when crude estimates $\hat{q}_x$ had to be graduated, and the graduations tested for goodness-of-fit. The calculation of the $\hat{q}_x$ themselves was not approached as a question of inference, given a probabilistic model capable of generating the data.
- The various rate intervals (life year, calendar year, policy year) encountered in actuarial work led to further complications in adjusting E_x^c. In fact the whole treatment tended to degenerate into immensely detailed formulae for initial exposed-to-risk.

It is hardly surprising that the calculation of initial exposed-to-risk has been a source of boundless confusion to students over the years. Ironically, the problems mostly concerned the adjustments to E_x^c, which could have been used directly to estimate a hazard rate.

We leave the final word to F. M. Redington, a leading British actuary of the mid twentieth century, famous for his clarity of thought. He was reported to say, in the discussion of Bailey and Haycocks (1947), that "In the ordinary census formula, $\frac{1}{2}P$ at the beginning of the year and $\frac{1}{2}P'$ at the end of the year were obvious and easily remembered; but when adjustments had to be made to give the deaths a full year's exposure the fog descended." (See Section 4.10 for the census formula referred to by Redington.)

17

Counting Process Models

17.1 Introduction

Here is a puzzle for the reader, alluded to before in Section 5.6. We observe that a woman survives from age 68 to 69, and then from age 69 to 70. We use that information in the exposed-to-risk in estimating $\hat{\mu}_{68.5}$ and $\hat{\mu}_{69.5}$. Using any of the methods discussed so far, we would assume, perhaps implicitly, that the estimation of $\hat{\mu}_{69.5}$ proceeds independently of the estimation of $\hat{\mu}_{68.5}$. But how can that be? We only got to observe the woman from age 69 to 70 because she did not die between age 68 and 69. These observations are not independent. How can any of the methods discussed so far be valid?

In Chapter 15 we found counting processes useful for estimating intensities in the Markov model. Their usefulness arose from their notational completeness, which let us write down likelihoods for any Markov model, and also from the introduction of the processes $Y_i^j(t)$ indicating the presence of the ith individual in state j just before time t. The latter greatly simplified the handling of left-truncation and right-censoring (we shall have more to say about this in Section 17.7.2).

In fact counting processes have much more to offer than some handy notation, including the answer to the puzzle above. Their introduction to survival modelling (Aalen, 1975; Aalen, 1978) led to a complete revolution in underlying theory, such that no book on the subject would be complete without mentioning it. Apart from anything else, it is the language in which most research in the subject is now written. Crowder (2001), while admitting the technical challenge of getting to grips with the counting process machinery, nevertheless saw it as the way forward: "The counting process approach has solid advantages: it provides a firm theoretical basis for modelling events unfolding over time and is particularly suited to models framed in terms of hazard rates and involving censoring."

307

The theory depends on the properties of certain martingales associated with a counting process. In recent years, actuaries have encountered martingales as part of the stochastic calculus used in financial economics. The technical apparatus needed for survival modelling is much less daunting than this; in particular we do not need the Itô integral.

In Section 17.2 we introduce the basic concepts needed in counting process models. This amounts to the notation necessary to answer the following questions:

- How can we represent the acquisition of knowledge by observing events as time passes?
- How does that knowledge change our expectations of future events?

A full (and excellent) treatment of this is Williams (1991). In Section 17.3 we define a certain integral, such that very many objects of interest in survival analysis can be expressed as integrals of this kind. With these concepts in place, we define martingales in Section 17.4.

We then (our ultimate goal) define certain martingales associated with the multivariate counting processes $N_i^{jk}(t)$ and the indicator functions $Y_i^j(t)$ defined in Chapter 15 (Section 17.5). We will omit all formal technicalities; full details can be found in Andersen et al. (1993) or Fleming and Harrington (1991).

Standard properties of these counting process martingales yield a harvest of insights and results which would otherwise be very difficult to obtain. These insights include:

- the answer to the puzzle at the start of this section;
- the reason why Poisson-like likelihoods pervade the whole subject; and
- the statistical properties of the very many quantities that have the integral representations mentioned above.

17.2 Basic Concepts and Notation for Stochastic Processes

Consider a stochastic process $X(t)$ in continuous time, $t \geq 0$, with $X(0) = 0$. We assume we are able to observe $X(t)$ completely over time, in the sense that there are no gaps in our knowledge.

At time t we have observed:

$$\{X(s) \,:\, s \leq t\}. \tag{17.1}$$

Or, immediately before time t, which we denote by t^-, we have observed:

$$\{X(s) \; : \; s < t\}. \tag{17.2}$$

We may also have knowledge obtained by other means. For example, if a person is underwritten, medically or otherwise, that is relevant knowledge, presumed known at time $t = 0$.

To allow for all such possibilities, we define $\mathcal{F}_t$ to be all that we know at time t, from all sources (and similarly $\mathcal{F}_{t^-}$ at time t^-). We say that $\mathcal{F}_t$ is *generated by* the knowledge at time t, and we use the symbol σ to denote knowledge generated by observations. For example, at time t we have:

$$\mathcal{F}_t = \sigma(\{X(s) \; : \; s \le t\}) \tag{17.3}$$

from observing the process only, or, if underwriting took place:

$$\mathcal{F}_t = \sigma(\{X(s) \; : \; s \le t\}, \text{ knowledge of risk factors}). \tag{17.4}$$

This knowledge will change our expectations about future values of $X(t)$. For example, if $X(t)$ is non-decreasing, $X(0) = 0$, $X(s) > 0$ at some $s > 0$, and $r > s$, then we know at time s that $X(r) \ge X(s) > 0$, which we did not know at time $t = 0$. This leads us to consider the expected value of $X(r)$, conditional on what we know at time t, for $t \le r$. We call this a *conditional expectation* and denote it by:

$$E[X(r) \mid \mathcal{F}_t]. \tag{17.5}$$

To illustrate these ideas we use a simple example. Suppose we toss a fair coin three times in succession, at times $t = 1, 2$ and 3. Define $X(t)$ to be the number of tosses that came up heads, up to and including time t. For $t > 0$ define $\Delta X(t) = X(t) - X(t-1)$, the *increments* of the process $X(t)$. So $\Delta X(t) = 1$ if the tth toss was "heads" and $\Delta X(t) = 0$ if the tth toss was "tails".

Clearly, $X(t)$ can be written as the sum of its increments: $X(t) = \sum_{i=1}^{t} \Delta X(i)$. (Compare this with the construction of a counting process $N_i^{jk}(t)$ as the integral of its increments $dN_i^{jk}(t)$ in Section 15.4.)

To give us a concrete example to work with, suppose we observe the three tosses to be head, tail, head in that order. This observation can be represented in two equivalent ways:

- We observe successively $X(0) = 0$, $X(1) = 1$, $X(2) = 1$ and $X(3) = 2$.
- We know that $X(0) = 0$, and then observe successively $\Delta X(1) = 1$, $\Delta X(2) = 0$ and $\Delta X(3) = 1$.

But in either case, we do not possess *all* this information until all three tosses have been observed. For example, at time $t = 1$ we have:

$$\mathcal{F}_1 = \sigma(X(0), X(1)) = \sigma(\Delta X(1)),\qquad(17.6)$$

but we do not yet know $X(2)$ or $X(3)$.

Suppose our main interest is in the final value $X(3)$. Though we will not finally know $X(3)$ until time $t = 3$, we certainly learn an increasing amount about $X(3)$ from our observations at $t = 1$ and $t = 2$. Having observed a head at time $t = 1$, we know that $X(3) > 0$, and having then observed a tail at time $t = 2$ we also know that $X(3) < 3$.

Our increasing knowledge about $X(3)$ may be summarised by its conditional expected value at times $0, 1, 2$ and 3. Let us work through this in detail:

- Before the first toss, $X(3)$ will take value 0 with probability 1/8, value 1 with probability 3/8, value 2 with probability 3/8 and value 3 with probability 1/8, and its expected value is 12/8.
- After the first toss is heads, we know that $X(3) = 1 + \Delta X(2) + \Delta(X3)$. The distribution function and expected value of $X(3)$ have changed. $X(3)$ now takes value 0 with probability 0, value 1 with probability 1/4, value 2 with probability 2/4 and value 3 with probability 1/4, and its expected value is 8/4.
- Then, when the second toss is a tail, we know that $X(3) = 1 + 0 + \Delta X(3)$. Now $X(3)$ will take values 0 or 3 with probability 0, value 1 with probability 1/2 and value 2 with probability 1/2, and its expected value is 3/2.
- Finally, when we know that the third toss is a head, we know that $X(3) = 2$.

In terms of classical conditioning on observed random variables we would express this as:

$$E[X(3)] = 12/8 \qquad(17.7)$$
$$E[X(3) \mid \Delta X(1)] = 8/4 \qquad(17.8)$$
$$E[X(3) \mid \Delta X(1), \Delta X(2)] = 3/2 \qquad(17.9)$$
$$E[X(3) \mid \Delta X(1), \Delta X(2), \Delta X(3)] = 2. \qquad(17.10)$$

The proper definition of conditional expectation is more subtle than portrayed above, one reason being to handle conditioning on events that have probability zero. See Williams (1991) for details. The conditional expectations

$E[X(3) \mid \mathcal{F}_0], E[X(3) \mid \mathcal{F}_1], E[X(3) \mid \mathcal{F}_2]$ and $E[X(3) \mid \mathcal{F}_3]$ defined as in equation (17.5) are not quite the same as the "classical" conditional expectations in equations (17.7) to (17.10) but we need not worry about that here.

Conditional expectations have two particularly important consequences:

* *The Tower law.* Suppose X is a random variable whose value will be known at time $T > 0$, and $s \le t \le T$. Then:

$$E[X \mid \mathcal{F}_s] = E[E[X \mid \mathcal{F}_t] \mid \mathcal{F}_s]. \tag{17.11}$$

In words, knowing $\mathcal{F}_s$, list all possible observations between time s and time t, hence every possible $\mathcal{F}_t$. For each of these, compute $E[X \mid \mathcal{F}_t]$. Then average over all of these, conditional on $\mathcal{F}_s$. The result is the same as computing $E[X \mid \mathcal{F}_s]$ in the first place. In our example, suppose we know $\mathcal{F}_1$. If $\Delta X(2) = 0$ then $E[X(3) \mid \mathcal{F}_2]$ will be $3/2$. If $\Delta X(2) = 1$ then $E[X(3) \mid \mathcal{F}_2]$ will be $5/2$. The mean of these possibilities is $8/4$, in agreement with equation (17.8).

* *Previsible processes.* In discrete time, a process $Z(t)$ is *previsible* (also called *predictable* if the value $Z(t)$ is already known at time $t - 1$, given $\mathcal{F}_{t-1}$. Then also:

$$E[Z(t) \mid \mathcal{F}_{t-1}] = Z(t). \tag{17.12}$$

In continuous time, a process $Z(t)$ is predictable if the value $Z(t)$ is already known at time t^-, given $\mathcal{F}_{t^-}$. Then also:

$$E[Z(t) \mid \mathcal{F}_{t^-}] = Z(t). \tag{17.13}$$

The prototype of a previsible process in continuous time is one whose sample paths are left-continuous. See the definition of $Y_i^j(t)$ in Chapter 15, and especially the discussion of the example life history in Section 15.3.

If $Z(t)$ is deterministic it is previsible trivially, but if $Z(t)$ is stochastic this property has important consequences.

Moreover, if $X(t)$ is the process generating the information $\mathcal{F}_t$ and $Z(t)$ is previsible, then:

$$E[Z(t) X(t) \mid \mathcal{F}_{t-1}] = Z(t) E[X(t) \mid \mathcal{F}_{t-1}] \quad \text{(discrete time)} \tag{17.14}$$

$$E[Z(t) X(t) \mid \mathcal{F}_{t^-}] = Z(t) E[X(t) \mid \mathcal{F}_{t^-}] \quad \text{(continuous time).} \tag{17.15}$$

Equations (17.14) and (17.15) express the key property of a previsible process. We have two main reasons for defining them:

- They are the processes that can be integrated in a stochastic integral; see Section 17.3. Equation (17.14) will be used in an essential way in equation (17.32).
- We often want to identify precisely the *innovations* obtained by observing $X(t)$. In our coin-tossing example, the innovation or new information at time t is the "unexpected" part of the increment, namely $\Delta X(t) - E[\Delta X(t) \mid \mathcal{F}_{t-1}]$. The process $X(t-1) + E[\Delta X(t) \mid \mathcal{F}_{t-1}]$ is previsible; both terms are known numbers at time $t-1$. It is the "best estimate" of $X(t)$ given $\mathcal{F}_{t-1}$. Writing:

$$X(t) = \underbrace{\left(X(t-1) + E[\Delta X(t) \mid \mathcal{F}_{t-1}]\right)}_{\text{Best estimate}} + \underbrace{\left(\Delta X(t) - E[\Delta X(t) \mid \mathcal{F}_{t-1}]\right)}_{\text{Innovation}} \quad (17.16)$$

decomposes $X(t)$ into the best estimate of $X(t)$ given all that was known at time $t-1$ – a previsible quantity – plus the new information arriving at time t. This is an important way to view a stochastic process; see Arjas (1989) and Arjas and Harra (1984).

Previsible processes in continuous time can be thought of as those with left-continuous paths; $Z(t)$ is known given $\mathcal{F}_{t^-} = \sigma(X(s) : s < t)$. When we discussed estimation of intensities in Markov models in Chapter 15, we defined the indicator of presence in state j, namely $Y_i^j(t)$, so that its value at time t indicated presence in state j at time t^-. This was because we need $Y_i^j(t)$ to be previsible in what follows.

With the observed information $\{\mathcal{F}_t\}_{t \geq 0}$ and the conditional expectations $E[X(s) \mid \mathcal{F}_t]$, we have answered the two questions posed in Section 17.1, namely:

- How can we represent the acquisition of knowledge by observing events as time passes?
- How does that knowledge change our expectations of future events?

17.3 Stochastic Integrals

We first say that the stochastic integrals described here are *not* the Itô integrals used in financial mathematics. They are much simpler. As a prototype, consider the expectation of a continuous random variable Y with distribution $F(t)$ and density $f(t)$. It would usually be written:

$$E[Y] = \int t \, f(t) \, dt. \quad (17.17)$$

Ignoring rigour (which can be supplied), we can write this also as:

$$E[Y] = \int t f(t) \, dt = \int t \frac{dF(t)}{dt} \, dt = \int t \, dF(t). \qquad (17.18)$$

In other words, the expected value is the weighted sum of the values of Y, the weights being the *increments* of the distribution function. More generally, if we have some function or process $X(t)$, many important quantities may often be written in the form:

$$\text{Quantity} = \int Y(t) \, dX(t), \qquad (17.19)$$

that is, as an integral of some function $Y(t)$ weighted by the *increments* of $X(t)$. Such an integral is called a Stieltjes or Riemann–Stieltjes integral. The function $Y(t)$ being integrated is called the *integrand* as usual, and the function $X(t)$ whose increments weight the integrand is called the *integrator*. One useful feature is that this formulation unifies continuous, discrete or mixed integrators $X(t)$.

- At a point t where $X(t)$ is continuous we picture the quantity under the integral sign as:

$$Y(t) \frac{dX(t)}{dt} \, dt. \qquad (17.20)$$

(It is possible that $X(t)$ may be continuous but not differentiable at some points. This can be accommodated; we write equation (17.20) purely to supply some intuition.)

- At a point t where $X(t)$ has a discrete jump $\Delta X(t) = X(t) - X(t^-)$ we suppose that the quantity under the integral sign is:

$$Y(t) \, \Delta X(t). \qquad (17.21)$$

As an example of the latter, suppose a person's lifetime is modelled by the two-state counting process $N_i^{01}(t)$, and a sum assured of S will be paid upon their dying at time $t \le T$. Then the present value at time 0 of the benefit (a random variable) at force of interest δ can be written as:

$$\text{Present value of sum paid} = \int_0^T e^{-\delta t} S \, dN_i^{01}(t), \qquad (17.22)$$

where the integrator is $N_i^{01}(t)$, which has at most one discrete jump.

This integral in equation (17.19) is a stochastic integral if the integrand, the integrator, or both, are stochastic processes. This is legitimate provided two conditions are met (they always will be in this chapter):

- The process $Y(t)$ must be previsible with respect to $X(t)$. In equation (17.22) the integrand was deterministic and continuous, hence previsible with respect to any process.
- In between any discrete jumps, the integrator $X(t)$ must be sufficiently smooth. This is where the Itô integral takes a different direction; there the integrators (Brownian motions) are so unsmooth that the Stieltjes integral cannot be defined.

The stochastic integral thus defined is a stochastic process, as a function of its upper limit of integration. It is known at time t given the information $\mathcal{F}_t$ generated by the integrator.

17.4 Martingales

Suppose a stochastic process $X(t)$ generates the information $\{\mathcal{F}_t\}_{t \geq 0}$ as it is observed. Suppose also that $E[|X(t)|] < \infty$ for all $t \geq 0$ (a technical condition).

Definition 17.1 $X(t)$ is a *martingale* with respect to $\{\mathcal{F}_t\}_{t \geq 0}$ if, for $s \leq t$:

$$E[X(t) \mid \mathcal{F}_s] = X(s). \tag{17.23}$$

In words, a martingale is a stochastic process whose conditional expected value at all future times is equal to its value now. The simplest example, in discrete time, is a random walk without drift: $\Delta X(t) = \pm 1$, each with probability $1/2$.

The information $\{\mathcal{F}_t\}_{t \geq 0}$ is an essential part of the definition, so much so that we usually call $X(t)$ an $\mathcal{F}_t$-martingale if it satisfies the conditions above.

A martingale may be multivariate, in which case each component process is a univariate martingale with respect to *all* the information $\{\mathcal{F}_t\}_{t \geq 0}$ (not just information gained from observing that single component).

Martingales have four key properties for our purposes.

- *Conditional expectations of a random variable form a martingale.* If Z is a random variable known given $\mathcal{F}_T$ for some $T > 0$ (that is, whose value will be known at time T but not necessarily before time T), then by the Tower law the process $X(t)$ defined for $0 \leq t \leq T$ by:

$$X(t) = E[Z \mid \mathcal{F}_t] \tag{17.24}$$

   is an $\mathcal{F}_t$-martingale.

- *The future increments of a martingale have mean zero.* This is simple to show in discrete time, but also true in continuous time. Let $X(t)$ be a discrete-time $\mathcal{F}_t$-martingale $X(t)$ and consider $E[X(t+1) \mid \mathcal{F}_t]$:

$$E[X(t+1) \mid \mathcal{F}_t] = E[X(t) + \Delta X(t+1) \mid \mathcal{F}_t] \tag{17.25}$$

$$= E[X(t) \mid \mathcal{F}_t] + E[\Delta X(t+1) \mid \mathcal{F}_t] \tag{17.26}$$

$$= X(t) + E[\Delta X(t+1) \mid \mathcal{F}_t]. \tag{17.27}$$

But, by the martingale property, $E[X(t+1) \mid \mathcal{F}_t] = E[X(t)]$, implying that $E[\Delta X(t+1) \mid \mathcal{F}_t] = 0$. This gives us an important way of characterising martingales, as processes whose future increments have mean zero, given all current information. A similar argument shows that this is true for the increments of $X(t)$ over any future time period, not just one step ahead.

The corresponding property in continuous time can be expressed as:

$$E[X(u) - X(t) \mid \mathcal{F}_t] = 0 \qquad \text{(for } u \geq t) \tag{17.28}$$

or, in terms of increments:

$$E[dX(t) \mid \mathcal{F}_{t^-}] = 0. \tag{17.29}$$

- *Martingales have uncorrelated future increments.* If $X(t)$ is an $\mathcal{F}_t$-martingale and $t \leq u < v \leq w < z$, then the increments of $X(t)$ on the non-overlapping intervals $[u, v]$ and $[w, z]$ are $X(u) - X(v)$ and $X(w) - X(z)$, respectively. It can be shown that:

$$E[(X(u) - X(v))(X(w) - X(z)) \mid \mathcal{F}_t] = 0. \tag{17.30}$$

- *When the integrator in a stochastic integral is a martingale, then the stochastic integral of a previsible process is also a martingale.* We can show this in discrete time, under appropriate conditions:

  – Let $X(t)$ be an $\mathcal{F}_t$-martingale.
  – Let $Y(t)$ be a process previsible with respect to $X(t)$ (or $\{\mathcal{F}_t\}_{t \geq 0}$). In discrete time this means that $Y(t)$ is known at time $t - 1$.
  – Define the following stochastic process as a function of time t:

$$Z(t) = Y(0) + \sum_{r=0}^{t-1} Y(r+1) \Delta X(r+1). \tag{17.31}$$

This is the discrete time analogue of the stochastic integral, namely the sum of the integrand weighted by the increments of the integrator.

Then $Z(t)$ is an $\mathcal{F}_t$-martingale. An outline proof is as follows. Consider $E[Z(t) \mid \mathcal{F}_s]$ for some $s < t$. This conditional expectation is then:

$$E[Z(t) \mid \mathcal{F}_s] = E[Y(0) \mid \mathcal{F}_s] + \sum_{r=0}^{t-1} E[Y(r+1)\,\Delta X(r+1) \mid \mathcal{F}_s]$$

$$= Y(0) + \sum_{r=0}^{s-1} E[Y(r+1)\,\Delta X(r+1) \mid \mathcal{F}_s]$$

$$+ \sum_{r=s}^{t-1} E[Y(r+1)\,\Delta(X(r+1) \mid \mathcal{F}_s]$$

$$= Y(0) + \sum_{r=0}^{s-1} Y(r+1)\,\Delta X(r+1)$$

$$+ \sum_{r=s}^{t-1} E[E[Y(r+1)\,\Delta X(r+1) \mid \mathcal{F}_r] \mid \mathcal{F}_s]$$

$$= Z(s) + \sum_{r=s}^{t-1} E[Y(r+1)\,E[\Delta X(r+1) \mid \mathcal{F}_r] \mid \mathcal{F}_s] \qquad (17.32)$$

$$= Z(s), \qquad (17.33)$$

where in the last two steps we have applied, successively, the Tower law, the previsibility of $Y(t)$ and the fact that the conditional expected values of the future increments of a martingale are zero.

The continuous time version of this last result is the statement that, under some technical conditions, the integral:

$$Z(t) = \int_0^t Y(s)\,dX(s) \qquad (17.34)$$

is an $\mathcal{F}_t$-martingale. Here, the previsibility of $Y(t)$ can be taken to mean that its paths are left-continuous. The most important example we shall see is when the integrand includes any of the processes $Y_i^j(t)$ indicating presence of individual i in state j at time t^-.

17.5 Martingales out of Counting Processes

Suppose $X(t)$ is a stochastic process which is non-decreasing over time (if it is multivariate, each of its component processes is non-decreasing over time). Counting processes are obvious examples. Let $\{\mathcal{F}_t\}_{t \geq 0}$ be the information generated by observing $X(t)$. Since it is non-decreasing, $E[X(s) \mid \mathcal{F}_t] \geq X(t)$ for $s \geq t$, so $X(t)$ is not an $\mathcal{F}_t$-martingale. But martingales have such nice properties that we ask: can we find any martingales associated with $X(t)$? The answer is yes, there exists a unique non-decreasing process, denoted by $A(t)$, previsible with respect to $\mathcal{F}_t$, such that $X(t) - A(t)$ is an $\mathcal{F}_t$-martingale. The process $A(t)$ is called the $\mathcal{F}_t$-*compensator* of $X(t)$. This result is known (in continuous time) as the Doob–Meyer decomposition. Its proof is formidably technical but its application is simple. (The corresponding result in discrete time, known as the Doob decomposition, is surprisingly simple; just rearrange equation (17.16).)

In Section 15.2 we defined counting processes, counting transitions between states. For now we consider the simplest two-state alive–dead model, with transition from state 0 to state 1 governed by transition intensity $\mu^{01}(t)$. In Section 15.2 we also introduced a process $Y_i^0(t)$, indicating *observed* presence of the ith individual in state 0 at time (or age) t^- (thus taking care of left-truncation and right-censoring). The counting process $N_i^{01}(t)$ generates the information $\{\mathcal{F}_{i,t}\}_{t \geq 0}$, where we have added a subscript i to identify observation of the ith individual. Until we meet multivariate counting processes, we will omit the 0 and 01 superscripts for brevity, and just write $N_i(t), Y_i(t), A_i(t)$ and $\mu(t)$.

Since $N_i(t)$ is non-decreasing, it has a compensator. We have already seen it, in equation (15.13). It is:

$$A_i(t) = \int_0^t Y_i(s)\mu(s)\,ds. \tag{17.35}$$

(It is this definition (or result) that made us require the processes $Y_i(t)$ to be previsible with respect to $\{\mathcal{F}_{i,t}\}_{t \geq 0}$.)

It is convenient to define a *counting process martingale* as follows:

$$M_i(t) = N_i(t) - A_i(t) = N_i(t) - \int_0^t Y_i(s)\mu(s)\,ds. \tag{17.36}$$

The martingale property can then be expressed compactly as:

$$E[dM_i(t) \mid \mathcal{F}_{i,t^-}] = 0. \tag{17.37}$$

This is interpreted in just the same way as equation (3.18), as follows. At time t^-, if the ith individual is under observation, $Y_i(t) = 1$, otherwise $Y_i(t) = 0$.

If $Y_i(t) = 1$, then $dN_i(t)$ is a random variable that takes the value 1 if there is a transition in time dt, and 0 otherwise. Its expected value is the probability that such a transition occurs, which is $Y_i(t)\mu(t)\,dt + o(dt)$, and in the limit $Y_i(t)\mu(t)\,dt = dA_i(t)$.

Most importantly for applications in survival analysis, the second-moment properties of these counting process martingales are simple. We have:

$$\text{Var}[dM_i(t) \mid \mathcal{F}_{i,t^-}] = \text{Var}[dN_i(t) - dA_i(t) \mid \mathcal{F}_{i,t^-}]. \tag{17.38}$$

But $dA_i(t)$ is known (constant) given $\mathcal{F}_{i,t^-}$, because $A_i(t)$ is previsible, and $dN_i(t)$ is a Bernoulli 0–1 random variable with parameter $dA_i(t)$, which in the limit is small, so:

$$\text{Var}[dM_i(t) \mid \mathcal{F}_{i,t^-}] = \text{Var}[dN_i(t) \mid \mathcal{F}_{i,t^-}] \tag{17.39}$$

$$= dA_i(t)(1 - dA_i(t)) \tag{17.40}$$

$$\approx dA_i(t). \tag{17.41}$$

A more rigorous treatment shows that this result is exact, not approximate. Equations (17.37) and (17.41) finally explain the pervasiveness of Poisson-like random variables and likelihoods in survival analysis. Heuristically, given $\mathcal{F}_{i,t^-}$, and when $Y_i(t) \neq 0$, the increment $dN_i(t)$ is a Bernoulli 0–1 random variable with mean and variance both equal to $\mu(t)\,dt$; that is, in the limit, proportional to a hazard rate. Thus, when $Y_i(t) \neq 0$, a counting process behaves exactly like a Poisson process. It can be shown that Poisson processes are precisely those counting processes that have continuous, deterministic compensators (Karr, 1991). So survival models are seen to be based on Poisson-like counting processes whose compensators are not continuous and deterministic because they are modified by events themselves, such as left-truncation, right-censoring and transitions between states.

If we have a multiple-state model with more than one possible transition, we define a multivariate counting process with components $N_i^{jk}(t)$, one for each pair of states $j \neq k$, governed by a transition intensity $\mu^{jk}(t)$, and previsible indicator processes $Y_i^j(t)$ indicating *observed* presence of the ith individual in state j at time t^-. For each pair of states, define the counting process martingale $M_i^{jk}(t)$ by:

$$M_i^{jk}(t) = N_i^{jk}(t) - A_i^{jk}(t) = N_i^{jk}(t) - \int_0^t Y_i^j(s)\mu^{jk}(s)\,ds, \tag{17.42}$$

and exact analogues of equation (17.37) (for first moments) and (17.41) (for second moments) will hold for each of them. Note that $E[dN_i^{jk}(t) \mid \mathcal{F}_{i,t-}]$ may be zero either because $Y_i^j(t) = 0$ (the ith individual is not in state j) or because $\mu_i^{jk}(t) = 0$ (transition from state j to state k is impossible), or both.

Further, recall our assumption (see Section 15.2) that no two components of the multivariate counting process can jump simultaneously. As a consequence, we have, for distinct transitions jk and $j'k'$ of the counting process, that:

$$E[dM_i^{jk}(t) \, dM_i^{j'k'}(t) \mid \mathcal{F}_{t-}] = 0, \qquad (17.43)$$

and the second-moment properties of the $M_i^{jk}(t)$ are complete. (Technically, equation (17.43) means that the components of this multivariate martingale are *orthogonal*.)

If we observe $n > 1$ individuals, we may aggregate all the counting (and associated) processes, $N^{jk}(t) = \sum_{i=0}^{n} N_i^{jk}(t)$ and so on and define $\{\mathcal{F}_t\}_{t \geq 0}$ to be the information generated by all the observations, and use these to define any quantities that appear in this book, such as crude mortality ratios (see Section 17.7.1) or non-parametric estimators (see Section 17.7.3).

None of the above is at all rigorous. See Andersen et al. (1993) or Fleming and Harrington (1991) for details. Kalbfleisch and Prentice (2002, Chapter 5), has a nice treatment intermediate between ours and the two books cited above.

17.6 Martingale Central Limit Theorems

A most important consequence of the "nice" first- and second-moment properties of martingales is the existence of *martingale central limit theorems*. Recall that the central limit theorem for univariate random variables states that if we have a series $X_1, X_2, \ldots$ of independent, identically distributed random variables meeting certain conditions, and if we define $S(n) = \sum_{i=1}^{n} X_i$, then:

$$\lim_{n \to \infty} \frac{S(n) - E[S(n)]}{\text{Var}[S(n)]^{1/2}} \sim \text{normal}(0, 1). \qquad (17.44)$$

(Note that $S(n) - E[S(n)]$ in the numerator above strongly resembles a martingale created out of a process and its compensator, if n were to play the role of time.) In the martingale version, aggregated quantities such as $N(t) = \sum_{i=0}^{n} N_i(t)$ (omitting jk superscripts here) play the part of the partial sums $S_n = \sum_{i=1}^{n} X_i$, and similarly the aggregated $M(t) = \sum_{i=0}^{n} M_i(t)$ play the part of the numerator in equation (17.44). The result is that, as the number of individuals grows large, the paths of a martingale, normalised as in equation (17.44), tend to the paths of a zero-mean Brownian motion with known variance process, which is the stochastic-process counterpart of the normal(0,1) distribution in equation (17.44).

Moreover, many important statistics arising in survival analysis can be written as a stochastic integral of the form:

$$X(t) = \sum_{i=1}^{n} \int_0^t Z_i(s)\, dN_i(s), \qquad (17.45)$$

where $Z_i(s)$ is some observed (previsible) quantity relating to the ith individual, which is summed over the jump times of $N_i(s)$ and then the resulting sums are summed again over all individuals. We will see examples in Section 17.7. Defining the counting process martingales $M_i(t) = N_i(t) - A_i(t)$ we see that:

$$\sum_{i=1}^{n} \int_0^t Z_i(s)\, dM_i(s) \qquad (17.46)$$

is also a martingale, with the same second-moment properties as $X(t)$ because compensators are previsible; that is:

$$\mathrm{Var}\left[\sum_{i=1}^{n} \int_0^t Z_i(s)\, dM_i(s)\right] = \mathrm{Var}\left[\sum_{i=1}^{n} \int_0^t Z_i(s)\, dN_i(s)\right] \qquad (17.47)$$

(see equation (17.41)). Then an application of the martingale central limit theorem will give us the asymptotic properties of the statistic $X(t)$.

We now have enough to survey the considerable harvest that can be gathered from this effort.

17.7 A Brief Outline of the Uses of Counting Process Models

Having pinpointed martingales as embodying all the information about counting processes and intensities in one package, we can use their remarkable properties to say rather a lot. The following list is by no means complete. For technical details, see Andersen et al. (1993), Fleming and Harrington (1991) or the short survey by Macdonald (1996).

The reason why counting processes unlock the range of topics below lies in the apparent puzzle about survival data, mentioned at the start of this chapter, which we now examine in more detail. Consider any statistic that involves data on the same individual life over some extended period, say a number of calendar years. For example, suppose we observe data in the form "time under observation" and "indicator of death" defined in Section 4.6, in respect of the calendar years 2010 and 2011 as follows: (D_i^{2010}, T_i^{2010}) and (D_i^{2011}, T_i^{2011}). The statistics of interest, in general, are then some function of these (and many other observations):

$$\text{Statistics of interest} = f(D_i^{2010}, T_i^{2010}), f(D_i^{2011}, T_i^{2011}), \dots \qquad (17.48)$$

Can we calculate the sampling properties of these statistics? An immediate problem is that (D_i^{2010}, T_i^{2010}) and (D_i^{2011}, T_i^{2011}) are not independent of each other, or even uncorrelated. If we observe $d_i^{2010} = 1$, for example, we know $(d_i^{2011}, t_i^{2011}) = (0, 0)$. So the second moments of these statistics ought, it seems, to contain non-zero correlations between the random variables representing different segments of the same individual's life history.

Another way of expressing the above, in the language of counting processes, is that the $N_i^{jk}(t)$ do not have uncorrelated increments. But the $M_i^{jk}(t)$ do, because they are martingales. The $M_i^{jk}(t)$ "carry" all the information of interest – the innovations of the $M_i^{jk}(t)$ are exactly the same as those of the $N_i^{jk}(t)$ (see equation (17.16)) – but have the "nice" properties of uncorrelated increments and orthogonality, making second moments of statistics based on life histories tractable.

17.7.1 Markov and Semi-Markov Models

The Markov and semi-Markov models described in earlier chapters, and used in applications such as disability insurance, fall into the counting process framework. In applying results from the theory of counting processes, including any relating to martingales, we need to use the full history $\{\mathcal{F}_t\}_{t \geq 0}$ of the process. However, when calculating probabilities of future events, the Markov or semi-Markov properties mean that we do not need the full history.

We saw in equation (15.21) that the intensities in the Markov model could be estimated separately because the likelihood factorises. The same is true if we write down the likelihood in the counting process model (which we shall omit). Moreover, the moment properties we need to describe the likelihood all follow from the martingale properties of the $M_i^{jk}(t)$. If we assume, for simplicity, that we are observing a single year of age over which all intensities are constant, we have the following:

- First moments follow from:

$$E\left[\sum_i N_i^{jk}(t)\right] = \mu^{jk} E\left[\sum_i \int_0^1 Y_i^j(t)\, dt\right], \qquad (17.49)$$

where, in the right-hand side, we can substitute the observed for the expected exposure.
- Asymptotic variances follow from equation (17.41).
- Covariances are asymptotically zero because the $M_i^{jk}(t)$ are orthogonal.

(These can also be obtained using the martingale central limit theorem.)

These results are not confined to Markov and semi-Markov models. They depend on the assumption of a constant hazard rate over a defined set of properties, which could include an interval of age, duration, gender, smoking status or anything else.

17.7.2 Right-Censoring

In Section 5.2 we discussed the role of right-censoring in a probabilistic model capable of generating the data. In almost all cases, we were forced to suppose that right-censoring was independent or non-informative and did not need to be modelled jointly alongside survival. Andersen et al. (1993) put it thus: "A fundamental problem that we have to face is that the right-censoring may *alter the intensities for the events of interest*" (italics in original). Here the idea of a "probabilistic model capable of generating the data" begins with a collection of processes $\tilde{N}_i^{jk}(t)$ and $\tilde{Y}_i^j(t)$, where the tilde denotes absence of left-truncation and right-censoring. An observation left-truncated at time (age) x and right-censored at time s is represented by the observed processes $N_i^{jk}(t)$ and $Y_i^j(t)$, setting $Y_i^j(t) = 0$ for all j and all $t < x$ and $t > s$.

Note that, although transition from one state to another is a form of right-censoring of observation in the first state, it is generated by an event that is part of the probabilistic model we wish to study. Here we are considering "true" right-censoring generated by a mechanism external to the model we wish to study, which would be a nuisance if we had to estimate anything about it.

The discussion in Section 5.2 carries over to counting process models. The information generated by the uncensored observations is (or would be):

$$\mathcal{G}_t = \sigma(\tilde{N}_i^{jk}(s), \tilde{Y}_i^j(s) : s \le t) \qquad (17.50)$$

and is smaller than that generated by the censored observations, because fewer events are possible (compare equations (5.4) and (5.5)), so $\mathcal{G}_t \subseteq \mathcal{F}_t$ for all t.

The uncensored processes $\tilde{N}_i^{jk}(t)$ would have compensators $\tilde{A}_i^{jk}(t)$ such that $\tilde{M}_i^{jk}(t) = \tilde{N}_i^{jk}(t) - \tilde{A}_i^{jk}(t)$ would be G_t-martingales (we see now why the information is an essential part of the definition of a martingale). The censoring is then independent if the compensator $A_i^{jk}(t)$ which makes $M_i^{jk}(t)$ an $\mathcal{F}_t$-martingale is "the same as" $\tilde{A}_i^{jk}(t)$ in the sense that, for the same hazard function $\mu^{jk}(t)$, we have:

$$A_i^{jk}(t) = \int_0^t Y_i^j(s)\,\mu^{jk}(s)\,ds \quad \text{and} \quad \tilde{A}_i^{jk}(t) = \int_0^t \tilde{Y}_i^j(s)\,\mu^{jk}(s)\,ds \qquad (17.51)$$

(see equation (17.42)). Formally, $\tilde{M}_i^{jk}(t)$ is an $\mathcal{F}_t$-martingale as well as a G_t-martingale. But, since we never observe G_t, this is nearly always an unverifiable assumption.

17.7.3 The Nelson–Aalen Estimator

Suppose we have n lives, observed until death or right-censoring. The ith individual's history is modelled by the two-state counting process $N_i^{01}(t)$ and indicator $Y_i^0(t)$, which have the associated martingale $M_i^{jk}(t) = N_i^{jk}(t) - A_i^{jk}(t)$, and as before we omit the "0" and "01" superscripts for brevity. Define:

$$N(t) = \sum_{i=1}^n N_i(t), \quad Y(t) = \sum_{i=1}^n Y_i(t) \quad \text{and} \quad M(t) = \sum_{i=1}^n M_i(t). \qquad (17.52)$$

Then form the stochastic integral:

$$J(t) = \int_0^t \frac{I_{\{Y(s)>0\}}}{Y(s)}\,dM(s), \qquad (17.53)$$

with the convention that the integrand is zero if $Y(s) = 0$. The numerator is therefore just a book-keeping item to take care of division by zero. Since the integrator is a martingale, so is $J(t)$. But then we have:

$$J(t) = \int_0^t \frac{I_{\{Y(s)>0\}}}{Y(s)}\,dN(s) - \int_0^t \frac{I_{\{Y(s)>0\}}}{Y(s)}\,dA(s) \qquad (17.54)$$

$$= \int_0^t \frac{I_{\{Y(s)>0\}}}{Y(s)}\,dN(s) - \int_0^t \frac{I_{\{Y(s)>0\}}}{Y(s)}\,Y(s)\,\mu(s)\,ds \qquad (17.55)$$

$$= \int_0^t \frac{I_{\{Y(s)>0\}}}{Y(s)}\,dN(s) - \int_0^t I_{\{Y(s)>0\}}\mu(s)\,ds. \qquad (17.56)$$

The first term on the right-hand side of equation (17.56) is the Nelson–Aalen estimator. The second term is the integrated hazard rate, allowing for the possibility of no lives being under observation. The fact that $J(t)$ is a martingale tells us at once that the Nelson–Aalen estimator is (nearly) an unbiased estimate of the integrated hazard function, at ages where the number of lives observed is not too small. The martingale central limit theorem leads to simple expressions for the variance and confidence intervals of the Nelson–Aalen estimator.

Moreover, the relationship between the non-parametric estimators discussed in Chapter 8 is clarified: the Nelson–Aalen estimator arises naturally in the counting process framework and the Kaplan–Meier estimator turns out to be the product integral of the Nelson–Aalen estimator (see Andersen et al., 1993 for details).

Suppose the $N_i^{jk}(t)$ are multivariate counting processes, say the paths of a multiple-state model. Then the collection of Nelson–Aalen estimators for each $j \neq k$ form the off-diagonal elements of a matrix-valued stochastic process, called the Aalen–Johansen estimator. This carries much the same properties as the Nelson–Aalen estimator.

17.7.4 Kernel Smoothing

Chapter 11 introduced some methods of smoothing survival data, other than by fitting a parametric function. Here we consider one such method based on counting processes, namely *kernel smoothing*, with particular reference to the Nelson–Aalen estimator.

A *kernel* is a function $K(t)$, zero outside $[-1, 1]$, integrating to unity. Examples are the uniform kernel $K(t) = 1/2$ and the Epanechnikov kernel $K(t) = 0.75(1 - t^2)$. Then define:

$$\hat{\mu}(t) = \frac{1}{b} \int K\left(\frac{t-s}{b}\right) \frac{I_{\{Y(s)>0\}}}{Y(s)} \, dN(s). \qquad (17.57)$$

The parameter b is the *bandwidth*. We see that $\hat{\mu}(t)$ is a weighted average of the Nelson–Aalen estimator on the range $[-b, b]$ (hence the usual choice of a symmetric kernel). We can show that this is an asymptotically consistent estimator of $\mu(t)$ based on the fact that:

$$\frac{1}{b} \int K\left(\frac{t-s}{b}\right) \frac{I_{\{Y(s)>0\}}}{Y(s)} \, dM(s) \qquad (17.58)$$

is a martingale, along similar lines to Section 17.7.3. This is a modern version of the well-known moving average graduation methods; see Chapter 11 and Benjamin and Pollard (1980). Special measures are needed at ages t that are within b years of the lowest and highest ages; see Andersen et al. (1993).

17.7.5 Parametric Survival Models

In Chapter 7 we considered features of each individual in the portfolio that could be represented as a vector z_i of covariates, the influence of which depended upon a vector ζ of regression parameters, which we had to estimate. As an example we fitted models with proportional hazards (see, for example, equation (7.5)), although nothing about this approach mandates the use of proportional hazards.

The likelihoood from equation (15.25), for the two-state alive–dead model (omitting superscripts and allowing for covariates) is:

$$L(\mu, \zeta) = \prod_{i=1}^{n} \exp\left(-\int_0^\infty Y_i(t)\mu(t, z_i, \zeta)\, dt\right) \prod_{t\in[0,\infty]} (\mu(t, z_i, \zeta))^{dN_i(t)}. \quad (17.59)$$

(Recall from Section 15.8 that the indicator $Y_i(t)$ takes care of left-truncation and right-censoring so we need not trouble with these in the limits of integration.) The log-likelihood is therefore:

$$\ell(\mu, \zeta) = \sum_{i=1}^{n} \int_0^\infty \log \mu(t, z_i, \zeta)\, dN_i(t) - \sum_{i=1}^{n} \int_0^\infty Y_i(t)\mu(t, z_i, \zeta)\, dt. \quad (17.60)$$

Apart from the counting process notation, equation (17.60) is the same as equations (5.13) and (15.25). However, it remains true if we replace the constant covariates by time-varying functions, or even stochastic processes, as long as they are previsible.

Supposing purely for notational simplicity that the parameter ζ is univariate, the score function is:

$$\frac{\partial \ell}{\partial \zeta} = \sum_{i=1}^{n} \int_{0}^{\infty} \frac{\partial}{\partial \zeta}\Big(\log \mu(t, z_i, \zeta)\Big) dN_i(t) - \sum_{i=1}^{n} \int_{0}^{\infty} Y_i(t) \frac{\partial}{\partial \zeta}\Big(\mu(t, z_i, \zeta)\Big) dt$$

$$= \sum_{i=1}^{n} \int_{0}^{\infty} \frac{\partial}{\partial \zeta}\Big(\log \mu(t, z_i, \zeta)\Big) dN_i(t)$$

$$\quad - \sum_{i=1}^{n} \int_{0}^{\infty} Y_i(t) \frac{\partial}{\partial \zeta}\Big(\mu(t, z_i, \zeta)\Big) \frac{\mu(t, z_i, \zeta)}{\mu(t, z_i, \zeta)} dt$$

$$= \sum_{i=1}^{n} \int_{0}^{\infty} \frac{\partial}{\partial \zeta}\Big(\log \mu(t, z_i, \zeta)\Big) dN_i(t)$$

$$\quad - \sum_{i=1}^{n} \int_{0}^{\infty} Y_i(t) \frac{\partial}{\partial \zeta}\Big(\log \mu(t, z_i, \zeta)\Big) \mu(t, z_i, \zeta) dt$$

$$= \sum_{i=1}^{n} \int_{0}^{\infty} \frac{\partial}{\partial \zeta}\Big(\log \mu(t, z_i, \zeta)\Big) dM_i(t), \tag{17.61}$$

and hence by equation (17.46) is a martingale. This result, and its straightforward generalisations when we restore all the details we stripped out, assure the actuarial user of the validity of likelihood estimation, under an extraordinarily wide range of models which historically have usually been treated separately. The sampling properties of the estimates can be read off from the martingale central limit theorem. The range of models includes ordinary survival models, multiple-state models, Markov and semi-Markov models, multiple-decrement models, models fitted to single years of age or to entire observed lifetimes, models with fixed, varying or even stochastic covariates, all subject to any of the forms of left-truncation and right-censoring likely to be encountered in actuarial practice. It may also be extended to semi-parametric models such as the Cox model, although we have not discussed these in detail.

These results can all be regarded, in a sense, as being distilled from the Poisson-process-like behaviour of survival data. No such unification arises from models based on a binomial distribution, as Chapter 16 showed.

17.7.6 Extensions to Life Insurance Models

We have not by any means exhausted the applications of counting processes to problems of inference. However, for an actuarial readership, we end this chapter with an application to life contingencies, which demonstrates many of the properties developed in the preceding sections. The approach we describe is based on Ramlau-Hansen (1988).

We use the two-state alive–dead counting process $N(t)$ and its associated processes. Suppose a life-contingent contract with term T years pays a benefit of $b(t)$ on death at time t, and an annuity benefit at rate $a(t)$ per annum is payable if alive at time t^-. Both may be previsible stochastic processes. The signs of $b(t)$ and $a(t)$ are not specified, so this covers most life insurance and annuity contracts, but we define outgo from the insurer to be positive. Then the cumulative outgo up to time t, denoted by $B(t)$, is:

$$B(t) = \int_0^t b(s)\,dN(s) + \int_0^t Y(s)\,a(s)\,ds. \qquad (17.62)$$

Therefore $dB(t)$ is the instantaneous cash-flow at time t.

Define $v(t)$ to be the discount factor from time t to time 0. It too may be a previsible stochastic process. Then the present value at time 0 of all cash-flows is the random variable V defined as:

$$V = \int_0^T v(s)\,dB(s). \qquad (17.63)$$

Then (see equation (17.24)) $E[V \mid \mathcal{F}_t]$ is an $\mathcal{F}_t$-martingale; denote this by $X(t)$. We can decompose $X(t)$ into what is known by time t and an expectation of what has yet to happen (see the discussion of the coin-tossing model following equation (17.6)):

$$X(t) = E[V \mid \mathcal{F}_t] = \int_0^t v(s)\,dB(s) + E\left[\int_t^T v(s)\,dB(s)\,\Big|\,\mathcal{F}_t\right] \qquad (17.64)$$

$$= \int_0^t v(s)\,dB(s) + v(t)\,E\left[\int_t^T \frac{v(s)}{v(t)}\,dB(s)\,\Big|\,\mathcal{F}_t\right], \qquad (17.65)$$

in which we recognise the last conditional expectation:

$$E\left[\int_t^T \frac{v(s)}{v(t)}\,dB(s)\,\Big|\,\mathcal{F}_t\right] \qquad (17.66)$$

to be the ordinary prospective policy value at time t, correctly taking the value 0 if death has already occurred. Denote this policy value by $V(t)$.

Now consider the loss in time interval $(r, t]$ discounted back to time 0. Denote this by $L(r, t)$. We start with the policy value at time r, discounted, namely $v(r)\,V(r)$. The discounted outgo between time r and time t is $\int_r^t v(s)\,dB(s)$. Finally we must set up the required policy value $V(t)$ at time t; its discounted value is $v(t)\,V(t)$. Hence:

$$L(r, t) = \int_r^t v(s)\, dB(s) + v(t)\, V(t) - v(r)\, V(r) \qquad (17.67)$$

$$= \left(\int_0^t v(s)\, dB(s) + v(t)\, V(t) \right) - \left(\int_0^r v(s)\, dB(s) + v(r)\, V(r) \right) \quad (17.68)$$

$$= X(t) - X(r). \qquad (17.69)$$

Therefore, the discounted loss over any time interval is the increment of the martingale $X(t)$. It follows that discounted losses over non-overlapping time intervals are uncorrelated. This is *Hattendorff's theorem*, which, remarkably, dates from 1868 (Hattendorff, 1868). It is easily extended to multiple-state models.

Appendix A
R Commands

A.1 Introduction

This book illustrates many of its algorithms with source code for R, the open-source statistical modelling software available at www.r-project.org. This appendix gives an overview of how to get R and lists some of the more useful commands.

R is a fourth-generation programming language (4GL), which means that the user can accomplish a lot with fewer commands than programming languages such as BASIC and C. R functions are collected in *packages*, which have to be downloaded and installed before use. To download a new R package, say *newpackage*, you would use the following at the R prompt:

```
>install.packages("newpackage")
```

whereupon you will be asked to select a mirror site from which to download the package. The usual rule is to select a site which is geographically close to you. You only need to call `install.packages("newpackage")` once on the system where you wish to use *newpackage*.

After installation the package will be stored locally on your computer or network, but it will not be automatically available for use every time you start R. Whenever you want to use functions within a package, you will have to have the following in your R script:

```
>library("newpackage")
```

You will have to call `library("newpackage")` for every R session in which you want to use a function in the package.

Tables A.1 to A.7 list various useful R functions by subject, together with the package where they can be found. If the package is given as *base*, *graphics*, *grDevices*, *stats* or *utils* then it is part of the basic R system and you do not need to call `install.packages()` or `library()`, as the functions in these packages are always available. Note that the descriptions in these tables are very basic and are intended merely to make the reader aware of the various functions. Fuller details on how to use the various functions are available by using the `help()` command in the R console.

A.2 Running R

R can be run in one of two modes: interactively or in batch. The interactive mode features a console for entering commands line-by-line. Alternatively, you can create the command lines in a separate text editor and paste in the lines you want to run.

For larger pieces of work it can be handy to run R in batch mode. The details of this depend on your operating system. For Windows computers you will need an MS-DOS prompt, i.e. the `cmd` command to open a command shell. For Unix computers you will need a terminal process. In both cases you can run an R script in batch mode using the `Rscript` command. For debugging complex scripts we find it handy to save objects prior to a problem and load them into an interactive R process; for this we find the `save()` and `load()` commands useful.

A.3 R Commands

All the graphs in this book have been generated in R, which has powerful options for plotting and image creation. An overview of some of the common commands is given in Table A.7.

A.4 Probability Distributions

R also has many functions for simulation and manipulating probability distributions. Typically four functions are available for each distribution, each prefixed according to the following scheme:

- d for the probability density function (continuous variables), or probability mass function (discrete variables)
- p for the cumulative distribution function
- q for the quantile function
- r for the simulation function.

Table A.1 *Basic R commands.*

R command	Package	Description
|	*base*	Logical OR operator.
&	*base*	Logical AND operator
a:b	*base*	Create a range of values from a to b in steps of 1
c()	*base*	Concatenate values into a vector
dim()	*base*	Find or set dimensions of an object
help("item")	*base*	Look up documentation for item
for()	*base*	Create a loop
function()	*base*	Define a new function
install.packages()	*base*	Download and install package
length()	*base*	Get length of vector
library("newpackage")	*base*	Load newpackage into your current R session
load()	*base*	Load previously saved R objects from a file
ls()	*base*	List all the current objects in the workspace
names()	*base*	List contents of an R object, or set their names
read.csv()	*utils*	Read in a CSV file
rep()	*base*	Replicate values
save()	*base*	Save R objects to a file
seq()	*base*	Create a range of values
source()	*base*	Read a file of R commands
summary()	*base*	Get summary information for an object
write.csv()	*utils*	Write data to a CSV file

Table A.2 *R mathematical functions.*

R command	Package	Description
abs()	*base*	Compute absolute values
cumsum()	*base*	Compute vector of cumulative sums
diff()	*base*	Compute lagged differences
exp()	*base*	Compute exponentials
kronecker	*base*	Form Kronecker product
log()	*base*	Compute logarithms
rnorm()	*stats*	Generate a sample from a normal distribution
sqrt()	*base*	Compute square root
var()	*stats*	Compute variance

The second part of the function name indicates the distribution concerned. Common distributions available include the following:

- norm for the normal distribution
- t for Student's *t* distribution (not to be confused with the t() function for transposing a matrix)

Table A.3 *R commands for model-fitting.*

R command	Package	Description
factor()	*base*	Define a factor, or convert values into factor levels
fit()	*survival*	Fit a Kaplan–Meier survival function
glm()	*stats*	Fit a generalised linear model
gnm()	*gnm*	Fit a generalised non-linear model
lm()	*stats*	Fit a linear model
Mort1Dsmooth()	*MortalitySmooth*	Compute P-spline smooth in 1-d
Mort2Dsmooth()	*MortalitySmooth*	Compute P-spline smooth in 2-d
predict()	*stats*	Predict results from a model fit
relevel()	*stats*	Redefine the levels of a factor to change the baseline
sarima()	*astsa*	Fit an ARIMA(p, q, d) model
sarima.for()	*astsa*	Forecast an ARIMA(p, d, q) model
summary()	*base*	Print summary of a model object, including coefficient values
survreg()	*survival*	Fit a parametric survival regression model
Surv()	*survival*	Create a survival object
vcov()	*stats*	Return covariance matrix for parameters of a model object

Table A.4 *R commands and operators for matrices.*

R command or operator	Package	Description
%*%	*base*	Matrix multiplication
apply()	*base*	Compute function on the margin of a matrix
cbind()	*base*	Create matrix from vectors
chol()	*base*	Calculate Cholesky factorisation of a matrix
diag()	*base*	Create a diagonal matrix
matrix()	*base*	Create a matrix
solve(A)	*base*	Find the inverse of matrix A
t(A)	*base*	Return the transpose of matrix A

- unif for the uniform distribution
- pois for the Poisson distribution.

Thus, the function dnorm() returns the density function for the normal distribution, while rpois() returns simulated values from the Poisson distribution. A full list of the distributions available in the *stats* package can be obtained by typing help(distributions) in the R console window.

Table A.5 *R commands for summarising data.*

R command	Package	Description
hdquantile()	*Hmisc*	Calculate the Harrell–Davis quantile estimate of a vector of values
kurtosis()	*moments*	Calculate the kurtosis of a vector of values
mean()	*base*	Calculate the mean of a vector of values
median()	*stats*	Calculate the median of a vector of values
qqnorm()	*stats*	Draw a normal quantile-quantile plot
quantile()	*stats*	Calculate the quantile of a vector of values
skewness()	*moments*	Calculate the skewness of a vector of values

Table A.6 *R commands for optimisation.*

R command	Package	Description
deriv()	*stats*	Analytically compute a derivative
integrate()	*stats*	Numerically integrate a function between a lower and an upper limit
nlm()	*stats*	Find the minimum function value in multiple dimensions
optim()	*stats*	Optimise a function value in multiple dimensions
optimise()	*stats*	Optimise a function value in one dimension

Table A.7 *R commands for graphs.*

R command	Package	Description
axis()	*graphics*	Set the axis details for a graph
dev.off()	*grDevices*	Close a graphics device and close the file
legend()	*graphics*	Draw a legend on a graph
lines()	*graphics*	Add data lines to a graph
matlines()	*graphics*	Add lines
matplot()	*graphics*	Plot columns of a matrix
matpoints()	*graphics*	Add points
par()	*graphics*	Control aspects of graph presentation
pdf()	*grDevices*	Direct graph commands to a portable document format file (PDF)
plot()	*graphics*	Plot a graph
png()	*grDevices*	Direct graph commands to a portable network graphic (PNG) file
points()	*graphics*	Add data points to a graph
postscript()	*grDevices*	Direct graph commands to a Postscript file
rect()	*graphics*	Draw a rectangle on a graph
setEPS()	*grDevices*	Set the Postscript type to "encapsulated"
text()	*graphics*	Write text on a graph

Appendix B

Basic Likelihood Theory

B.1 Scalar Parameter Models: Theory

Suppose $X' = (X_1, X_2, \ldots, X_n)$ is a vector of random variables, where the X_i are independent (not necessarily identical) with probability functions $f_i(x, \theta)$, each of which depends on the same scalar parameter θ (which does not determine the range of any of the X_i).

We regard θ as a variable ranging over a parameter space, which is a set of possible values of θ usually fixed by the model under consideration. For example, if the parameter is interpreted as a death probability q in the model, the parameter space may be the interval $[0, 1]$. Or, if the parameter is interpreted as a hazard rate μ in the model, the parameter space may be the half-line $[0, \infty)$.

The value of the parameter θ determines the probability functions $f_i(x, \theta)$ and therefore any operations involving the probability functions, such as the calculation of moments. So we need to include the parameter value explicitly in means, variances and so on; we will write $E_\theta[X]$, $Var_\theta[X]$ and so on as required.

Definition B.1 (Score function)   Let:

$$\ell = \ell(\theta) = \ell(\theta, X) = \sum_{i=1}^{i=n} \log f_i(X_i, \theta) \tag{B.1}$$

be the log-likelihood function (viewed as a random variable). Then:

$$U = U(\theta) = \frac{\partial \ell(\theta)}{\partial \theta} \tag{B.2}$$

is called the *score function*.

Definition B.2 (Fisher's information function)   *Fisher's information function* (or just the *information function*) is:

$$I(\theta) = -\mathrm{E}_\theta \left[\frac{\partial^2 \ell(\theta)}{\partial \theta^2} \right] = -\mathrm{E}_\theta \left[\frac{\partial U(\theta)}{\partial \theta} \right] = \mathrm{Var}_\theta[U(\theta)]. \tag{B.3}$$

Note a slight abuse of notation in the partial derivatives here (and later). For example:

$$-\mathrm{E}_\theta \left[\frac{\partial U(\theta)}{\partial \theta} \right] \quad \text{strictly means} \quad -\mathrm{E}_\theta \left[\frac{\partial U(\theta')}{\partial \theta'} \Big|_{\theta'=\theta} \right]. \tag{B.4}$$

The score function has some very simple and general properties.

Proposition B.3 *For all values of θ in the parameter space, the score function satisfies:*

(a) $\mathrm{E}_\theta[U(\theta)] = 0$

(b) $\mathrm{Var}_\theta[U(\theta)] = -\mathrm{E}_\theta \left[\dfrac{\partial U(\theta)}{\partial \theta} \right] = -\mathrm{E}_\theta \left[\dfrac{\partial^2 \ell(\theta)}{\partial \theta^2} \right]$ *(from equation (B.3)).*

Proof:

(a) For any possible θ, consider the identity:

$$\int f_i(x_i, \theta) \, dx_i = 1,$$

where the integration is over the range of X_i. Differentiate this with respect to θ:

$$\int \frac{\partial f_i(x_i, \theta)}{\partial \theta} \, dx_i = 0$$

$$\Rightarrow \int \left(\frac{\partial}{\partial \theta} \log f_i(x_i, \theta) \right) f_i(x_i, \theta) \, dx_i = 0 \tag{B.5}$$

$$\Rightarrow \sum_{i=1}^{i=n} \int \left(\frac{\partial}{\partial \theta} \log f_i(x_i, \theta) \right) f_i(x_i, \theta) \, dx_i = 0. \tag{B.6}$$

The left-hand side of this equation is $\mathrm{E}_\theta[U]$, so $\mathrm{E}_\theta[U] = 0$ as required.

(b) The proof uses the same trick. Differentiate equation (B.6) with respect to θ (we abbreviate $f_i(x_i, \theta)$ as f_i to make the following clearer):

$$\sum_{i=1}^{i=n} \int \left\{ \frac{\partial}{\partial \theta} \left(\frac{\partial}{\partial \theta} \log f_i \right) f_i + \left(\frac{\partial}{\partial \theta} \log f_i \right) \frac{\partial f_i}{\partial \theta} \right\} \, dx_i = 0$$

$$\Rightarrow \sum_{i=1}^{i=n} \int \frac{\partial}{\partial \theta} \left(\frac{\partial}{\partial \theta} \log f_i \right) f_i \, dx_i + \int \left(\frac{\partial}{\partial \theta} \log f_i \right) \left(\frac{\partial}{\partial \theta} \log f_i \right) f_i \, dx_i = 0$$

$$\Rightarrow E_\theta \left[\frac{\partial U(\theta)}{\partial \theta} \right] + E_\theta [U(\theta)^2] = 0$$

$$\Rightarrow E_\theta [U(\theta)^2] = -E_\theta \left[\frac{\partial U(\theta)}{\partial \theta} \right]$$

$$\Rightarrow \text{Var}_\theta [U(\theta)] = -E_\theta \left[\frac{\partial U(\theta)}{\partial \theta} \right] \quad \text{since } E_\theta[U] = 0$$

$$\text{or} \quad \text{Var}_\theta [U(\theta)] = -E_\theta \left[\frac{\partial^2 \ell}{\partial \theta^2} \right],$$

and (b) is proved. □

The score function $U(\theta)$ is a fundamental quantity in statistics. Among other things, it gives us properties of the MLE of θ:

- The MLE satisfies $\partial \ell(\theta)/\partial \theta = 0$ i.e. $U(\theta) = 0$.
- The asymptotic variance of the MLE is found from $\text{Var}_\theta[U(\theta)]$.

We are now ready to state (without proof) the *Maximum likelihood theorem*:

Proposition B.4 (Maximum likelihood theorem) *Suppose $\hat{\theta}$ satisfies $U(\theta) = 0$ and θ_0 is the true (but unknown) value of θ. If $\tilde{\theta}$ is the random variable (the estimator) corresponding to $\hat{\theta}$ (the estimate), then $\tilde{\theta}$ is:*

(i) asymptotically unbiased, that is, as $n \to \infty$, $E_{\theta_0}[\tilde{\theta}] \to \theta_0$

(ii) asymptotically efficient, that is, it achieves the Cramer–Rao lower bound
$\text{Var}_{\theta_0}[\tilde{\theta}] \to I(\theta_0)^{-1}$

(iii) asymptotically normal. □

We can summarise this as $\tilde{\theta} \approx N(\theta_0, I(\theta_0)^{-1})$. Of course, θ_0 is unknown, so how do we calculate $I(\theta_0)^{-1}$? There are two main approximations:

- If the expectation in equation (B.3) can be written down explicitly, simply replace θ_0 with $\hat{\theta}$, that is:

$$I(\theta_0) \approx I(\hat{\theta}). \tag{B.7}$$

- If we cannot work out the expectation in equation (B.3) then try:

$$I(\theta_0) \approx -\left(\frac{\partial^2 \ell}{\partial \theta^2}\right)\bigg|_{\theta=\hat{\theta}}. \tag{B.8}$$

B.2 The Single-Decrement Model

Suppose we have a single-decrement model with a constant hazard rate μ, and data consisting of total time V spent exposed to risk and D observed deaths. From equation (5.19), the log-likelihood is $\ell(\mu) = -\mu V + D \log \mu$, so the score is:

$$U(\mu) = -V + \frac{D}{\mu}, \tag{B.9}$$

leading to estimator $\tilde{\mu} = D/V$. Fisher's information function is:

$$I(\mu) = -E_\mu\left[\frac{\partial^2 \ell(\mu)}{\partial \mu^2}\right] = -E_\mu\left[\frac{\partial U(\mu)}{\partial \mu}\right] = \text{Var}_\mu[U(\mu)]. \tag{B.10}$$

Applying Proposition B.3(a) to equation (B.9), we get:

$$E_\mu[D/\mu] = E_\mu[V] \Rightarrow E_\mu[D] = \mu E_\mu[V]. \tag{B.11}$$

Now we apply the maximum likelihood theorem. Suppose that μ_0 is the true, but unknown, value of the hazard rate.

(i) First, $\tilde{\mu}$ is asymptotically unbiased, so $E_{\mu_0}[\tilde{\mu}] \to \mu_0$.
(ii) Second, $\tilde{\mu}$ is asymptotically efficient, so:

$$\text{Var}_{\mu_0}[\tilde{\mu}] \to I(\mu_0)^{-1} = \left\{-E\left[\frac{\partial U}{\partial \mu}\right]\right\}^{-1}\bigg|_{\mu=\mu_0}. \tag{B.12}$$

Differentiating equation (B.9), this becomes:

$$-\left\{E_{\mu_0}\left[-\frac{D}{\mu_0^2}\right]\right\}^{-1} = \left(\frac{E_{\mu_0}[D]}{\mu_0^2}\right)^{-1} = \frac{\mu_0^2}{E_{\mu_0}[D]} = \frac{\mu_0}{E_{\mu_0}[V]}, \tag{B.13}$$

where the last step follows from equation (B.11).

(iii) Third, $\tilde{\mu}$ is asymptotically normal: $\tilde{\mu} \approx N\left(\mu_0, \dfrac{\mu_0}{E_{\mu_0}[V]}\right)$. The true values of μ_0 and $E_{\mu_0}[V]$ are unknown, so in practice we substitute the estimate $\hat{\mu}$ and the observed value v, finally obtaining the approximate result we can use in practice:

$$\tilde{\mu} \approx N\left(\hat{\mu}, \frac{\hat{\mu}}{v}\right). \tag{B.14}$$

B.3 Multivariate Parameter Models: Theory

Definition B.5 (Score function)   Suppose $X' = (X_1, X_2, \ldots, X_n)$ is a vector of random variables, where the X_i are independent with probability functions $f_i(x, \theta)$, not necessarily identical, that depend on the same parameter $\theta' = (\theta_1, \ldots, \theta_r)$ with r components (which do not determine the range of any of the X_i, $i = 1, 2, \ldots, n$). Let:

$$\ell = \ell(\theta) = \ell(\theta, X) = \sum_1^n \log f_i(X_i, \theta) \tag{B.15}$$

be the log-likelihood function (viewed as a random variable). Let:

$$U_i = U_i(\theta) = \frac{\partial \ell}{\partial \theta_i}, \quad i = 1, \ldots r. \tag{B.16}$$

Then $U' = (U_1, \ldots, U_r) = (U_1(\theta), \ldots, U_r(\theta))$ is called the *score function*.

Proposition B.6   *For any possible value θ of the parameter, the score function satisfies*:

(a) $E_\theta(U(\theta)) = \mathbf{0}$

(b) $Var_\theta(U(\theta)) = -E_\theta\left(\dfrac{\partial U(\theta)}{\partial \theta}\right) = (v_{ij})_{r \times r}$, *where* $v_{ij} = -E_\theta\left(\dfrac{\partial^2 \ell(\theta)}{\partial \theta_i \partial \theta_j}\right)$.

Proof: The proof is very similar to the one-parameter case:

(a) For any possible θ, and $k = 1, 2, \ldots, n$, consider the identity:

$$\int f_k(x_k, \theta)\, dx_k = 1,$$

where the integration is over the range of X_k. Differentiate this with respect to θ_i:

$$\int \frac{\partial f_k(x_k, \boldsymbol{\theta})}{\partial \theta_i} \, dx_k = 0$$

$$\Rightarrow \int \left(\frac{\partial}{\partial \theta_i} \log f_k(x_k, \boldsymbol{\theta}) \right) f_k(k_i, \boldsymbol{\theta}) \, dx_k = 0 \tag{B.17}$$

$$\Rightarrow \sum_{k=1}^{k=n} \int \left(\frac{\partial}{\partial \theta_i} \log f_k(x_k, \boldsymbol{\theta}) \right) f_k(x_k, \boldsymbol{\theta}) \, dx_k = 0. \tag{B.18}$$

The left-hand side of this equation is $E_{\boldsymbol{\theta}}[U_i]$, so $E_{\boldsymbol{\theta}}[U] = 0$ as required.

(b) The proof uses the same trick. Differentiate equation (B.18) with respect to θ_j:

$$\sum_{i=1}^{i=n} \int \left\{ \frac{\partial}{\partial \theta_j} \left(\frac{\partial}{\partial \theta_i} \log f_k \right) f_k + \left(\frac{\partial}{\partial \theta_i} \log f_k \right) \frac{\partial f_k}{\partial \theta_j} \right\} \, dx_k = 0$$

$$\Rightarrow \sum_{i=1}^{i=n} \int \frac{\partial}{\partial \theta_j} \left(\frac{\partial}{\partial \theta_i} \log f_k \right) f_k \, dx_k + \int \left(\frac{\partial}{\partial \theta_i} \log f_k \right) \left(\frac{\partial}{\partial \theta_j} \log f_k \right) f_k \, dx_k = 0$$

$$\Rightarrow \sum_{i=1}^{i=n} \int \frac{\partial U_i(\boldsymbol{\theta})}{\partial \theta_j} f_k \, dx_k + \int U_i(\boldsymbol{\theta}) U_j(\boldsymbol{\theta}) f_k \, dx_k$$

$$\Rightarrow E_{\boldsymbol{\theta}}[U_i(\boldsymbol{\theta}) U_j(\boldsymbol{\theta})] = -E_{\boldsymbol{\theta}} \left[\frac{\partial U_i(\boldsymbol{\theta})}{\partial \theta_j} \right]$$

$$\Rightarrow \mathrm{Var}_{\boldsymbol{\theta}}[U_i(\boldsymbol{\theta}) U_j(\boldsymbol{\theta})] = -E_{\boldsymbol{\theta}} \left[\frac{\partial^2 \ell(\boldsymbol{\theta})}{\partial \theta_i \theta_j} \right],$$

and (b) is proved. □

The information function $I(\boldsymbol{\theta})$ extends as follows:

Definition B.7 (Fisher's information function) $I(\boldsymbol{\theta})$ is:

$$I(\boldsymbol{\theta}) = -E_{\boldsymbol{\theta}} \left[\frac{\partial^2 \ell(\boldsymbol{\theta})}{\partial \theta_i \theta_j} \right] = -E_{\boldsymbol{\theta}} \left[\frac{\partial U}{\partial \boldsymbol{\theta}} \right] = \mathrm{Var}_{\boldsymbol{\theta}} U(\boldsymbol{\theta}). \tag{B.19}$$

As before we usually call $I(\boldsymbol{\theta})$ the *information function*. We are now ready to state (without proof) the multi-parameter version of the maximum likelihood theorem.

Proposition B.8 (Maximum likelihood theorem) *Suppose $\hat{\boldsymbol{\theta}}$ satisfies $U(\boldsymbol{\theta}) = 0$ and $\boldsymbol{\theta}_0$ is the true (but unknown) value of $\boldsymbol{\theta}$. If $\tilde{\boldsymbol{\theta}}$ is the random variable corresponding to $\hat{\boldsymbol{\theta}}$, then $\tilde{\boldsymbol{\theta}}$ is:*

(i) *asymptotically unbiased, that is, as* $n \to \infty, \mathrm{E}_{\theta_0}[\tilde{\theta}] \to \theta_0;$

(ii) *asymptotically efficient, that is, it achieves the Cramer–Rao lower bound, i.e.* $\mathrm{Var}_{\theta_0}[\tilde{\theta}] \to I(\theta_0)^{-1};$

(iii) *asymptotically multivariate normal.*   □

We can summarise this as $\tilde{\theta} \approx \mathcal{N}(\theta_0, I(\theta_0)^{-1}).$

For a more detailed theoretical treatment of maximum likelihood, see for example Cox and Hinkley (1974).

Appendix C

Conversion to Published Tables

C.1 Reasons to Use Published Tables

It is one thing to be able to fit a bespoke model with multiple risk factors. However, actuaries also need to express their bespoke bases in terms of a published table for communication purposes. Examples include reserving bases, communication with third parties (such as regulators and auditors) and of course circumstances where one wants to keep one's bespoke basis private (as in competitive pricing).

C.2 Equivalent-Reserve Method

A good way to convert mortality bases is by equating reserves, the so-called *equivalent-annuity method* when applied to pensions and annuities; see Willets (1999), Richards et al. (2013) and Richards (2016) for examples. This involves solving an equation where the only difference on each side is the mortality basis being used. For example, two mortality bases B and T would be deemed equivalent for a given annuity portfolio if the following held true:

$$\sum_{i=1}^{n} w_i \ddot{a}_{x_i}^{T} = \sum_{i=1}^{n} w_i \ddot{a}_{x_i}^{B}, \tag{C.1}$$

where w_i is the annual pension paid to life i and $\ddot{a}_{x_i}$ values an immediate level lifetime annuity paid to a life aged x_i. Without loss of generality, we can think of B as the bespoke, multi-factor basis and T as the much simpler one based on the published table. In the case of gender-differentiated rates in T, which is usually the norm, the summation in equation (C.1) is generally performed separately for males and for females.

Alternatively, the actuary can solve for a target reserve value:

$$\sum_{i=1}^{n} w_i \ddot{a}_{x_i}^T = S^p. \tag{C.2}$$

S^p could be an arbitrary target value, or it could be the appropriate quantile of the set of valuations carried out with similar $\ddot{a}$ functions. For example, Richards (2016) derives an approximate 95% confidence interval for the best-estimate basis for a pension scheme by solving for $S^{2.5\%}$ and $S^{97.5\%}$, where S is a set of valuations assessing mis-estimation risk (see Section 9.6). Note that the confidence interval need not be symmetric around the best-estimate percentages. One important point to note is that for gender-differentiated rates we would have to solve equation (C.2) separately for males and for females. Alternatively, we could solve for the male and female reserves combined and use the same percentage of table for both genders.

Another potential use of equation (C.2) is to express a regulatory stress. For example, a 99.5% mis-estimation stress could be expressed in terms of a standard table by using $S^{99.5\%}$ in equation (C.2). One reason for doing this might be if one needed a rough-and-ready approximation for daily stress-testing.

An emerging application of equation (C.2) is to compare the strength of mortality projection bases, i.e. where the current rates of mortality are the same under bases B and T and where the only difference lies in the projection. For a single portfolio this can be done simply enough by valuing the portfolio under each projection and comparing the reserve values. However, this is inadequate for comparing the bases between portfolios, especially for a regulator or analyst wanting to compare the bases used by different insurers or pension schemes. Mortality projections are two-dimensional surfaces in age and time, and the financial impact varies according to the liability distribution in each portfolio. Also, different portfolios have different effective discount rates depending on their backing assets and whether the liabilities are fixed, indexed or escalating. One approach is to solve equation (C.2) for an equivalent constant rate of mortality improvement. A regulator or analyst could then plot the equivalent constant rate of assumed improvement against the liability-weighted average age and thus compare the strength of different companies' mortality projections.

C.3 Algorithms for Solving for Percentage of Published Table

Equations like (C.1) and (C.2) can be non-linear, which points to the need for a robust method for root-solving. One example is the bisection method. This involves first establishing two percentages of the published table which respectively over- and under-shoot the target value f_1 and f_2, say then calculating the reserves produced by the mid-point percentage, $f_3 = (f_2 + f_2)/2$. Depending on whether the resulting reserve value is an over- or underestimate, f_3 replaces one of the two values so that (f_1, f_2) forms an ever-narrower interval and iteration proceeds until the desired degree of accuracy is achieved.

However, in a number of instances it is possible for a simple bisection algorithm for the equivalent-reserve calculation to be caught oscillating between two values. We have found that a practical solution to this is stochastic bisection: instead of always trying the mid-point of the two percentages, $f_3 = (f_1 + f_2)/2$, one uses a randomly weighted average $f_3 = (Uf_1 + (1 - U)f_2)/2$, where U is a randomly generated value from the uniform $(0,1)$ distribution. Although stochastic bisection is generally slower than deterministic bisection, we have found it to be more reliable in solving for roots for the equivalent-reserve method.

Appendix D
Numerical Integration

D.1 Introduction

A key task in evaluating the likes of equation (5.26) is the evaluation of the integrated hazard function, $\Lambda_{x,t}$ (see equation (3.25)). Many of the most useful mortality laws for actuaries have closed-form expressions for $\Lambda_{x,t}$; examples for the most common are given in Table 5.1, while further examples for less common mortality laws are tabulated in Richards (2012).

However, there may well be circumstances when the analyst needs to use a mortality law for which there is no closed-form expression. Under such circumstances, $\Lambda_{x,t}$ will have to be evaluated numerically. R offers a simple function for numerical integration: `integrate()`. This is illustrated in Figure D.1 for the evaluation of the Gompertz integrated hazard between ages 60 and 70, and where $\alpha = -10$ and $\beta = 0.1$. The exact integrated hazard is $\Lambda_{60,10} = 0.3147143$, while the numerical approximation returns 0.3147143 with a stated absolute error of less than 3.5×10^{-15}.

D.2 Implementation Issues

There are two important practical aspects to equation (5.26):

(i) Summation of a large number of small values into a large one (the log-likelihood). Mathematically this is not an issue, but readers should be aware of the potential problems due to computer representation of real numbers. Where n is large in equation (5.26), floating-point underflow can occur during summation. The underflow issue is described by Kahan (1965), who also gives a simple algorithm to guard against it.

(ii) Evaluation of expressions with problematic intermediate values. Consider

344

```
gdAlpha <- -10.0
gdBeta  <- 0.1

# Define Gompertz hazard function:
mu <- function(x)
{
  exp(gdAlpha+gdBeta*x)
}

# Numerically integrate the hazard between 60 and 70:
integrate(mu, 60, 70)

# Define closed-form Gompertz integrated hazard function:
Lambda <- function(x, t)
{
  (exp(gdBeta*t)-1) / 0.1 * exp(gdAlpha+gdBeta*x)
}

# Evaluate exact integrated hazard function:
Lambda(60, 10)
```

Figure D.1 R source illustrating numerical approximation of the integrated hazard function for a Gompertz hazard with $\alpha = -10$ and $\beta = 0.1$.

the expression $e^x/(1 + e^x)$, variants of which appear in a number of mortality laws. Clearly, as $x \to \infty$ then $e^x/(1 + e^x) \to 1$. However, for large values of x the limitations of floating-point computer arithmetic can lead to the intermediate values overflowing. Richards (2012) lists some commonly encountered problematic expressions and describes safe ways of programming them to avoid under- or overflow during calculation.

D.3 Numerical Integration over a Grid of Fixed Points

The R code in Figure D.1 works when one wishes to integrate a function that can be evaluated at any point. However, in actuarial work it is common to have to calculate a life expectancy – or an annuity factor – where the survival function, ${}_tp_x$, is only available for integer values of t. How can we accurately integrate a function where the values of the function are only available at fixed points? One approach is to use some results from the history of mathematics.

For two points separated by one year, we use the trapezoidal rule:

$$\int_a^{a+1} f(x) \approx \frac{1}{2}\left[f(a) + f(a + 1) \right]. \tag{D.1}$$

Table D.1 *Numerical integration over a fixed grid.*

Age range	Rule applied	$\Lambda_{x,t}$
60–64	Boole's Rule (equation D.4)	0.090081
64–68	Boole's Rule	0.134385
68–70	Simpson's 3/8 Rule (equation D.3)	0.090249
60–70		0.314714

For three points spaced one year apart, we use Simpson's rule:

$$\int_a^{a+2} f(x) \approx \frac{1}{3} \left[f(a) + 4f(a+1) + f(a+2) \right]. \tag{D.2}$$

For four points spaced one year apart, we use Simpson's 3/8 rule:

$$\int_a^{a+3} f(x) \approx \frac{3}{8} \left[f(a) + 3f(a+1) + 3f(a+2) + f(a+3) \right]. \tag{D.3}$$

For five points spaced one year apart, we use Boole's rule:

$$\int_a^{a+4} f(x) \approx \frac{2}{45} \left[7f(a) + 32f(a+1) + 12f(a+2) + 32f(a+3) + 7f(a+4) \right]. \tag{D.4}$$

To integrate over n points at yearly intervals we would first apply Boole's rule as many times as possible, then Simpson's 3/8 rule, Simpson's rule or the trapezoidal rule for any remaining points at the highest ages. How accurate would this simple approach be? We illustrate by applying the above rules to our Gompertz example in Section D.1. We evaluate the Gompertz hazard μ_{60+t} for $t = 0, 1, 2, \ldots, 10$, then we apply the rules as shown in Table D.1 to get an acceptable approximation of the integrated hazard. The final value of 0.314714 agrees with the known theoretical value for the first six significant figures.

Appendix E
Mean and Variance-Covariance of a Vector

We generalise the familiar notions of the mean and variance of a random variable. Let X be a random vector, i.e. a vector whose components are random variables. In general, these components need not be independent, thus $X = (X_1, \ldots, X_n)'$ where n is the number of components. Let $E[X_i] = \mu_i$, $i = 1, \ldots, n$. We define the expectation of X, denoted by $E[X]$, by:

$$E[X] = E\begin{bmatrix} X_1 \\ X_2 \\ \vdots \\ X_n \end{bmatrix} = \begin{bmatrix} E[X_1] \\ E[X_2] \\ \vdots \\ E[X_n] \end{bmatrix} = \begin{bmatrix} \mu_1 \\ \mu_2 \\ \vdots \\ \mu_n \end{bmatrix} = \mu, \tag{E.1}$$

where $\mu = (\mu_1, \ldots, \mu_n)'$.

In the same way, we define the expectation of a random matrix $M = (M_{i,j})$ to be the matrix whose (i, j)th element is $E[M_{i,j}]$. For a random vector X, let:

$$V = E[(X - \mu)(X - \mu)'], \tag{E.2}$$

which gives

$$v_{i,i} = \text{Var}[X_{i,i}], \quad v_{i,j} = \text{Cov}[X_{i,j}], \quad i \neq j. \tag{E.3}$$

We refer to V as the *variance-covariance matrix* of X and write $V = \text{Var}[X]$.

The following results on the mean and variance of a linear transformation of a random vector are useful.

Proposition E.1 *Let X be a random vector with $E[X] = \mu$ and $\text{Var}[X] = V$. Let A be a matrix of constants such that AX is defined. Then:*

(a) $E[AX] = A\mu$;

(b) $\text{Var}[AX] = AVA'$.

Proof: Let $Y = AX$. Then we have:

$$E[Y] = E[AX] = AE[X] = A\mu. \qquad (E.4)$$

In a similar way we have:

$$\text{Var}[Y] = E\left[\{Y - E[Y]\}\{Y - E[Y]\}'\right] = E\left[A(X - \mu)(X - \mu)'A'\right] = AVA'. $$
$$(E.5)$$

$\square$

Appendix F
Differentiation with Respect to a Vector

We shall have occasion to differentiate functions of the form $a'x$ and $x'Ax$ with respect to x, where x is a vector; here a is a vector of constants and A is a symmetric matrix of constants.

Definition F.1 Let $x' = (x_1, \ldots, x_n)$ be a vector and $f(x)$ be a scalar function of x. Then we define:

$$\frac{\partial f}{\partial x} = \begin{pmatrix} \dfrac{\partial f}{\partial x_1} \\ \vdots \\ \dfrac{\partial f}{\partial x_n} \end{pmatrix}. \tag{F.1}$$

The following results are easily proved by appealing to the definition and elementary calculus.

Proposition F.2 *Let a be a vector and A a symmetric matrix, neither depending on x. Then:*

(a) $\dfrac{\partial a'x}{\partial x} = a$

(b) $\dfrac{\partial x'Ax}{\partial x} = 2Ax.$

Proof: These follow directly from the definition. □

We make two remarks: first, these generalise the familiar results of elementary calculus; second, $\dfrac{\partial x'x}{\partial x} = 2x$ is a special case of (b).

Appendix G

Kronecker Product of Two Matrices

Suppose we have two matrices $A = (a_{i,j})$, $m \times n$, and $B = (b_{i,j})$, $p \times q$. Then the Kronecker product of B by A, written $B \otimes A$, is the $mp \times nq$ block matrix:

$$
B \otimes A = \begin{pmatrix} b_{1,1} A & \cdots & b_{1,q} A \\ \vdots & \ddots & \vdots \\ b_{p,1} A & \cdots & b_{p,q} A \end{pmatrix}. \tag{G.1}
$$

Notice that $B \otimes A \neq A \otimes B$ in general.

The Kronecker product arises naturally in regression models when the data are in the form of a matrix, as in mortality data; here the rows of the data matrix are indexed by age and the columns by calendar year. When the model has a row-and-column structure the model matrix can usually be expressed compactly using the Kronecker product.

Example Let I_3 be the identity matrix of size 3, and $x = (x_1, \ldots, x_m)'$ be a vector of length m. Then:

$$
I_3 \otimes x = \begin{pmatrix} x & 0 & 0 \\ 0 & x & 0 \\ 0 & 0 & x \end{pmatrix}, \tag{G.2}
$$

where 0 is a vector of 0s of length m. This example would arise in the Cairns–Blake–Dowd model if there were m ages and three calendar years; see Section 12.3. The Kronecker product also arises in the smooth two-dimensional models of Section 12.4.

Appendix H
R Functions and Programs

In this Appendix we describe the R programs provided through the book's online website, www.cambridge.org/9781107045415. These programs can be used to produce many of the results and graphs in the book. File references in these programs may be changed to those of the reader. The section titles below correspond to the titles of Chapters 4, 6, 10, 11, 12 and 13, where these programs are introduced.

H.1 Statistical Inference with Mortality Data

Function: `splitExperienceByAge()`

Function to split up the individual time lived into single years of age. Note that age zero is excluded, i.e. experience is only split up over the age range from 1 to `uiMaxAge`. Returns deaths (dx), central exposures (Ecx) and the amounts-weighted equivalents (dxa and Ecxa).

H.2 Model Comparison and Tests of Fit

Function: `calculateDevianceResiduals()`

This function calculates the integrated hazard over single years of age. Note that age zero is excluded, i.e. experience is only split up over the age range from 1 to `uiMaxAge`. Returns deaths (dx), integrated hazards (Hx) and deviance residuals (DevRes). This is model-dependent since it uses the hazard function; the example given is for the Gompertz model.

351

Function: `bootstrap()`

This function calculates bootstrap test statistics for the integrated hazard function $H()$, which is presumed defined in the same R workspace and has access to the current values of the parameters. See Section 6.7.

H.3 Methods of Graduation I: Regression Models

Program: `Read_HMD.r`

This program defines the R function `Read.HMD()`, which takes a single argument, the name of a downloaded HMD file, say `HMD_UK_Deaths.txt` or `HMD_UK_Exposures.txt`, i.e. HMD files of deaths or central exposures. A typical sequence of commands would be:

```
source("Read_HMD.r")
Deaths = Read.HMD("Data/HMD_UK_Deaths.txt")
Exposures = Read.HMD("Data/HMD_UK_Exposures.txt")
```

Here we have arranged that all data files are stored in the sub-directory `Data` of the directory that R is running in.

Program: `Gompertz.r`

This program uses UK data to 2009 to illustrate the Gompertz law. Figure 10.1 is produced, the Gompertz model is fitted, various formulae in Section 10.3 are illustrated and the Poisson model in Section 10.8 is fitted.

Program: `Time_Models.r`

This program uses US data on males and females to illustrate the fitting of quadratic and cubic models of mortality. The models are fitted with the `glm()` function. Figure 10.3 is produced.

H.4 Methods of Graduation II: Smooth Models

Program: `Whittaker_Smooth.r`

This program defines the R function `WhittakerSmooth()`, which takes four arguments: the ages, number of deaths, central exposures and smoothing parameter. A typical call would be:

```
WhittakerSmooth(Age, Dth, Exp, 10)
```

Program: `Whittaker.r`

This program calls the function `WhittakerSmooth()`, performs Whittaker smoothing and outputs a graph of the result. The value of the smoothing parameter, here set to 10, can be varied. A figure similar to Figure 11.1 is produced.

Program: `Bspline.r`

This program defines the R function `bspline()`; see the function listing for its arguments. The function requires the splines library. A typical call would look like:

```
source("Bspline.r")
library(splines)
Out = bspline(Year, min(Year), max(Year), 6, 3)
```

Program: `Basis.r`

This program calculates a *B*-spline basis and plots it, as in Figure 11.3.

Program: `Pspline_Regression.r`

This program performs *P*-spline smoothing in two ways, plots the data, the smooths and the regression coefficients. See Figure 11.7.

Program: `Pspline.r`

This program uses the function `Mort1Dsmooth()` to investigate effective dimension, deviance and dealing with overdispersion. Table 11.1 is a table of effective dimension, deviance, AIC and BIC. Figure 11.8 is an AIC plot and Figure 11.10 shows the fitted smooth curve after allowing for overdispersion.

H.5 Methods of Graduation III: Two-Dimensional Models

Program: `Lee_Carter.r`

This program fits the Lee–Carter model to UK male data with the gnm() function, estimates α, β and κ, and converts these to estimates satisfying the standard constraints. Figures 12.1 and 12.2 are obtained.

Program: CBD.r

This program fits the Cairns–Blake–Dowd model to UK male data with the glm() function and estimates κ_1 and κ_2. Figure 12.3 is obtained.

Program: 2d_Pspline.r

This program fits the two-dimensional *P*-spline model to UK male data with the Mort2Dsmooth() function in the package *MortalitySmooth*. Figures 12.6 and 12.7 are produced.

H.6 Methods of Graduation IV: Forecasting

Program: Pspline_Forecast.r

This program fits the one-dimensional *P*-spline model to USA female data and forecasts for 25 years with the Mort1Dsmooth() function in the package *MortalitySmooth*. Figures 13.6 and 13.7 are produced.

Program: Forecast_LC.r

This program illustrates forecasting with the Lee–Carter model. The model is fitted with the gnm() function in the *gnm* package. Forecasting is done with the sarima.for() function in the *astsa* package. Figures 13.9 and 13.10 are produced.

Program: Simulation.r

This program illustrates the simulation of time series. We simulate future values of κ in the Lee–Carter model with the drift model. Simulation is discussed in Section 13.5 and Figures 13.11, 13.12 and 13.13 are produced.

Program: Forecast_CBD.r

This program illustrates forecasting of the CBD model with the bivariate model with drift, as in Section 13.6. The code follows this section closely and Figure 13.15 is produced.

Program: 2d_Pspline_Forecast.r

This program illustrates forecasting of the 2-d *P*-spline model, as in Section 13.7. Figures 13.17 and 13.18 are produced.

References

Aalen, O. O. 1975. Statistical inference for a family of counting processes. Ph.D. thesis, University of California, Berkeley.

Aalen, O. O. 1978. Non-parametric inference for a family of counting processes. *The Annals of Statistics*, **6**, 701–726.

Aalen, O. O. 1987. Dynamic modelling and causality. *Scandinavian Actuarial Journal*, **1987**, 177–190.

Akaike, H. 1973. Information theory and an extension of the maximum likelihood principle. Pages 267–281 of: Petrov, B. and Cźaki, F. (eds), *2nd International Symposium on Information Theory*. Akademiai Kiadó, Budapest.

Akaike, H. 1987. Factor analysis and AIC. *Psychometrica*, **52**, 317–333.

Andersen, P. K., Borgan, Ø., Gill, R. D. and Keiding, N. 1993. *Statistical Models Based on Counting Processes*. Springer, New York.

Arjas, E. 1989. Survival models and martingale dynamics. *Scandinavian Journal of Statistics*, **16**, 177–225.

Arjas, E. and Harra, P. 1984. A marked point process approach to censored failure data with complicated covariates. *Scandinavian Journal of Statistics*, **11**, 193–209.

Bailey, W. G. and Haycocks, H. W. 1947. A synthesis of methods of deriving measures of decrement from observed data. *Journal of the Institute of Actuaries*, **73**, 179–212 (with discussion).

Beard, R. E. 1959. Note on some mathematical mortality models. Pages 302–311 of: Wolstenholme, G. E. W. and O'Connor, M. (eds), *The Lifespan of Animals*. Little, Brown, Boston.

Benjamin, B. and Pollard, J.H. 1980. *The Analysis of Mortality and Other Actuarial Statistics*. Heinemann, London.

Bielecki, T. R. and Rutkowski, M. 2002. *Credit Risk: Modeling, Valuation and Hedging*. Springer, Berlin, Heidelberg.

Booth, H. and Tickle, L. 2008. Mortality modelling and forecasting: A review of methods. *Annals of Actuarial Science*, **3(I/II)**, 3–44.

Bowers, N. L., Gerber, H. U., Hickman, J. C., Jones, D. A. and Nesbitt, C. J. 1986. *Actuarial Mathematics*. Society of Actuaries, Itasca, IL.

Brouhns, N., Denuit, M. and Vermunt, J. K. 2002. A Poisson log-bilinear approach to the construction of projected lifetables. *Insurance: Mathematics and Economics*, **31(3)**, 373–393.

355

Cairns, A. J. G., Blake, D. and Dowd, K. 2006. A two-factor model for stochastic mortality with parameter uncertainty: Theory and calibration. *Journal of Risk and Insurance*, **73**, 687–718.

Cairns, A. J. G., Blake, D., Dowd, K., Coughlan, G. D., Epstein, D., Ong, A. and Balevich, I. 2009. A quantitative comparison of stochastic mortality models using data from England and Wales and the United States. *North American Actuarial Journal*, **13(1)**, 1–35.

Camarda, C. G. 2012. MortalitySmooth: An R package for smoothing Poisson counts with *P*-splines. *Journal of Statistical Software*, **50**, 1–24.

Carstairs, V. and Morris, R. 1991. *Deprivation and Health in Scotland*. Aberdeen University Press, Aberdeen.

Collett, D. 2003. *Modelling Survival Data in Medical Research*, second edn. Chapman & Hall/CRC, Boca Raton, FL.

Conte, S. D. and de Boor, C. 1981. *Elementary Numerical Analysis: An Algorithmic Approach*, third edn. McGraw-Hill, New York.

Continuous Mortality Investigation. 1991. *Continuous Mortality Investigation Report No. 12*. Institute of Actuaries and Faculty of Actuaries, London.

Continuous Mortality Investigation. 2007. *Working Paper 26: Extensions to Younger Ages of the "00" Series Pensioner Tables of Mortality*. Institute of Actuaries and Faculty of Actuaries, London.

Cox, D. R. 1972. Regression models and life tables. *Journal of the Royal Statistical Society: Series B*, **24**, 187–220 (with discussion).

Cox, D. R. and Hinkley, D. V. 1974. *Theoretical Statistics*. Chapman & Hall, London.

Cox, D. R. and Miller, H. D. 1987. *The Theory of Stochastic Processes*. Science Paperbacks, vol. 134. Chapman & Hall, London.

Crowder, M. 1991. On the identifiability crisis in competing risks analysis. *Scandinavian Journal of Statistics*, **18**, 222–233.

Crowder, M. 2001. *Classical Competing Risks*. Chapman & Hall/CRC, Boca Raton, FL.

Currie, I. D. 2013. Smoothing constrained generalized linear models with an application to the Lee–Carter model. *Statistical Modelling*, **13**, 69–93.

Currie, I. D. 2016. On fitting generalized linear and non-linear models of mortality. *Scandinavian Actuarial Journal*, **2016**, 356–383.

Currie, I. D., Durban, M. and Eilers, P. H. C. 2004. Smoothing and forecasting mortality rates. *Statistical Modelling*, **4**, 279–298.

Delwarde, A., Denuit, M. and Eilers, P. H. C. 2007. Smoothing the Lee–Carter and Poisson log-bilinear models for mortality forecasting: A penalized likelihood approach. *Statistical Modelling*, **7**, 29–48.

Dickson, D. C. M., Hardy, M. R. and Waters, H. R. 2013. *Actuarial Mathematics for Life Contingent Risks*, second edn. Cambridge University Press, Cambridge.

Djeundje, V. A. B. and Currie, I. D. 2011. Smoothing dispersed counts with applications to mortality data. *Annals of Actuarial Science*, **5(I)**, 33–52.

Dobson, A. J. 2002. *An Introduction to Statistical Modelling*. Chapman & Hall, London.

Durbin, J. and Watson, G. S. 1971. Testing for serial correlation in least squares regression, III. *Biometrika*, **58(1)**, 1–19.

Efron, B. and Tibshirani, R. J. 1993. *An Introduction to the Bootstrap*, first edn. Monographs on Statistics and Applied Probability, vol. 57. Chapman & Hall, London.

Eilers, P. H. C. and Marx, B. D. 1996. Flexible smoothing with *B*-splines and penalties. *Statistical Science*, **11**, 89–121.

Feller, W. 1950. *An Introduction to Probability and its Applications*, third edn. Vol. 1. John Wiley & Sons, New York.

Fleming, T. R. and Harrington, D. P. 1991. *Counting Processes and Survival Analysis*. John Wiley & Sons, New York.

Forfar, D. O., McCutcheon, J. J. and Wilkie, A. D. 1988. On graduation by mathematical formula. *Journal of the Institute of Actuaries*, **115**, 1–149.

Gerber, H.U. 1990. *Life Insurance Mathematics*. Springer, Berlin, and the Swiss Association of Actuaries, Zurich.

Gini, C. 1921. Measurement of inequality of incomes. *The Economic Journal*, **31(121)**, 124–126.

Girosi, F. and King, G. 2008. *Demographic Forecasting*. Princeton University Press, Princeton, NJ.

Gompertz, B. 1825. The nature of the function expressive of the law of human mortality. *Philosophical Transactions of the Royal Society*, **115**, 513–585.

Green, P. J. and Silverman, B. W. 1994. *Nonparametric Regression and Generalized Linear Models: A Roughness Penalty Approach*. Chapman & Hall, London.

Greenwood, M. 1926. The natural duration of cancer. *Reports on Public Health and Medical Subjects*, **33**, 1–26.

Hannan, E. J. and Quinn, B. G. 1979. The determination of the order of an autoregression. *Journal of the Royal Statistical Society: Series B*, **41**, 190–195.

Hardy, M. R. 2006. *An Introduction to Risk Measures for Actuarial Applications*. Construction and Evaluation of Actuarial Models Study Note. Society of Actuaries, Schaumburg, IL, and Casualty Actuarial Society, Arlington, VA. Available online at www.casact.org/library/studynotes/hardy4.pdf.

Harrell, F. E. and Davis, C. E. 1982. A new distribution-free quantile estimator. *Biometrika*, **69**, 635–640.

Hastie, T. J. and Tibshirani, R. J. 1990. *Generalized Additive Models*. Chapman & Hall, London.

Hattendorff, K. 1868. Das Risiko bei der Lebensversicherung. *Masius Rundschau der Versicherungen*, **18**, 169–183.

Haycocks, H. W. and Perks, W. 1955. *Mortality and Other Investigations*. Vol. 1. Cambridge University Press, Cambridge.

Hoem, J. M. 1969. Markov chain models in life insurance. *Blätter der Deutschen Gesellschaft für Versicherungsmathematik*, **9**, 91–107.

Hoem, J. M. 1988. The versatility of the Markov chain as a tool in the mathematics of life insurance. *Transactions of the 23rd International Congress of Actuaries, Helsinki*, **S**, 171–202.

Hoem, J. M. and Aalen, O. O. 1978. Actuarial values of payment streams. *Scandinavian Actuarial Journal*, **1978**, 38–47.

Hurvich, C. M. and Tsai, C. L. 1989. Regression and time series model selection in small samples. *Biometrika*, **76(2)**, 297–307.

Hyndman, R.J. and Fan, Y. 1996. Sample quantiles in statistical packages. *American Statistician*, **50(4)**, 361–365.

Kahan, W. 1965. Further remarks on reducing truncation errors. *Communications of the ACM*, **8(I)**, 40.

Kaishev, V. K., Haberman, S. and Dimitrova, S. 2009. *Spline Graduation of Crude Mortality Rates for the English Life Table 16*. Office for National Statistics, London. Pages 14–24.

Kalbfleisch, J. D. and Prentice, R. L. 2002. *The Statistical Analysis of Failure Time Data*, second edn. John Wiley & Sons, Hoboken, NJ.

Kaplan, E. L. and Meier, P. 1958. Nonparametric estimation from incomplete observations. *Journal of the American Statistical Association*, **53**, 457–481.

Karr, A. F. 1991. *Point Processes and their Statistical Inference*, second edn. Marcel Dekker, New York, Basel.

Kendall, M. G. and Stuart, A. 1973. *The Advanced Theory of Statistics*, third edn. Vol. 2. Griffin, London.

Kleinow, T. and Richards, S. J. 2016. Parameter risk in time-series mortality forecasts. *Scandinavian Actuarial Journal*, **2017(9)**, 804–828.

Lawless, J. F. 1987. Negative binomial and mixed Poisson regression. *Canadian Journal of Statistics*, **15**, 209–225.

Lee, R. D. and Carter, L. 1992. Modeling and forecasting US mortality. *Journal of the American Statistical Association*, **87**, 659–671.

Li, J. S. H., Hardy, M. R. and Tan, K. S. 2009. Uncertainty in mortality forecasting: An extension to the classic Lee–Carter approach. *Astin Bulletin*, **39**, 137–164.

Macdonald, A. S. 1996. An actuarial survey of statistical models for decrement and transition data, III: Counting process models. *British Actuarial Journal*, **2**, 703–726.

Madrigal, A., Matthews, F., Patel, D., Gaches, A. and Baxter, S. 2011. What longevity predictors should be allowed for when valuing pension scheme liabilities? *British Actuarial Journal*, **16(I)**, 1–62 (with discussion).

Makeham, W. M. 1860. On the law of mortality and the construction of annuity tables. *Journal of the Institute of Actuaries and Assurance Magazine*, **8**, 301–310.

McCullagh, P. and Nelder, J. A. 1989. *Generalized Linear Models*, second edn. Monographs on Statistics and Applied Probability, vol. 37. Chapman & Hall, London.

McCutcheon, J. J. 1985. Experiments in graduating the data for the English Life Tables (No. 14). *Transactions of the Faculty of Actuaries*, **40**, 135–147.

McCutcheon, J. J. and Eilbeck, J. C. 1975. Experiments in the graduation of the English Life Tables (No. 13) data. *Transactions of the Faculty of Actuaries*, **35**, 281–296.

McLoone, P. 2000. *Carstairs Scores for Scottish Postcode Sectors from the 1991 Census*. Public Health Research Unit, University of Glasgow, Glasgow.

Neill, A. 1986. *Life Contingencies*. Heinemann, London.

Nelder, J. A. and Wedderburn, R. W. M. 1972. Generalized linear models. *Journal of the Royal Statistical Society: Series A*, **135, Part 3**, 370–384.

Nelson, W. 1958. Theory and applications of hazard plotting for censored failure times. *Technometrics*, **14**, 945–965.

Pawitan, Y. 2001. *In All Likelihood: Statistical Modelling and Inference Using Likelihood*. Oxford University Press, Oxford.

Perks, W. 1932. On some experiments in the graduation of mortality statistics. *Journal of the Institute of Actuaries*, **63**, 12–40.

Perperoglou, A. and Eilers, P. H. C. 2010. Penalized regression with individual deviance effects. *Computational Statistics*, **25**, 341–361.

Philips, L. 1990. Hanging on the metaphone. *Computer Language*, **7(12)**, 39–43.

Prentice, R. L., Kalbfleisch, J. D., Peterson, A. V., Jr., Flournoy, N. S., Farewell, V. T. and Breslow, N. E. 1978. The analysis of failure times in the presence of competing risks. *Biometrics*, **34**, 541–554.

Press, W. H., Teukolsky, S. A., Vetterling, W. T. and Flannery, B. P. 1986. *Numerical Recipes in C++: The Art of Scientific Computing*, second edn. Cambridge University Press, New York.

Ramlau-Hansen, H. 1988. Hattendorff's theorem: A Markov chain and counting process approach. *Scandinavian Actuarial Journal*, **1988**, 143–156.

Renshaw, A. E. and Haberman, S. 2006. A cohort-based extension to the Lee–Carter model for mortality reduction factors. *Insurance: Mathematics and Economics*, **38**, 556–570.

Richards, S. J. 2008. Applying survival models to pensioner mortality data. *British Actuarial Journal*, **14(II)**, 257–326 (with discussion).

Richards, S. J. 2009. Selected issues in modelling mortality by cause and in small populations. *British Actuarial Journal*, **15 (supplement)**, 267–283.

Richards, S. J. 2012. A handbook of parametric survival models for actuarial use. *Scandinavian Actuarial Journal*, **2012 (4)**, 233–257.

Richards, S. J. 2016. Mis-estimation risk: Measurement and impact. *British Actuarial Journal*, **21(3)**, 429–457.

Richards, S. J. and Currie, I. D. 2009. Longevity risk and annuity pricing with the Lee–Carter model. *British Actuarial Journal*, **15(II) No. 65**, 317–365 (with discussion).

Richards, S. J. and Jones, G. L. 2004. *Financial Aspects of Longevity Risk*. Staple Inn Actuarial Society (SIAS), London.

Richards, S. J., Kirkby, J. G. and Currie, I. D. 2006. The importance of year of birth in two-dimensional mortality data. *British Actuarial Journal*, **12(I)**, 5–61 (with discussion).

Richards, S. J., Kaufhold, K. and Rosenbusch, S. 2013. Creating portfolio-specific mortality tables: A case study. *European Actuarial Journal*, **3 (2)**, 295–319.

Richards, S. J., Currie, I. D. and Ritchie, G. P. 2014. A value-at-risk framework for longevity trend risk. *British Actuarial Journal*, **19(1)**, 116–167.

Schwarz, G. E. 1978. Estimating the dimension of a model. *The Annals of Statistics*, **6** (2), 461–464.

Shumway, R. H. and Stoffer, D. S. 2010. *Time Series Analysis and its Applications*, third edn. Springer, London.

Spencer, J. 1904. On the graduation of the rates of sickness and mortality. *Journal of the Institute of Actuaries*, **38**, 334–343.

Sverdrup, E. 1965. Estimates and test procedures in connection with stochastic models for deaths, recoveries and transfers between states of health. *Skandinavisk Aktuaritidskrift*, **48**, 184–211.

Thatcher, A.R., Kannisto, V. and Vaupel, J.W. 1998. *The Force of Mortality at Ages 80 to 100*. Odense University Press, Odense.

The Economist. 2012. The ferment of finance. Special report on financial innovation. **25 February 2012**, 8.

Thurston, S. W., Wand, M. P. and Wiencke, J. K. 2000. Negative binomial additive models. *Biometrics*, **56**, 139–144.

Tsiatis, A. A. 1975. A nonidentifiability aspect of the problem of competing risks. *Proceedings of the National Academy of Sciences of the United States of America*, **72**, 20–22.

Turner, H. and Firth, D. 2012. Generalized non-linear models in R: An overview of the gnm package (R package version 1.0-6). Available online at http://CRAN.R-project.org/package=gnm.

Waters, H. R. 1984. An approach to the study of multiple state models. *Journal of the Institute of Actuaries*, **111**, 363–374.

Wedderburn, R. W. M. 1974. Quasi-likelihood functions, generalized linear models, and the Gauss–Newton method. *Biometrika*, **61**, 439–447.

Whittaker, E. T. 1923. On a new method of graduation. *Proceedings of the Edinburgh Mathematical Society*, **41**, 63–75.

Willets, R. C. 1999. *Mortality in the Next Millennium*. Staple Inn Actuarial Society (SIAS), London.

Willets, R. C. 2004. The cohort effect: Insights and explanations. *British Actuarial Journal*, **10**, 833–877.

Williams, D. 1991. *Probability with Martingales*. Cambridge University Press, Cambridge.

Wood, S. N. 2006. *Generalized Additive Models: An Introduction with R*. Chapman & Hall, London.

Author Index

361

Index

363